THE POLITICAL SPECTRUM

THOMAS WINSLOW HAZLETT

The Political Spectrum

THE TUMULTUOUS LIBERATION OF WIRELESS TECHNOLOGY,

FROM HERBERT HOOVER TO THE SMARTPHONE

Yale

UNIVERSITY PRESS

NEW HAVEN & LONDON

Published with assistance from the foundation established in memory of Calvin Chapin of the Class of 1788, Yale College.

Yale University Press books may be purchased in quantity for educational, business, or promotional use. For information, please e-mail sales.press@yale.edu (U.S. office) or sales@yaleup.co.uk (U.K. office).

Set in Scala type by Integrated Publishing Solutions, Grand Rapids, Michigan.
Printed in the United States of America.

Library of Congress Control Number: 2016950652
ISBN 978-0-300-21050-7 (hardcover : alk. paper)

A catalogue record for this book is available from the British Library.

This paper meets the requirements of ANSI/NISO Z39.48-1992 (Permanence of Paper).

10 9 8 7 6 5 4 3 2 1

For my loving mother, Marilyn Kay Hazlett, 1926–1992,
and her joie de vivre.
And for my friend, Giancarlo Ibárgüen, 1963–2016,
a remarkable visionary.

CONTENTS

ACKNOWLEDGMENTS

THE AUTHOR STANDS ON THE SHOULDERS of giants—and hopes not to crush his mentors. These scholarly heroes include Ronald Coase, Harold Demsetz, Alfred Kahn, Leo Herzel, Armen Alchian, William Allen, and Harvey J. Levin. I have been exceedingly fortunate to speak with or formally interview many experts on subjects pertaining to this book, and gratefully acknowledge the grace, wisdom, and patience of Henry Geller, Brian Lamb, Sol Schildhause, Lionel van Deerlin, Henry Manne, Robert Hahn, Marty Cooper, Michael Marcus, Peter Huber, Pablo Spiller, Harold Furtchgott-Roth, Evan Kwerel, Jonathan Levy, Florence Setzer, John Williams, Richard Bennett, Lex Felker, Charla Rath, Janice Obuchowski, Gerald Faulhaber, Adam Clayton Powell III, Gregory Rosston, Greg Sidak, Robert Pepper, Michael Chartier, Thomas Welter, Dale Hatfield, Chris Sacca, Andrew Kreig, Glen O. Robinson, Marius Schwartz, Judith Mariscal, Jon Leibowitz, Jonas Wessel, Rune Ostgard, Ted Hearn, Gary Libecap, Karen Wrege, Brett Tarnutzer, Robert Corn-Revere, Chris DeMuth, Joan Marsh, Tom Whitehead, Blair Levin, Andrew McLaughlin, Dean Brenner, Hal Varian, Jim Snider, Frank Washington, Dorothy Robyn, Shane Greenstein, Charles Plott, Ted Frech, and Dennis Patrick.

Peter Pitsch read an early draft of this book and offered deep insights as well as practical suggestions, golden nuggets for a free-riding author. The guidance and encouragement of George Mason University Law School Dean Daniel Polsby, a remarkable man, were instrumental to my efforts, as was the Information Economy Project at George Mason University Law School. Clemson University has generously continued the Project and extended its support, for which I am indebted to Skip Sauer, recently chairman of the Department of Economics, and Robert McCormick, dean of the College of Business. James Montgomery, a bold editor at the *Financial Times,* created the New Technology Policy Forum in 2002; until 2011 it provided a lively space for me to spar with big thinkers Richard Epstein, Jamie Boyle, Eli Noam, and Lawrence Lessig, developing many of the ideas in this book. Thomas Donlan, an editor at *Barron's* who paid attention to tech policy before tech policy was cool, has also been supportive. Charlie Firestone has regularly invited me to Aspen Institute discussions of spectrum policy; I must commend Aspen for organizing such stimulating events and thank Charlie for so often including me. Dewayne Hendricks's dedication to a technology article e-list, and Ken Robinson's remarkable one-man show, the weekly *Telecommunications Policy Review,* have proven valuable educational tools for me, as well.

My academic coauthors have been a bountiful source of knowledge, and I have learned vast amounts from Arthur Havenner, Giancarlo Ibárgüen, Dennis Weisman, Robert Crandall, Roberto Muñoz, Diego Avanzini, Evan Leo, Babette Boliek, Scott Wallsten, George Bittlingmayer, David Teece, Len Waverman, Sarah Oh, Robert J. Michaels, Zhiqiang Leng, David Sosa, Coleman Bazelon, George Ford, Wayne Leighton, Brent Skorup, Bruno Viani, Drew Clark, Ralitza Grigorova, Michael Honig, and Joshua D. Wright. Vernon Smith and David Porter, academic collaborators par excellence, are particularly notable in this regard, as are their leadership and bountiful insights on projects conducted by our boutique consulting firm, Arlington Economics.

Individuals who helped me in the production of this manuscript include Laura Lieberman, Ryan Tacher, James Spurlock, Haobin Fan, Emily Wright, Cesar Castellon, Ben Schwall, Jonathan Earnest, and Lynn Evans. Special gratitude is owed to my intrepid friend and editor Martin Morse Wooster, and to Charles Jackson, whose excellent radio engineering tutorials I imperfectly absorbed. Teresa Hartnett, my agent, was a strong source of enthusiastic support. My hope is that some fraction of that vital input was justified. William Frucht is the ultimate professional, and I have been most fortunate to have him as my editor at Yale University Press. He offered invaluable editorial assistance reflecting his well-developed literary acumen and an impressive dedication to the understanding of history, economics, and public policy. He selected two anonymous referees for the initial manuscript, each of whom offered sage criticisms and suggestions. Bill's assistant, Karen Olson, gave me guidance in organizing the manuscript and tracking down the illustration permissions, and manuscript editor Dan Heaton helped polish what I had hammered. The finished volume owes much to each of these contributions. All errors and omissions remain, sadly, the responsibility of the author.

It is customary to end such a list with a tribute to one's family for the great sacrifice they have made, while pleading for forgiveness for time spent writing. This tradition seems beside the point in the instant case. The fact of the matter is that my amazing wife, Alexandra, and our teenage daughters, Marilyn and Lauren, made certain—through this project and all else— that our best life continued to happen. I was unfamiliar with the country tune about a dad who spends time with his daughter, "She Thinks We're Just Fishin'." But my girls made sure I heard that music, and so much else. And now, ladies, it's time to find where they're biting.

THE POLITICAL SPECTRUM

Introduction: Magic and Cacophony

How easily are base and selfish measures masked by pretexts of public good and apparent efficacy.

—James Madison

NO NATURAL RESOURCE IS LIKELY TO be more critical to human progress in the twenty-first century than radio spectrum. Invisible, odorless, and ubiquitous, it is the space through which electronic communications travel. We have been grappling with its possibilities for just over a century. While children today see wireless technology as a standard social amenity, only a short time ago it appeared as something of a magic trick. One description, commonly if erroneously attributed to Albert Einstein, was, "The wireless telegraph is not difficult to understand. The ordinary telegraph is like a very long cat. You pull the tail in New York, and it meows in Los Angeles. The wireless is the same, only without the cat."[1]

When television broadcasting debuted at the New York World's Fair in 1939, exhibitors created a special glass-encased TV set for the demonstration. It was an attempt to preempt any claims by skeptics that tiny performers were being squeezed onto a compact stage.

That light rippled through the atmosphere in the form of waves was shown by English physicist Thomas Young's famous "double-slit" experiment in 1801. After other waves—unseen, unheard, unfelt—were discovered, the theory developed that some of these oscillating signals could be tamed. The first such effort was the telegraph, developed in the 1830s: it worked by turning an electrical circuit over a wire on and off. When Samuel F. B. Morse developed a code, information flowed. In 1874, Alexander Graham Bell pumped conversations through those wires, translating human speech into electrical pulses and back again. This grand achievement illustrated the potential of "spectrum in a tube."

In 1895, Guglielmo Marconi removed the tube. Setting up two devices on either bank of a river in England, he created a voice conversation, no phone cord needed.

The long cat had meowed.

But there were no apps. No one, save Marconi, owned a radio. When such units became available, they were sold in pairs to provide point-to-point communications. A radio in St. Louis could "talk" to a radio in Kansas City via a beam transmitted between them. Conflicts were unlikely and easily fixed. Soon, Marconi and others figured out how to multiply traffic in the same spot by adopting different wavelengths for different channels—"frequency division." (Frequency is inversely proportional to wavelength; the shorter the wavelength, the more frequent are its cycles.)

Science continued to march. "Geographic division" would reuse frequencies from area to area. "Time division" would slice up a given channel, interspersing extra conversations during short quiet times, while "code division" would allow a receiver to decipher a seeming jumble of overlapping signals. Such techniques can dramatically increase communications capacity. Progress has been relentless, with wireless traffic doubling roughly every two years—upholding Cooper's Law.[2] This produces about a millionfold increase every fifty years. We today enjoy one trillion times the wireless capacity of networks a century ago.

This furious pace is not slackening. The Apple iPhone, introduced in 2007, triggered a smartphone revolution. From 2008 through 2011, global mobile data flows increased not by the long-term trend of 4x but by 30x—more than doubling each year.[3] U.S. regulators call this the "mobile data tsunami." Policy makers in Europe dub spectrum the "economic oxygen" powering society.

At the dawn of wireless, radio kits became a craze among America's

youth. Hobbyists cobbled together crude transmitters and receivers, playing and experimenting. Then tragedy intervened. In April 1912, radio operators at the British Marconi Company wireless station atop Wanamaker's
Department Store in New York City heard and relayed distress calls (and
news of individual survivors) from the sinking *Titanic* and the rescuing
Carpathia; the much closer *Californian* was not listening. It was thought
that the lack of attention had contributed to the disaster's toll. The Radio
Act became law the following August, mandating that seagoing vessels give
"absolute priority to signals and radiograms relating to ships in distress."[4]

The Radio Act of 1912 also set out the first rules on potential airwave
conflicts. It required radio transmitters to be licensed by the secretary of
labor and commerce, and for the secretary to issue such licenses so as to
"minimize interference."[5]

This grant of authority was not contentious. Radios were few, traffic was
limited, and conflicts inconsequential. Then, on November 2, 1920, radio
station KDKA in Pittsburgh transmitted election returns in the presidential
contest between Republican Warren G. Harding and Democrat James M.
Cox. The event launched a new business model: broadcast. The term is
agricultural. A "broad cast" tosses seeds not narrowly but in all directions.
Radio as a one-way, point-to-*multipoint* service was a disruptive innovation
revealing the possibility of a mass market for electronic news, information,
and entertainment.

Game on. By December 1922, there were more than five hundred broadcasting stations across the United States. Marketing wizards proclaimed
1924's holiday season "Radio Christmas," and customers flocked to buy
consoles. By 1926, four million households had radios, and by July of that
year spouses were complaining that their mates were spending too much
time playing with their gadgets.

The technology was transformative. Citizens far and wide could hear and
know—in an instant—of events across the world. In 1865 the famous as-
sassin of an American president could outrun, on horseback, the news of
his crime—escaping the nation's capital after exchanging pleasantries with
Union soldiers guarding a bridge, some thirty minutes after the murder on
a public stage just blocks from the White House.[6] Now, however, a newly
connected world was born. Information traveled at the speed of light, no

1. *Radio News*, July 1926. Broadcasting developed into a wildly popular social phenomenon well before the 1927 Radio Act.

wires needed. Sports promoters, commercial sponsors, and politicians were quick to seize on the emerging platform. By the 1940s, when moving pictures were added to the audio streams, the disruption seemed complete.

But broadcasting was just the beginning. In the 1980s, governments began distributing licenses for cellular telephone service. Less than three decades later, there are more than six billion wireless phone users. More people today own cell phones than own toothbrushes or have access to working toilets.[7]

And this sensational progress, too, will prove simply a glimmer of the future. One hundred years from now, wireless devices will be loaded with faster, better cameras, sensors, monitors, and other, not yet invented technologies to produce as yet unimagined services. A few are already obvious: telemetry to steer driverless cars, crime detection applications to enhance personal security, biometric screenings to continuously check our vital signs. Countless more dots are forming on the horizon. The conventional wisdom is that 2020 will be a world of fifty billion connected devices, most wireless. Machine-to-machine (m2m) radios—used, for example, when a stolen car broadcasts its location via Lojack, a working pacemaker is reprogrammed without surgery, or a vending machine signals the warehouse to refill the Oreos but not the granola bars—are already standard.

The trillionfold increase in capacity has not delivered the "end of spectrum scarcity." Instead, it has triggered a torrent of new devices, applications, content, and businesses to gobble up the emergent opportunities. The cry is heard, *more bandwidth, please!* Every advance has spurred commensurate leaps in demand.

Entrepreneurs in garages continue to discover killer apps to fill the air. Were mobile phone service today that of 1990—analog cellular voice service—the tenfold increase in mobile bandwidth allotments that has occurred since then would simply provide for excellent phone calls. But your father's mobile market is gone. Younger subscribers—between texts, emails, Facebook posts, tweets, Spotify tunes, YouTube videos, Netflix movies, Instagrams and Snapchats—may be surprised to find "human interactive voice" an available app.

Wireless devices, once "car phones" or "handsets," are now computers, netbooks, smartphones, tablets, personal hotspots, watches, wearables, or embedded m2m devices. Slices of radio spectrum serve them all. Economic interests jockey to push (or keep) frequency allocations favoring their competitive niche—broadcasters to protect the TV allocation table of 1952; cellular phone carriers to repurpose TV band spectrum for mobile licenses; cable TV operators for set-asides dedicated to "unlicensed" use; car makers for other bands optimized specifically for vehicle informatics. Meanwhile, regulators and communications law firms unite to defend the labor-intensive bureaucratic labyrinth handling all such requests under a nineteenth-century policy technology, pioneered in the Interstate Commerce Act of 1887 and codified in the Radio Act of 1927.

Even as generations of PANS (pretty amazing new stuff) have blown away POTS (plain old telephone service), that traditional structure for spectrum allocation has stood tall. It has not stopped all progress. Yet each leap forward has come only following a process too long, too painful. In March 2010, in its National Broadband Plan, the Federal Communications Commission agreed. It decried the pervasive delays—estimated by the FCC to take from six to thirteen years—before new wireless technologies could be authorized. But the bold new allocations it advocated are now themselves taking six to thirteen years to implement.

We are swimming in underutilized frequency spaces. Bands that lie mostly fallow could be intensively deployed, serving pent-up demands and unleashing transformative social progress. Depriving markets of available spectrum saps entrepreneurial energy. For all that wireless has delivered, it might deliver orders of magnitude more. Insiders blame technical complexity, echoing Soviet commissars who attributed seventy years of poor harvests to bad weather. Both outcomes result from system design defects.

The goodwill of regulators is not at issue. They are often conscientious and pro-consumer in their outlook, as frustrated by the shortcomings of spectrum allocation as private sector innovators or academic critics. The structure of their administrative apparatus is the devil. As a rule, to access unused frequencies wireless investors must file a plan with the government, detail their prospective business, document their technology, and then prove that such operations will serve "public interest, convenience, or necessity."

Six to thirteen years is best case. Most requests are killed outright or simply fail to be filed due to the high expense and low probability of regulatory success. The screen forms an "attractive nuisance," in the apt words of former FCC Chief Legal Adviser Ken Robinson, wherein interested parties— often, powerful industry incumbents—are invited to protest innovations as threatening existing regulatory practices and bargains. These claims are unduly persuasive.

The political spectrum becomes littered with regulatory gridlock, what Michael Heller calls a "tragedy of the anticommons."[8] This results in a situation where cooperation can yield healthy benefits, but property rights are divvied up among so many parties that deals become impossible. In 2012, the tragedy struck anew, as a $14 billion national network (using state-of-the-art 4G technology) being built by upstart mobile operator LightSquared was derailed. The company was to deploy airwaves set aside for satellite phone service, hosting almost no traffic and providing essentially zero economic value. Gains from the new (terrestrial) network promised to be huge, running to more than $100 billion.[9] But when opponents alleged interference in the neighboring GPS band, spillovers that could be fixed for (at most) a few hundred million dollars, the regulatory process broke down. The FCC revoked LightSquared's mobile licenses, and Americans lost a competitive option they never knew they had.

Hopeless deadlock in one band borders innovation beyond compare. If we can realign the tragic failures of spectrum allocation to coincide with the spectacular successes of wireless markets, we can triple-down on innovation. The momentum for such a course correction is surprisingly sluggish. Many interest groups favor existing inefficiencies, and take great care to obscure our policy choices. They are assisted by the concentrated benefits that flow to well-informed incumbent licensees,[10] and by the seemingly obscure subject matter.

For instance, the term *technical reasons* is commonly deployed as a weapon. Wielded as a conversation killer, it is a blunt instrument preempting simple questions on the grounds that only select experts may grasp the tropospheric complexities involved. In September 1993, while visiting the Czech Republic, I read that the country's first private TV broadcasting license had just been authorized. I asked my academic hosts why, of twenty-seven applicants, only one had been issued a license. They did not know but offered to introduce me to two members of the Broadcasting Commission, whom I could ask directly. (Their offer included an English-Czech translator.) In accepting, and without studying up on the local situation, I proffered a prediction: the rationale for monopoly licensing would be *technical reasons*.

At the appointed time I put my query to the commissioners. *I understand that some twenty-seven applications for a TV license were received, but just one license was issued by the government. Why not allow more competition?* The translator restated this, and one of the commissioners began answering in Czech. As I heard *technika, technika,* I was unable to suppress a chuckle. My unfortunate mirth took the Czech commission members aback; they had been told I spoke only English. In fact, we had stumbled into the universal language of anticompetitive spectrum allocation.

I restated my question, in even less politic terms. *There is room for many TV broadcasts in an area. Today, in Rome, one can tune into some forty broadcast stations. I have not undertaken any atmospheric studies of the Czech Republic, but if it is like the rest of the planet, it is possible to squeeze in another broadcast or two.*

The answer was twofold. First, the commission's chief engineer was attending a conference in London. Thus the available commissioners could not directly address the "technical issues." Second, even were there room for more stations, they had made a decision to encourage Czech-language programming. With competition, they asserted, no station would have the financial resources to create local programming.

In the political spectrum, regulators speak various tongues but explain their actions in common code. Sharp restrictions on market rivalry and free speech are said to be embedded in nature. Artificial policy choices are transformed into necessities. But those limits are then seamlessly transformed into further, less natural, constraints. Profits flow from these man-made barriers that punish consumers—and often, as in broadcasting, free speech— in the name of the "public interest." Thus a monopoly operator is declared a social benefit.

That such gains are illusory, achievable without the extravagantly wasteful expense of monopolistic distortions, is studiously ignored. And extensive real-world evidence certifies that the ostensible *quid pro quo* always delivers the *quid* but rarely the *quo*. Judge Richard A. Posner raised this possibility when he noted the recurring practice of "taxation by regulation."[11] A more lively formulation is posited by Chilkoot Charlie's, a popular watering hole in Anchorage, Alaska, which proudly displays its motto: *We Cheat the Other Guy and Pass the Savings On To You!*[12]

Were the infirmities of frequency regulation solely a matter of unjust enrichment, they would be awful. But the extant system is worse: we have seen the future but refuse to embrace it. In 1983, the late MIT political scientist Ithiel Pool presciently forecast the emergence of computer networks and mobile technologies, already charting their path to social centrality. A sage observer, Pool grasped the bountiful promise of innovation. But he also saw how sharply the emergent "technologies of freedom" conflicted with existing rules and political incentives for control, as had historically played out in the information revolution ignited by the printing press. "The onus is on us," wrote Pool, "to determine whether free societies in the twenty-first century will conduct electronic communications under the conditions of freedom established for the domain of print through centuries of struggle, or whether that great achievement will become lost in a confusion about new technologies."[13]

PART ONE

WELCOME TO THE JUNGLE

There is a great deal of spectrum . . . but of course not an unlimited amount. The limited supplies . . . can be squandered or used well. They can be rationed out by a market or rationed out in a political way by the authoritarian practices of government.

—Ithiel de Sola Pool

WIRELESS COMMUNICATIONS HAVE disrupted economies and transformed cultures. But the enormous gains have been sharply truncated by regulatory structures even older than the technology itself. This history is rich with irony and filled with implication for the modern world.

1

Dances with Regulators

We fight for and against not men and things as they are, but for and against the caricatures we make of them.

—Joseph A. Schumpeter

THE MOST FAMOUS SPEECH EVER GIVEN by an American regulator was delivered in Las Vegas on May 9, 1961. Federal Communications Commission Chairman Newton Minow was the keynote speaker at the annual convention of the National Association of Broadcasters. Citing the substantial profits TV station owners were enjoying, Minow told the assembled executives, "I have confidence in your health. But not in your product." He proceeded to lecture the industry elite as though they were schoolboys.

> I have seen a great many television programs that seemed to me eminently worthwhile. . . . But when television is bad, nothing is worse. I invite you to sit in front of your television set. . . .
>
> You will see a procession of game shows, violence, audience participation shows, formula comedies about totally unbelievable families, blood and thunder, mayhem, violence, sadism, murder, Western badmen, Western goodmen, private eyes, gangsters, more violence and cartoons. And, endlessly, commercials—many screaming, cajoling and offending. And most of all, boredom.[1]

He then dismissed their product as "a vast wasteland."

Inside the hall, stunned silence. Outside, universal applause. Newspaper headlines heralded a tough new sheriff riding into town. "F.C.C. Head Bids TV Men Reform 'Vast Wasteland'—Minow Charges Failure in Public Duty—Threatens to Use License Power" was how the *New York Times* framed it. Wisconsin Senator William Proxmire called the speech "courageous and provocative" and inserted it into the *Congressional Record*.[2] The FCC chairman was awarded a Peabody, a prestigious prize for excellence in journalism.[3] Three years later, after Minow left the FCC, the Washington Post Company (a TV station group owner) published an expanded version of the speech as a book, *Equal Time: The Private Broadcaster and the Public Interest.*

Broadcasting executives may have sneered (the stranded boat on *Gilligan's Island*, a CBS show debuting in 1964, was mockingly named the *Minnow*), but the words *vast wasteland*, coming from the leader of the agency that licensed the offending actors, resonated with the press and the public. Citing a reporter who had written that "the New Frontier is going to be one of the toughest FCC's in the history of broadcast regulation," Minow leaped to confirm it. "Let me make it perfectly clear that he is right."[4]

In the speech, Minow lectured broadcasters that they were "squandering our airwaves," a "lavish waste of [a] precious natural resource." The government did not lack tools to correct the situation. "Clearly, at the heart of the FCC's authority lies its power to license." Yet that formal authority was widely seen as toothless, exercised by a bureaucracy captured by the industry it regulated. "I say to you now," Minow announced, that broadcast TV license "renewals will not be pro forma in the future. . . . I intend to take the job of Chairman of the FCC very seriously."[5]

The FCC would combat what Minow called "the dictatorship of numbers"—viewer ratings. To renew their licenses, stations would have to show that they had supplied "the informational programs that exemplify public service." Strict enforcement was the key. "As you know, we are readying for use new forms by which broadcast stations will report their programming to the Commission. . . . Special attention will be paid in these reports to public service programming."[6]

With that, the regulator returned to Washington and went to work.

But Minow's FCC never drafted new programming rules of any consequence.[7] The "vast wasteland" continued unabated, a conclusion offered by Minow in his 1995 book *Abandoned in the Wasteland*. By then he had upped the metaphoric ante, describing TV broadcasters as "drug dealers."[8] His "new forms" threat had failed to produce either the forms or better TV. In 1963, the chairman left government service, returning to his law practice and later joining the Board of Directors of CBS.

It would be unfair, however, to say that Minow accomplished little at the FCC. During his tenure, U.S. television policy was fundamentally redirected. Almost before the ink was dry on the headlines from Las Vegas, Minow was crafting ambitious new regulatory barriers to shield TV stations, and the "vast wasteland" they aired, from an emerging competitive threat.

In the 1950s, the federal government had seen no reason to be concerned about TV programs sent via wires—"community antenna television" (CATV), or "cable." The product was local, not interstate, and was not delivered to customers via radio spectrum. In a 1959 report,[9] the Commission concluded it had no reason—or authority—to regulate.

Cable slowly made inroads. At first, operators wired areas where households were unable to receive over-the-air TV signals. Extending Philadelphia TV programs to viewers in Allentown, Pennsylvania, enhanced stations' audiences. Broadcasters had no objection, and neither did the FCC.

But then some cable TV systems began wiring areas with excellent broadcast reception. In 1961, Cox Cable offered subscription TV to San Diego residents with a twelve-channel system, creating a set of video choices well beyond the existing "2½ stations" (as Minow described the CBS-NBC-ABC triopoly, discounting ABC). To fill the dial, "entrepreneurs erected a sophisticated antenna, capable of picking up Los Angeles channels from 100-odd miles away." For $5.50 a month, the cable retransmitted the three network signals plus "four independent stations that served Los Angeles with sports, old moving pictures and reruns of network shows."[10]

This was war. Incumbent broadcast licensees launched volleys of lobbyists. The chairman who had roared in Las Vegas, lecturing industry leaders for their shoddy product, was now more receptive to their interests. In a "startling reversal," as economists Stan Besen and Robert Crandall put it, the Commission repudiated its policy of benign neglect toward cable and aggressively intervened on the triopoly's behalf.[11]

The new policy was signaled in a 1962 FCC ruling, *Carter Mountain*.[12] Whereas the 1959 report mocked TV broadcasters' request for anticompetitive rules as a "logical absurdity . . . [that] requires no elaboration,"[13] in 1962 the Commission provided just that—a creative legal analysis expanding the FCC's authority to regulate (and preempt) cable. Overturning a routine FCC staff award of a microwave license to Carter Mountain Transmission Co., which planned to beam video programming to a cable operator in Wyoming, the agency voted five to one (Chairman Minow in the majority) to revoke the authorization and deny the license.

The Communications Act of 1934 specified that microwave licensees were "common carriers" prohibited from discriminating among customers. Yet the regulators now blocked a point-to-point radio link precisely because an economic rival objected to the competitive implications of the content being transmitted. To protect the financial health of KWRB-TV—the Casper, Wyoming, station whose owners had filed the protest—the FCC asserted broad new regulatory authority and denied Carter Mountain's license.

The FCC could not openly state that it acted from naked protectionism. It needed a "public interest" justification. Alas, these are manufactured domestically. So in 1962 and in a string of orders in 1965, 1966, 1968, and 1970, the FCC opined about the benefits of suppressing cable television.

The argument began with a market prediction. Cable TV would never compete as a national distribution platform, and could do no more than supplement broadcast TV here and there. Over-the-air stations would inevitably remain Americans' primary means of accessing news and public affairs. Allowing cable systems to "siphon" their audiences could threaten stations' financial viability, causing some stations to go dark and leaving the public less informed. The Commission actively precluded this dire outcome by nipping rivals in the bud.

In these anticable orders, the FCC seemed untroubled by the "vast wasteland." Instead, it claimed that the "public interest" demanded that the wasteland be guarded against competition. At the very instant the Commission was reaping sensational publicity for holding broadcasters' feet to the fire and sending "shockwaves among television executives nationwide,"[14] it was carrying water for the purveyors of game shows, sadism, murder, endless commercials, and boredom.

The FCC found many reasons why cable should not compete with broadcast—to protect localism, diversity of expression, and public affairs programming, as well as to enhance competition from fledgling UHF stations against the stronger VHF stations. On each of these asserted "public interests," the FCC was wrong. The benefits claimed for the licensing system were fiction. Policy makers later discovered that the "public interest" lay in *reversing* the anticable policy. In the deregulation wave of the Ford and Carter years, buttressed by a newly skeptical D.C. Court of Appeals, a reform-minded Commission stripped away protectionist barriers. By 1980 America was being wired for video competition.

News and public affairs programs, touted as the heart of the "public interest" in broadcast regulation, were minuscule in the wasteland. In the early 1960s the three broadcast TV networks programmed just fifteen minutes of news per day. As late as 1973, CBS aired just ninety-two hours of news programs for the entire year. The conventional wisdom was that the American viewer simply was not interested in news or documentaries.

But with the deregulation of cable, innovative 24/7 news and public affairs networks emerged, scheduling scores of hours per *week*. The political, social, and cultural diversity of content vastly increased. Specialized programming networks radically altered the "lowest common denominator"[15] approach Minow had attributed to broadcaster greed and sloth.

Localism also took root, despite the FCC's predictions, as cable offered new local news channels, public access, and the mixed blessing of televised city council meetings. Contrary to the FCC's considered premise, UHF stations were not threatened but saved by cable, which eliminated their reception disadvantage versus VHF. The struggling "U's" had, in hindsight, been used as poster boys for the powerful "V's," whose lobbyists funded the study by an eminent MIT economist predicting the demise of UHF with the rise of cable[16]—swallowed whole by the FCC[17] but debunked by subsequent academic studies and real-world experience.[18]

In the political spectrum, perverse regulatory consequences mean never having to say you're sorry. In *Abandoned in the Wasteland* (1995), Minow continued to condemn television for mindless content, laying the blame on the 1934 Communications Act. "Because the act did not define what the public interest meant, Congress, the courts, and the FCC have spent sixty frustrating years struggling to figure it out."[19] But he simultaneously gushed about what the market had delivered: "The FCC objective in the early 60's, to expand choice, has been fulfilled—beyond all expectations."[20] To which law professors Thomas Krattenmaker and Lucas A. Powe, Jr. responded: "No kidding." Citing the emergence of CNN, C-SPAN, C-SPAN2, Discovery, the Learning Channel, MTV, Nickelodeon, A&E, Bravo, AMC, religious programming, BET, and Univision, they concluded, "Everything that anyone wanted in the 1960s . . . is available at the flip of a switch."[21]

Minow claimed credit. His FCC, he said, had rescued UHF TV stations from oblivion by lobbying Congress to pass the All-Channel Receiver Act of 1962 and then working with the Kennedy administration to launch satellite communications. The fiftieth anniversary of the "vast wasteland" speech was yet another forum for such assertions, and journalists eagerly repeated Minow's boasts.[22]

In fact, the FCC—under Minow and for years afterward—did not promote cable but aggressively moved to block it. The All-Channel Receiver Act did not produce net benefits for consumers and did not rescue UHF. Only after cable deregulation reversed Minow's policies in the late 1970s and early 1980s was a fourth broadcast network, Fox, able to form, in 1986. That was precisely because so many UHF stations were made viable via cable retransmission. The Spanish-language services Univision and Telemundo also became successful national networks as cable, unleashed, delivered new life for UHF.

Nor did the 1961 FCC launch satellite TV. Under Kennedy administration policy, satellite communications were bottled up in a monopoly enterprise dubbed Comsat, a consortium involving AT&T and the U.S. government. Its performance was poor and its prices astronomically high. When its exclusive franchise was ended, in the 1972 "Open Skies" policy,[23] Western Union, Hughes, RCA, GTE, and others rushed in to compete. Prices to transmit voice, data, and video plummeted. Whole new business models appeared. Brian Lamb, founder of C-SPAN, the first basic cable TV network, testifies that the market for new national TV programming opened because the new competition lowered video distribution costs by more than 95 percent.[24]

Here lurks a template for the creation of communications policy. Government identifies a problem and takes measures to address it. The market failure is misdiagnosed, and the regulatory "fix" reflects political bargains with powerful industry incumbents. Barriers to entry are created. The "public interest" is asserted. When the purported solution is finally abandoned, more robust market forces assert themselves. The government proudly claims that its policy was a success. The cycle repeats.

In an age when "memory is free," there is a haunting lack of institutional recall. FCC personnel pursue what they believe to be good policy. Many are quite knowledgeable about rules, radio engineering, or economics. But the administrative apparatus of the FCC reliably thwarts new and improved wireless technologies. Insiders dominate; incumbents prosper; entrants are stymied; outside observers are overwhelmed.

The pro-consumer bureaucrat receives little help. The press is as likely to advance an anticompetitive special pleader as to expose him. Deciphering the impact of entry barriers, which lobbyists are paid handsomely to drape in "syrupy talk about [licensees] acting in the public interest,"[25] is almost impossible for the full-time reporter. As longtime FCC official Sol Schildhause once commented about a policy matter of intense concern: "With all this talk about the 'public interest,' there must be a helluva lot of money at stake for someone."

In the 1970s, FCC Commissioner Glen O. Robinson colorfully described the way broadcast station license awards, for instance, wandered from due process to farce:

> There is less to such "contests" than meets the eye.... In fact it is ... a pretend game played between the Commission and the public. The outcome of the game is predetermined.... It rather resembles a professional wrestling match in which the contestants' grappling, throwing, thumping —with attendant grunts and groans—are mere dramatic conventions having little impact on the final result. Of course, wrestling fans know the result is fixed.[26]

Not all spectrum policy battles are as formulaic as broadcast license decisions. Some areas have seen distinct improvements: under a long-sought reform, finally authorized in 1993, the FCC now assigns wireless licenses via competitive bidding. In the two decades since Congress acted, wireless license auctions have raised (depending on pending sales) more than $100 billion for the U.S. Treasury and, more important, have cleansed and stream-lined the process for distributing spectrum rights. But traditional spectrum allocation continues, with regulators determining how bandwidth is used according to a legal apparatus that emerged nearly a century ago.

Things were (literally) quiet until "broadcasting" spawned hundreds of competing stations in the 1920s. Now transmitters needed to coordinate. Frequency rights were established on a first come, first served basis, echo-ing common law.

Almost instantly, however, incumbent broadcasters saw greater economic opportunity via rules to block additional competition. At the same time, policy makers like Secretary of Commerce Herbert Hoover, seeing that the emerging media would become a powerful platform for social influence, moved to assert control over stations and their programming. A bargain was struck. A coalition of broadcasters and policy makers, supported by a coterie of public advocates (ironically, perhaps, including the American Civil Liberties Union), crafted a new regime around the National Associa-tion of Broadcasters' 1925 argument that broadcasting was a privilege, not a right, and should be awarded according to the "public interest" as deter-mined by an administrative process.[27]

The reforms this coalition championed were enacted in the Radio Act, signed by President Coolidge on February 23, 1927. Broadcasters were re-quired to waive ownership rights in the "ether," and a new regulatory agency, the Federal Radio Commission (FRC), was created. Its five mem-bers, nominated by the president and confirmed by the Senate, would enjoy broad powers to define how radio waves could be used and who could use them. It would intimately manage a large and growing sector of the econ-omy and hence become the focus of intense political interest. In 1932, a Brookings Institution scholar wrote that "probably no quasi-judicial body was ever subject to so much Congressional pressure as the Federal Radio Commission."[28]

The FRC was an activist agency before being replaced in 1934 by the newly constituted Federal Communications Commission. The FRC's major contribution was to liquidate about a fourth of radio outlets, stations owned almost entirely by small businesses or nonprofit organizations like schools or labor unions. Their exit was hailed by regulators and major commercial broadcasters as a rationalization of the industry. That it carried ominous implications for consumer choice and First Amendment freedoms was barely noticed by the mainstream press or the general public.

No matter its infirmities, the spectrum allocation scheme was locked in. That these old institutions still today determine the bandwidth that wireless markets will have to work with is a crime against science. The 1927 regime is hugely overrestrictive, needlessly limiting both economic efficiency and free speech. But it lives on. Some scholars call this path dependency. Incumbents and policy makers call it the public interest. Communications lawyers call it a living.

1981's Toasters

In 1961, Newton Minow's "vast wasteland" paradigm established a vision of spectrum allocation that excites considerable enthusiasm to this day. Yet during the "deregulation wave" of the 1970s and 1980s, an opposing perspective was developed by economists like Alfred Kahn, Harvey Levin, Roger Noll, and Bruce Owen. In just two decades, the official view of regulation had spun around.

Odds are good that the most harshly condemned statement ever made by an FCC official appeared in a 1981 *Reason* magazine interview.[29] Mark Fowler, appointed to chair the Commission by President Ronald Reagan, laid out his vision.

> Why is it that we now single out one form—over-air television—and imbue it with specific social duties when we don't do the same for film, for example? Why is there this national obsession to tamper with this box of transistors and tubes when we don't do the same for *Time* magazine? Why don't we have a "fairness doctrine" for *Time* or the *Washington Post*, when we have one *Washington Post* in the city and seven television stations? . . . The television is just another appliance—it's a toaster with pictures.[30]

Condemnations were both swift and long-lasting. Fowler was a "knee jerk First Amendment purist,"[31] a "cynical corrupter."[32] His approach demonstrated "the brooding arrogance of capitalism."[33] More than thirty years later, the popular TV quiz show *Jeopardy!* featured the category "Mad Men." Under the $800 panel, the answer was: IN THE '80S GROWING BROADCAST EM- PIRES HAD HELP FROM THE "MAD MONK OF DEREGULATION," MARK FOWLER OF THIS COMMISSION. The correct response: "What is the FCC?"[34] In 2012, law profes- sor Susan Crawford introduced a discussion of mobile licenses by referring to Fowler's analogy, warning that his view had triggered a "deregulatory trend" that led to markets where "the consumer is toast."[35]

New York Times writer Peter J. Boyer summarized Fowler's tenure as a time when the "F.C.C. treated TV as commerce," noting that many experts "saw Mr. Fowler's vision as an abdication of the commission's responsibil- ity for the public airwaves."[36] The *Times* report was accurate: regulators, scholars, pundits, activists, industry lawyers, and many licensees were outraged at the idea that spectrum-based services were not unique. "Once everyone heard from Chairman Mark S. Fowler of the Reagan FCC that television was nothing more than a 'toaster with pictures,' there was little 'special' left."[37] Fowler trumpeted regularization. This presaged a radical pivot: wireless markets should be treated as standard economic institutions and competitive forces given broad rein.

To chart this reformist path it is important to review the road to spec- trum's special status. The idea that radio services did not conform to legal or economic norms was formalized in law under Justice Felix Frankfurter's key 1943 Supreme Court opinion in *NBC v. United States*.[38] The case re- solved a legal challenge to the FCC's Chain Broadcasting Rules brought by the largest such network, the National Broadcasting Company (NBC). Reg- ulators had imposed various limits on the agreements made between radio stations and networks that sold them programs. NBC argued that this ex- ceeded the statutory authority of the Commission, which was to police air- wave interference while leaving broadcasters free to create (or buy) their content. Moreover, to regulate content would abridge free speech, violating the First Amendment's prohibition.

NBC lost. It was a turning point in U.S. jurisprudence, pushing back constitutional limits and awarding wide discretion to government regula- tors to engineer markets.[39] *Technical reasons* formed the premise. Observing "certain basic facts" about radio stations, the Court wrote that "the radio spectrum is simply not large enough to accommodate everybody" who would like to transmit.[40] Government controls were thus not policy choices but imperatives dictated by science. "There is a fixed natural limitation upon the number of stations that can operate without interfering with one another."[41] To prevent pandemonium and to create the opportunity for *any* communications, policy makers had to privilege certain speakers and silence others. In this context it was found appropriate for the government to judge content: how else "could the Commission choose between two applicants for the same facilities?"[42]

Frankfurter posed the question rhetorically, but the answer was simple: by auctioning rights to the highest bidder. Scarcity, which occurs when not all demands can be met for free, is not unique to radio spectrum but is a general feature of valuable opportunities. When anarchy would lead to chaos, societies commonly limit resource access to mitigate conflicts. In successful modern economies, these restrictions give rise to rights that can be exchanged, allowing them to flow toward employments that satisfy the most pressing demands. Ownership accomplishes this by channeling competitive forces. Instead of "open access" to a pasture, yielding to a race to the bottom as ranchers graze their cattle until the last blade of grass is devoured,[43] property owners tend to conserve the resource to protect their wealth. Why not sell frequency rights and allow markets, unleashing competitive efficiencies, to mitigate conflicts?

There was extreme hostility directed toward those who advanced this simple logic in the 1950s and 1960s. Yet over time, even communications policy makers got the point. Markets could operate where spectrum use rights were defined and assigned. At no point did rules to police traffic require the government to control the content of broadcast stations, the technologies they deployed or the business models they used. "No one besides the Supreme Court actually believes the scarcity rationale," wrote law professor Jim Chen in 2005.[44]

But the fiction that spectrum was unique had allowed policy makers, and the judges who passed judgment over their actions, to set aside legal constraints. Ownership of radio waves was prohibited. Instead, valuable spectrum use rights were distributed by regulators, under the watchful eye of Congress and the White House, to private firms. The licenses were awarded without charge, and even better, they were restricted in number, protecting awardees from competition. Rivalry between licensees was also limited, as the state specified what services, technologies, and business models could be deployed.

Such governmental largesse for favored interests has not gone unrewarded. Licensees tend to comply not only with the explicit rules of the license but also with policy makers' implicit demands—"regulation by raised eyebrow." This has allowed officials to sidestep even the weak form of the First Amendment applied to broadcasting—and be happy to do so, as it enhances their clout. When, in 1969, *The Smothers Brothers Comedy Hour* joked about drug use and criticized U.S. military involvement in Vietnam, it drew criticism. But when the comics singled out for mockery Senator John O. Pastore (D-RI), the powerful head of the Senate Commerce Committee (overseeing the FCC)—he was awarded a "Flying Fickle Finger of Fate" by special guest Dan Rowan—the show never aired. Despite the series' popularity, CBS abruptly canceled it.[45] No government ruling ended the program. It did not have to.

Henry Geller, the FCC's general counsel in the 1960s, who became one of the top public advocates in communications,[46] by 1987 came to bemoan the fact that "the public trustee scheme" had failed to achieve its "goals after forty years of trying."[47] At that point, Geller proposed that in exchange for being released from the Fairness Doctrine, the Equal Time Rule, and requirements to provide educational programming for children, TV and radio broadcasters be required to pay an annual fee of 1 percent of station revenues, which would fund Public Broadcasting System and National Public Radio programs. While the Fairness Doctrine and the Equal Time Rule were effectively undone by subsequent reforms, Geller's more pointed deregulation went nowhere. Licensees opposed it, just as did regulatory activists. "A broadcaster loves to be considered a public interest figure," Geller observed. "Broadcasters generally want the economic benefits of being a public fiduciary without having to meet the burden of adhering to public interest content regulation. And they have great clout with Congress."[48] The same outcome obtained in 1998 when (to usher in new digital TV broadcasting) the ill-fated Gore Commission rejected Geller's reformist agenda, maintaining that the government should ambitiously promote "the public interest obligations of digital broadcasters." Alas, in 2008, members of that commission conceded that Geller had been right, that the highly touted "public interest obligations" had amounted to nothing.[49]

Dueling Metaphors

Minow and Fowler frame spectrum policy. Minow's "vast wasteland" is the paradigmatic embrace of administrative planning in the "public interest." Fowler's "toaster" is meant to relax mandates and release competitive rivalry. Minow engaged the challenge directly, calling the toaster model "ill-considered and uncaring" in 1995. "Fowler," he wrote, "inadvertently put his finger on America's greatest failing: no thinking adult would leave a small child unattended to play with a plugged in toaster for hours on end. . . . Yet because we have neglected to take seriously the Communications Act's 'public interest' requirement, leaving an unattended child in the company of a television set is just as dangerous."[50]

Setting aside the fact that leaving a toddler to watch television is not "just as dangerous" as a play date with a hot toaster, it is Minow who "inadvertently" clarifies the debate. Society relies on parents and guardians to keep children from hurting themselves with toasters. Government regulation to fully childproof all heat-bearing surfaces has not been the policy. Nor has regulation been effective in making television safe—as Minow himself complains. He argues that kids might as well watch the regulated product as play with a "plugged in toaster."

Alfred Kahn, the late dean of U.S. regulatory economists and a force of nature at Cornell University for more than six decades, learned that regulatory agencies are complex organizations and that the markets they seek to govern are often inscrutable. While chalkboard strategies cleanly chart straight lines from interventions to outcomes, actual results may vary. "Society's choices are always between or among imperfect systems," Kahn wrote. But given his experience in economics and government, he found that markets typically generate an irreplaceable dynamism. "Wherever it seems likely to be effective, even very imperfect competition is preferable to regulation."[51]

Let's cut to the chase: ordinary toasters have outperformed the special "public interest." Two historical comparisons paint the picture.

Television receivers v. cellular radios. In the two decades following enactment of the analog TV standard in 1941, imposed on the market by regulators, receiver quality advanced at a snail's pace. Yet in the two decades
following the issuance of cellular telephone licenses, which permitted technologies, devices, applications, and content to be selected by competing
networks,[52] cell phone design was shaken by waves of disruptive innovation
—from the "brick" car phones that sold for two thousand dollars in the mid-
1980s, to clamshell handsets in the mid-1990s, to BlackBerries, iPhones,
and Android-based smartphones in the 2000s, to the tablet market then
launched with the iPad in 2010.

The arc of progress across the differently regulated sectors has been
markedly distinct. Under the "public interest" standard, 1945 television receivers got a little bigger and a little better by 1965. Black and white became
color, and remote controls were introduced. Over a similar span, however,
mobile handsets exploded in functionality. The heavyset "car phone" became a mobile device carried in a shirt pocket, 1987 to 2007, delivering not
only voice calls but emails, texts, multimedia messages, tweets, electronic
books, music, video, machine-to-machine services, and broadband Internet
access. "Phones" were packed with alarm clocks, cameras, calculators, maps,
audio recorders and scanners. Thousands of wireless applications were developed, billions of copies downloaded.

In a now infamous 1980 forecast, consultants for AT&T predicted that
fewer than 1 million cell phones would be in use in the United States by
2000; the actual number exceeded 100 million.[53] By 2011, that number had
tripled: mobile phones were ubiquitous, and households were dumping
landline service. At the end of 2013, 41 percent of U.S. households were
wireless-only.[54] The toasters were smoking the competition.

Programming in broadcasting v. cable TV. Similar distinctions in progress
have emerged under rival regimes in television programming. Americans'
viewing choices over the first twenty years of "public interest" TV regulation actually declined, from four commercial networks in 1950 to just three
in 1970, after the demise of the DuMont Network in 1955. Once the government's anticable (and pro–monopoly satellite) rules were overturned in the
1970s, however, unregulated programming networks sprang to life. Video
choices expanded, bringing a cornucopia of new content into American liv-

ing rooms. From 1980, with the deregulation of cable, to 2000, nearly three hundred national TV networks were created. More than nine hundred existed in the United States in 2014. Toasters triumphant.

So if "public interest" spectrum regulation has gone the way of rabbit-ear antennas, why not sit back, tweet a radical political statement, and relax with a romantic comedy delivered wirelessly to your ninety-inch flat-screen via HBO or Netflix? Here's the hitch: while the political spectrum's mission has collapsed, the 1927 Radio Act continues to govern the allocation of radio spectrum. Save for finely targeted liberalizations that have worked to enable PANS, the vast bulk of wireless bandwidth remains walled off, dedicated to legacy services planned decades ago, often moribund, never efficient.

Yet the PANS must compete for bandwidth. Fertile frequency spaces remain virtually empty, while savvy incumbents and political operatives cook up new rationales for enhancing their rights by imposing artificial scarcity on newcomers. The process of obtaining spectrum rights forms a barrier to experimentation and entrepreneurship.

Regulators see this. "The current spectrum policy framework," wrote the FCC in 2010, "sometimes impedes the free flow of spectrum to its most highly valued uses."[55] Indeed, "sometimes" is "always" whenever (a) a traditional allocation is made, and (b) things change with respect to technology, market innovation, or consumer demand. Since the times are *always* changing, traditional rules are often problems. The FCC concludes: "Flexibility should be the norm."[56]

Flexible spectrum use allows competitors to innovate, no permission needed. It is the Toaster Model. Yet the Commission must explicitly grant flexibility under the old "public interest" standard; rigidity is still the default rule. Firms trying to escape the straitjacket confront a bureaucratic labyrinth, a well-lawyered gauntlet where parties wedded to the status quo are invited to raise questions regarding how the "public interest" might be affected by more robust rivalry. Questions are free for opponents to raise, but they are costly and time-consuming for innovators to answer. With defense easier than offense, the game is rigged. Pressure groups benefiting from existing allocations drag out proceedings. An industry publication, without irony, trumpeted a 2002 FCC decision in the assignment of 700 MHz licenses (where scheduled auctions were postponed seven times): "Spectrum Auction Delay Hits Fast Track."[57] The one regulatory action that can be implemented quickly is *in*action.

Spectrum allocation by political means is the devil's playground. It begins with a valuable resource that no one owns. Competing private interests organize to divert the asset to their benefit. Administrators don't know what, or whom, to believe. After all, the allocation decisions are complex and numerous—a myriad of choices exists within each choice. A wireless service may be supplied from a vast possible mix of inputs. More or less frequency space may be used, depending on what services are delivered and what radios, antennas, software, or processors are deployed. Network architectures vary. Communications options are relentlessly expanded by advances in frequency division, geographic division, time division, code division, and orthogonal frequency division multiplexing techniques. Upgrade technology or add expensive new cells, and a given network needs less bandwidth to deliver the same service. Radio frequency engineers, computer scientists, and software engineers are constantly enlarging, and thus complicating, the range of options. How to rank them depends on what the inputs cost and how the outcomes satisfy consumers. And because the investments made today create network services used years hence, there are deep uncertainties over future demands and evolving technologies.

This is a problem of economic optimization. Markets deal with such complexity via the trial-and-error revealed when decentralized enterprises compete for financial reward. Profit seekers reach out for all the help they can get: operations research, strategic business partners, state-of-the-art scientists, experienced marketers, savvy investors, entrepreneurial innovators, skilled workers, production chain experts. Risks are taken, payoffs received, bad gambles punished. The successful grow, and the failures go away.

Government administrators, meanwhile, stay put. They hold hearings and read comments. They ponder the choices before them, studiously in some cases, but do not own the spectrum allocations they make. If great products result and millions of consumers benefit, the gains accrue to others. A civil servant who expedites a spectrum allocation may improve economic efficiency and even advance free speech. But he or she cannot expect to personally capture economic returns. In fact, the hope is to realize *zero*, as decisions that threaten to shake the established order are those most often penalized, perhaps with banishment to a bureaucratic Siberia. The lesson is not lost in translation—which is why even FCC officials call the regime "command and control"[58] or "GOSPLAN."[59]

Marshall McLuhan's bon mot was "the medium is the message."[60] The transformative power of television on the culture was less about video content than about the way viewing habits changed social life. In U.S. spectrum allocation, the administrative structure is the verdict. Placing the burden of a "public interest" showing on competitive entrants and wireless disrupters surrounds the status quo with a piranha-infested moat. The system lets entrepreneurs gain access to radio spectrum only after they have submitted their business plans and technologies for federal review, revealing their plans to rivals, in a quest to gain approval of disinterested agency actors. These latter public servants will know far more about the services offered by incumbent firms, whose employees they enjoy mixing with at industry conferences, where they recall the fun they had while working together at the agency in years past.

It is not an entirely unintended consequence that organized interests will largely prevail, that newcomers will encounter stiff barriers, that progress will often be thwarted—and that not one citizen in a million will know the extent to which this has happened in the name of the "public interest."

Liberalization—Proof of Concept

Remarkably, productive spectrum reallocations have occasionally survived the gauntlet. Cellular licenses, after decades of protests from incumbent TV stations, were finally issued by lottery between 1984 and 1989. The new licenses allocated (some of the) bandwidth previously set aside for UHF TV channels 70–83. The TV stations on those frequencies were moved to vacant slots on lower channels (2–69). Through a spasmodic deregulatory process involving key FCC rulings in 1988[61] and 1994,[62] with a 1993 congressional statute tossed in (authorizing license auctions),[63] cellular operators were granted considerable scope to determine technologies, services, prices, network architectures, base station locations, and business models. Policies under both Republican and Democratic administrations eschewed either traditional broadcasting or "common carrier" rules, relying instead on market rivalry to serve the "public interest."

The reforms proved powerful. Advanced networks quickly evolved; innovative devices were developed; price and service experiments were tried. The 1998 Digital One Rate plan introduced by AT&T Wireless—instantly copied by its rivals—sent per-minute phone call prices plummeting; at more than forty cents in 1996, they were four cents by 2010. By then, of course, texts were replacing voice: the average U.S. teenage girl was sending more than four thousand a month. *OMG!!!* With the Apple App Store, Google Play, Twitter, Facebook, and a stream of others, exciting new ecosystems emerged, perched atop wireless networks. By 2008, the annual U.S. consumer gain for mobile voice alone (over and above the $140 billion paid in service fees) was estimated to be at least $260 billion[64]—just as the value of mobile data and video services was rapidly eclipsing that of voice.

This departure from traditional licenses to liberal frequency rights proved enormously successful. The FCC notes that "spectrum [is] the great enabler,"[65] which is both correct and reason for pause. The rules in place give the impression that airwaves are everywhere accounted for, that "the great enabler" is maxed out. This is an optical illusion created by a legal regime that continues to widely preempt market competition with regulatory fiat.

So stultifying are these institutional chains that in 1990 the *National Journal* reported that the FCC had just "handed out the last remaining substantial portion of prime radio waves."[66] The allocation was for a 6 MHz band set aside for airplane phone calls. The conventional wisdom, even among experts at the Commission, was that there would be no more bandwidth for anything "substantial." The "prime radio waves" were all taken.

In fact, even at the FCC's sluggish pace, "prime radio waves" available for mobile services increased from 50 MHz in 1990—when two cellular licenses in each market were allotted 25 MHz each—to 547 MHz by 2006. Impressive social gains resulted. But far more progress could have resulted had the transitions been smooth, routine, and more expansive. Reforms exist, tried and tested, that can achieve this, but they require switching out the comforting confines of the political spectrum, where decadelong food fights are the norm and many get paid by the scrum.

When the FCC proposed, in 2010, that 300 MHz more be allocated to flexible-use licenses and auctioned by 2015, with another 200 MHz made available by 2020, it aimed in the right direction. Yet it elected to manage the transition internally, using tools that had proven slow and cumbersome. Within two years, Blair Levin, who headed the FCC's National Broadband Plan Task Force, remarked that the U.S. was "moving backwards, not forwards" in spectrum. "It's clear we aren't going to get close to the 300 MHz by 2015 goal laid out in the Plan, never mind the longer-term goal of 500 MHz by 2020."[67] Time has proven him correct.

Alas, the FCC's Five-Year Plans have been hardly more successful than those in other command economies. In 1999, for instance, the FCC proposed doubling the spectrum available for mobile carriers, then about 170 MHz, over five years. The actual result: about 0 MHz.[68] The FCC still endeavors to make each allocation case-by-case, trusting technocrats to sort through claims and counterclaims about proposed uses, superior technologies, perceived airwave conflicts, band plans, and alternative programs.

What emerges is something like the present TV band, which sits on the mother lode of prime radio frequencies. Under the TV allocation table of 1952, forty-nine television channels continue to be set aside for over-the-air viewing all across America. But Elvis has left the building. Some 91 percent of households—and a higher percentage of actual TV watchers—pay monthly subscription fees to a cable or satellite TV provider to escape "free TV." Today, migration to the next platform is well under way "over the top," where consumers access video via broadband networks. The TV band stands as a historical landmark, celebrating a world two generations gone.

U.S. spectrum policy is reluctant to catch up. Economists long ago observed that, as a 1974 text put it, "Perhaps the most significant event in the history of television regulation was the creation of an artificial scarcity of VHF-TV licenses."[69] Decades later, the most significant event is the protection of an artificial *abundance* of broadcast TV licenses. Either way, innovators and consumers lose.

If this were the only way to organize airwave use, we would be stuck with such rigidities as a cost of going wireless. Lacking alternatives to regulatory fiat, the existing spectrum allocation regime would be efficient, the best of all possible worlds. But daring experiments have been launched in the United States and abroad. A better world exists.

As we saw, reforms in cellular have downloaded complex economic organization decisions to the market by issuing broad, unrestricted, generic
commercial mobile radio service (CMRS) licenses. CMRS authorizations
leave spectrum allocation choices to firms, a test of the thesis that chaos
was the natural consequence of the private marketplace. Instead, these rights
support billions of dollars in network infrastructure by operators, which
expand their spectrum footprints by taking on myriad partners via roaming
agreements and wholesale access deals, enabling networks to supply coverage across the United States and around the globe. These, in turn, support
the emergence of dense innovation ecosystems and a torrent of pretty amazing new stuff—smartphones, tablets, phablets, e-readers, mHealth monitors, MyFi hotspots, multimedia messaging, video streaming, broadband
data access, m2m, and mobile app stores—all without bureaucratic authorization. Social media have sprung up spontaneously. Networks have made
seamless transitions from 1G to 2G to 3G to 4G and are now embracing
5G—fifth-generation technology. New architectures have evolved, including small cells (covering just a household or small business), wi-fi offload
(integrating local fixed networks with mobile services), or those using array
antennas (dramatically increasing spectrum reuse within cells), with no
FCC reallocation required.

Yet at best, just 15 percent of the prime spectrum (below 4 GHz)[70] is allocated to such liberal licenses in the United States, which is among the more
liberal regimes worldwide. Most of the remaining bandwidth (and even
more in the frequencies above) lies idle, locked into decades-old services
or technologies dictated by long-dead lobbyists and bureaucrats.

Despite the unrivaled success of the toaster model, diversions lurk everywhere. For two decades, the prolific writer George Gilder has excited policy
makers with a vision of the end of "spectrum scarcity." In 1994, promising
abundance through smarter radios that no longer needed the coordination
of either government or entrepreneurs, Gilder called for a halt to the "spectrum auctions" then just getting under way. He touted the tiny Steinbrecher
Corporation, creator of a nifty little radio that used not just one frequency
but hundreds, crossing bands, sniffing out unoccupied spectrum. This, said
Gilder, was the wave of the future—literally—and it obviated the need for
lines denoting bands or their owners. Open access—use whatever frequency

space you like, no harm done—was the efficient paradigm. "This entire auction concept is tied to thousands of exclusive frequency licenses. It has no place for broadband radios that treat all frequencies alike and offer bandwidth on demand. . . . By the time the FCC gets around to selling its 1,500 shards of air, the air will have been radically changed by new technology." Gilder saw auctions as rooted in an obsolete "real estate paradigm for the spectrum," which lets a licensee obtain "a stretch of beach and build a wall" with which to exclude others. "The Steinbrecher system, by contrast, suggests a model not of a beach but of an ocean. You can no more lease electromagnetic waves than you can lease ocean waves. Enabled by new technology, this new model is suitable for an information superhighway in the sky."[71]

Actually, the U.S. government goes one better today, assigning offshore wind power rights by competitive bidding.[72] And alas, Steinbrecher—no longer in business—did not end spectrum scarcity. But another firm has arisen to host the "information superhighway in the sky." Qualcomm has prospered by developing a "spread spectrum" technology known as code division multiple access (CDMA). The technology is widely used in modern cellular networks, including those formed with the very mobile licenses assigned in the 1994 auction denounced by Gilder. CDMA leverages bandwidth to create additional communications capacity, powering the radios found in billions of 3G and 4G cell phones. That success is valuable owing to the scarcity of spectrum access, not its abundance. Decades after Gilder's pronouncement that bandwidth would become free, U.S. mobile licenses—which traded for 62¢ per person per MHz in 1995—sold for more: some $1.28 in 2008 and $2.21 in January 2015. Investors betting against the Steinbrecher vision are writing (or cashing) large checks.

It always seems to come as a surprise that great new technological opportunities beget fantastic new demands. The regulators of the 1980s, taking their cue from telecommunications industry experts, believed "car phones" would be a tiny niche market. "Commissioner Robert E. Lee," wrote cellular pioneer James Murray, Jr., "opposed a large allocation of radio spectrum, calling it a 'frivolous use' of the public airwaves to provide 'each automobile owner another status symbol—a telephone for the family car.' "[73] Carriers started out selling phones and giving away the service.[74] But the service quickly gravitated to mass-market adoption, and the operators ended up doing the reverse. Permitting markets to search for a better path upended predictions.

Ithiel de Sola Pool observed that "the main reason for the misuse of spectrum is that it is given away for free."[75] Close, but it is not the initial "free" assignment, per se, that results in waste. Rather, waste results from the pro forma restrictions such awards carry. When use of spectrum is locked in, rigid and unchangeable, that inflexibility eliminates opportunities beyond those envisioned by regulators. In this, the allocation eliminates opportunity costs—making the precious resource free to waste. A TV broadcaster uses spectrum for low-value purposes because its license prohibits it from doing other things. If it owned the resource and could divert it to more useful employments, it would behave differently. Broadcasters are accused of squandering "their spectrum," but the squandering occurs because it is not theirs to use as they determine. Just as they distribute the vast bulk of their video programming (like ABC-owned ESPN, ESPN2, and ESPN3) not via their TV stations but over cable, telco, satellite, or broadband, they would continue to make smart choices with the assets they own, freeing up valuable airwaves to capture greater value.

At the heart of the spectrum mismanagement solution is not a hapless regulator who should be disciplined but an institutional error that can be fixed. MIT economist Nancy Rose, channeling Fred Kahn in the *American Economic Review*, notes: "The policy tradeoff is not between imperfect markets and perfect regulation, but which imperfection—market or regulatory—is less costly. This conclusion . . . is stunningly overlooked in discussions that presume one simply needs 'better regulation' or 'better-intentioned' regulators to costlessly correct market failures."[76]

Regulators are confined to a system featuring a design defect. Given latitude to craft any policy in the "public interest," they are tied in knots by interest group politics, each constituency (and its agents) striving for favorable economic outcomes. This is not a new phenomenon, a conspiracy of the Baby Bells, a failure by TV broadcasters to face new realities, or an outcome of the polarization in Congress. It is the result of a political structure crafted by Republican Commerce Secretary Herbert C. Hoover and Democratic U.S. Senator Clarence C. Dill in 1927. It was designed to increase administrative discretion, and it did. That jurisdiction was designed to accommodate political realities, most often flowing through Congress, and it did. Newton Minow once remarked, "When I was Chairman, I heard from the Congress about as frequently as television commercials flash across the television screen."[77] Members of Congress may not know how to successfully allocate radio spectrum, but they are very good at allocating their time.

While episodes of liberalization have chipped away at the regulatory edifice, the core of the case-by-case spectrum allocation process remains. It has produced a quiet crisis in public policy. At the moment when tentative deregulatory actions have produced bountiful results, and the imagination of entrepreneurs promises far further great leaps forward, radio spectrum constraints remain fixed in the work of century-old champions of the administrative state. Markets are keen to turn bandwidth into gold, but each frequency nugget takes perhaps a decade for regulators to make ready—counting just the successful attempts. To remedy the bottleneck in radio spectrum, we must modernize the technology of control.

2
Etheric Bedlam

The United States has for one hundred years led the world in the quality of electronic communications services enjoyed by the public. That might suggest that in the big scheme of things our regulatory policies have done well. Before you rush to that conclusion, though, you might want to consider Otto von Bismarck's reputed quip that "God has a special providence for fools, drunks, and the United States of America," which suggests not that regulation has done well, but that with God's help we don't need the FCC's.
—Former FCC member Glen O. Robinson

ACCORDING TO STANDARD TEXTS, the allocation of radio spectrum mandated by the Radio Act of 1927 was a matter of simple necessity. "Before 1927," wrote the U.S. Supreme Court, "the allocation of frequencies was left entirely to the private sector . . . and the result was chaos." The physics of radio frequencies and the dire consequences of interference in early broadcasts made an ordinary marketplace impossible, and radio regulation under central administrative direction was the only feasible path. "Without government control, the medium would be of little use because of the cacaphony [sic] of competing voices."[1]

This narrative has enabled the state to pervasively manage wireless markets, directing not only technology choices and business decisions but licensees' speech. Yet it is not just the spelling of *cacophony* that the Supreme

Court got wrong. Each of its assertions about the origins of broadcast regulation is demonstrably false.

For nearly seven decades this reasoning blocked market-oriented reforms. The FCC awarded transmission rights inefficiently, unduly constraining competition while encouraging socially wasteful and democratically dubious lobbying. When former FCC Chairman Charles Ferris testified to Congress in 1987 in support of the Fairness Doctrine, regulating news programs on radio and TV stations, he made handy use of the "etheric bedlam"[2] argument: "The public interest standard and the concept of broadcasters as public trustees date back to the origins of broadcasting. Back then, anyone who could put up a transmitter could broadcast, and the result was chaos."[3]

Testifying at the same hearing, Newton Minow chimed in:

Charlie Ferris said something very important and I want to amplify it. We all forget history. Why is broadcasting regulated in the first place? It started out unregulated, and then when broadcasters realized that they had to have the exclusive right to a channel because all the public was getting was static, then broadcasters came to Washington, and they went to the then Secretary of Commerce, Herbert Hoover, and they said: Mr. Hoover, you have got to do something to regulate us, you have got to do something so the public can hear the radio.[4]

When, in 1995, a suggestion to abolish the FCC was floated following the Republican takeover of Congress, the reflexive response came quickly from an attorney for a law firm representing broadcast companies:

Critics of the FCC . . . argue that virtually all spectrum should be auctioned to the highest bidders who will use it for its "best" purpose . . . Some immediate problems come to mind that even Calvin Coolidge and Herbert Hoover (not noted champions of big government) foresaw when they launched government spectrum management. Who would set and police interference curbs . . . ?[5]

Perhaps Friedrich Nietzsche was exaggerating just a tad when he postulated: "what everybody believes is never true."[6] But it is correct that the popularity of an argument does not constitute evidence. The notion that wireless markets in the 1920s were governed by anarchy, and that they proved unmanageable without the tight administrative controls imposed by Congress in 1927, is close to pure folklore.

The 1927 Radio Act, which still governs U.S. spectrum allocation, was the vision of Secretary of Commerce Herbert C. Hoover. The future President needed several years, a strategic use of state power, and broad support from incumbent radio broadcasters to push the statute through. The chaos and confusion that supposedly made strict regulation necessary were limited to a specific interval—July 9, 1926, to February 23, 1927. They were triggered by Hoover's own actions and formed a key part of his legislative quest. In effect, he created a problem in order to solve it, with new tools that afforded policy makers far broader regulatory discretion than was needed to restore order.

Radio broadcasting began its meteoric rise in 1920–1926 under common-law property rules that went by several names, including *priority in use* and *right of user*. These rights were defined and enforced by the U.S. Department of Commerce, operating under the Radio Act of 1912.[7] They supported the creation of hundreds of stations, encouraged millions of households to buy (or build) expensive radio receivers, and ushered in a national craze over the world's first electronic mass media technology.

While experimental broadcasts had appeared sporadically since at least 1908, AM radio broadcasting is generally said to have been launched on Tuesday, November 2, 1920, when Pittsburgh station KDKA began regular transmission of a daily program lineup. The station was owned by Westinghouse, and its business purpose was to generate demand for Westinghouse radios. Both the broadcast transmissions and the customer hardware spread like wildfire. Between November 1920 and December 1922, more than five hundred broadcast stations sprang up.

The Radio Act gave the Commerce Department the power to issue permits to wireless stations so as to "minimize interference." In 1920 such conflicts had been modest. Radio communications were point-to-point, connecting two radios in two different places. This worked the way microwave links do today, say, when an offshore drilling rig signals back and forth with a base station on the mainland. With wireless applications confined to one-on-one exchanges, there were few users and they generated relatively little traffic. Radio deployments encountered few interference problems.

The rise of five hundred broadcasters changed that. These operations blasted high wattage signals from elevated towers to enable reception far and wide. One station's broadcast could puzzle the receivers of listeners tuned to another station's signal. The Commerce Department responded by setting limits on access to the "ether."

This it did in at least two ways. The first was to designate bands for radio broadcasting. From 1920 to 1923, broadcasters were limited to two AM frequencies. Time-sharing agreements were common, and emission levels were restricted. In 1923, the Commerce Department expanded the number of frequencies to seventy, and in 1924, to eighty-nine channels between 550 KHz and 1500 KHz—essentially today's AM dial.

The second policy was a priority-in-use rule for license assignments. The Commerce Department gave preference to stations that had been broadcasting the longest. This reflected a well-established principle of common law, that when a resource is effectively utilized in a socially useful way, a right is acquired against those who might later encroach.[8]

These rules of the road allowed an impressive new industry to form. Millions of dollars were invested to construct stations; programs were developed; enormous audiences formed despite the relatively high cost of receivers.

It was not the case that "anyone who could put up a transmitter could broadcast." Rights to broadcast were sufficiently clear that a brisk secondary market developed, of which government officials were well aware. In Senate hearings on February 26–27, 1926, the solicitor general of the Department of Commerce, Stephen Davis, testified about the issue.

SENATOR BURTON WHEELER: I want to get that clear. Supposing I have a wavelength and sell it to you, I do not sell you my permit. They have got to come to the department and get their permit or else their permit is not any good to me.
SENATOR ROBERT HOWELL: Yes, but the practice is to transfer that permit with the apparatus.
SENATOR WHEELER: Of course, they are not bound to do that.
SENATOR HOWELL: No, but that is the practice.
MR. DAVIS: The practical situation is as the Senator says—the wave-lengths today are taken and used and occupied. . . . The Senator is correct in say-

ing that we have . . . recognized transfers of that sort. In other words, we recognize the purchaser as stepping into the shoes of the licensee.[9]

In 1924 the *Chicago Tribune* bought a radio license for $50,000 and changed its call letters to WGN ("World's Greatest Newspaper").[10] Two years later, market prices for station permits in Chicago had climbed to $250,000.[11] No investor would pay such serious money in a market open to interlopers without limit.

Perhaps the most colorful proof came 2,669 miles from Washington, DC. The Reverend Aimee Semple McPherson, who preached out of the Four Square Gospel Church in Los Angeles,[12] created the spectacular Angelus Temple in 1923, seating 5,300[13] and costing $1.2 million. (One observer noted that McPherson "put the *cost* in Pentecost.")[14] She pioneered the use of airwaves for Christian evangelism. Upon learning that there were 200,000 radio receivers within 100 miles of downtown Los Angeles, she had her church purchase a local radio station for $25,000, renaming it KFSG—Kall Four Square Gospel. The station took to the air in February 1924 with the song "Give the Winds a Mighty Voice, Jesus Saves!"[15]

The winds did indeed prove mighty, pushing the station's signal from its assigned frequency. This prompted Secretary Hoover, after several warnings, to order KFSG off the air. He was soon greeted by a telegram:

TO SECRETARY OF COMMERCE HERBERT HOOVER:
PLEASE ORDER YOUR MINIONS OF SATAN TO LEAVE MY STATION ALONE. STOP. YOU CANNOT EXPECT THE ALMIGHTY TO ABIDE BY YOUR WAVE LENGTH NONSENSE. STOP. WHEN I OFFER MY PRAYERS TO HIM I MUST FIT INTO HIS RECEPTION. STOP. OPEN THE STATION AT ONCE. STOP.
                    AIMEE SEMPLE MCPHERSON[16]

The station was eventually allowed to resume broadcasting on its assigned channel. For the next two decades, until her death in 1944, Sister Aimee continued to defy the minions of Satan, making national headlines with a thirty-six-day disappearance in 1926 (which ended with a tale of abduction by Mexican bandits), accusations of an extramarital affair, and two inconclusive L.A. prosecutions. But, of moment for the history of spectrum allocation, her radio transmissions were constrained by law *prior* to the Radio Act of 1927.

The system worked to encourage industry growth. But Hoover, writing in early 1923, saw "two distinct problems." The first was "a question of traffic control." That function was being achieved via the authority already granted. Hence Hoover's plea of an "urgent need for radio legislation" depended on the second governmental interest: "the determination of who shall use the traffic channels and under what conditions."[17]

Hoover sought to leverage the government's traffic cop role to obtain political control. He did not, at first, succeed. Radio was booming, the technology craze of the 1920s. "One of our troubles in getting legislation," he wrote, "was the very success of the voluntary system we had created. Members of the Congressional committees kept saying, 'it is working well, so why bother?' A long period of delay ensued."[18]

Hoover's reference to a "voluntary system" was rhetorical garnish. As members of the Four Square Gospel Church could attest, the Commerce Department had the power to enforce priority-in-use broadcasting rights. There was surely ample cooperation between the "radio men" and Commerce in crafting these rules; annual radio conferences were held by the department from 1922 to 1925 to foster coordination. As Hoover said in greeting the 1925 gathering, "We have . . . developed, in these conferences, traffic systems by which a vastly increasing number of messages are kept upon the air without destroying each other."[19]

Hoover was, however, dissatisfied with this equilibrium and invited a court challenge to his regulatory powers. Zenith, a broadcasting station, obliged, transmitting its signal over a frequency reserved for Canadian broadcasting; the Commerce Department dutifully enjoined the action. When Zenith refused to stand down, a federal district court found in April 1926 that the department was powerless to enforce its order.[20] The ruling clearly conflicted with *Hoover v. Intercity Radio Co.,* a 1923 federal court verdict that upheld Commerce's authority to enforce frequency and time separation limits.[21]

In July 1926, however, in a stunning move, Hoover announced that he would not embrace the *Intercity* ruling or appeal *Zenith,* but would instead abandon Commerce's powers. The strategic position was framed by a ruling on July 8 by a fellow cabinet official, Acting Attorney General (and Hoover protégé) William J. Donovan. "Wild Bill" Donovan, who became famous in World War II as director of the Office of Strategic Services, precursor of the

Central Intelligence Agency, issued a legal opinion that the *Zenith* ruling was correct and *Intercity* faulty. The opinion had been requested by Hoover and could not have surprised him. The following day, Commerce issued a well-publicized statement that it could no longer police the airwaves.

This breathed new life into Hoover's agenda. The roughly 550 stations on the air were soon joined by 200 more. Many jumped channels. Conflicts spread, annoying listeners. Meanwhile, Commerce did nothing. Annual radio conferences had been held by the department since 1922; there was none in 1926. Senator Clarence C. Dill of Washington noted that Hoover's actions seemed "almost like an invitation to broadcasters to do their worst."[22] The press commented that the policy "tended to fulfill the Secretary's gloomy prophecy about chaos."[23] Calls for additional legislation to bring "trespassers," "pirates," and "wave-jumpers" to heel became urgent. Then, in November, the broadcast regulation question was set afire in a Chicago courtroom.

In *Tribune Company v. Oak Leaves Broadcasting Station*[24] the *Chicago Tribune* asked a state court to issue an injunction against a station that had jumped to within 40 kilocycles of WGN's frequency in September 1926. Chancellor Francis S. Wilson, a Cook County magistrate, considered the request "new and novel" but not entirely lacking in precedent. He found that wireless activity had commenced under property rules: "Wave lengths have been bought and sold . . . stations have contracted with each other so as to broadcast without conflicting. . . . The public has itself become educated . . . to obtain certain particular items of news, speeches, or programs over its own particular sets."[25]

The WGN complaint was that radio interference from the Oak Leaves station trespassed, encroaching upon its property. Chancellor Wilson agreed, finding that WGN had "carved out for itself a particular right . . . in considerable cost in money and considerable time in pioneering." The judge ordered the Oak Leaves station to respect WGN's place on the dial, locating no closer than 50 kilocycles. While one state court case could not definitively decide the spectrum property issue nationwide, no matter the claims of a triumphant front-page headline in the *Tribune,* the first such courtroom encounter heralded a solution: enforcing priority-in-use rules. No less than the solicitor general of the Commerce Department "contended that a ruling following up this decision in a higher court would protect businessmen against wavelength piracy."[26] The statement was both an obvious recital of the facts and a pointed warning to policy makers: time for a political solution was running out.

The case did, indeed, catch the attention of Congress. Senator Dill inserted the Cook County decision in the *Congressional Record* on December 10, 1926. (The Illinois court did not publish the opinion. It is available to students of the law due to Senator Dill's action.) Congressman William H. White of Maine, who had carried Hoover's preferred radio legislation in the House since 1922, described the effect: "Some of us have . . . believed that in the absence of legislation by Congress it was inevitable that the courts of the country sooner or later . . . would determine, as they have determined, that priority in point of time in the use of a wave length established a priority of right. That was precisely what was determined in this famous Chicago Tribune case."[27]

Now Congress acted. An emergency measure, introduced and passed just days after the ruling, mandated that all wireless operators immediately waive any vested rights in frequencies; failure to do so would result in license termination. This provision was then included in the Radio Act,[28] a compromise being cobbled together from separate House and Senate bills. It provided for allocation of wireless licenses according to "public interest, convenience or necessity" by a newly constituted Federal Radio Commission, and was signed into law by President Calvin Coolidge on February 23, 1927.

"We all forget history," Minow warns. Indeed. While the traditional story is that radio broadcasting began wholly "unregulated," developing a burgeoning "chaos" that required licensing in the "public interest," the actual events were markedly distinct. The radio industry launched under de facto property rules, and was disrupted when the agency enforcing these rules withdrew. The predictable confusion over rights could have been remedied by renewed Commerce Department enforcement or by the courts; either path could have been supported by legislation. Instead, a coalition formed to advance a more political system featuring far greater government control. The resulting barriers to entry created a windfall for existing station owners, while policy makers gained power over emerging media. Preempting private ownership of spectrum was key to the bargain. Senator Dill, a Democrat and the author of the 1927 Radio Act, explained this in his 1938 book *Radio Law*:

> Why Congress Became Aroused on Subject
>
> The development of these claims of vested rights in radio frequencies has caused many members of Congress to fear that this one and only remaining public domain in the form of free radio communication might soon be lost unless Congress protected it by legislation. It caused renewed demand for the assertion of full sovereignty over radio by Congress. . . .
>
> The purpose of Congress from the beginning of consideration of legislation concerning broadcasting was to prevent private ownership of wave lengths or vested rights of any kind in the use of radio transmitting apparatus.[29]

The broadcast license bargain of 1927 did not suffer the misunderstandings common among later commentators. The official U.S. government version, per the First Annual Report of the Federal Radio Commission, explains nicely:

> We have had about six years of radio broadcasting. It was in 1921 that the first station (KDKA) started operating, and soon grew in popularity, sales mounted, and a great new industry was in the making. Then something happened.
>
> In July, 1926, just 10 months ago, the Attorney General of the United States rendered his famous opinion that the Secretary of Commerce, under the radio law of 1912, was without power to control the broadcasting situation or to assign wave lengths. Thus, after five years of orderly development, control was off. Beginning with August, 1926, anarchy reigned in the ether.[30]

Federal policy makers understood that radio broadcasting had well established itself years "before 1927 . . . and the result," despite what a Supreme Court majority thought forty years later, was not chaos.[31] There existed a "breakdown of the law," as it was called at the time,[32] caused by the end of legal enforcement within a system of "orderly development." There was no existential collapse of a system that could not be salvaged except by a "comprehensive scheme of control over radio communication."[33] In fact, Senator

Dill described the move from property rules to administrative allocation as a departure from standard practice:

> It is interesting to note that some of the long established principles of law were not applied to radio. In fact, the ratio statute specifically denies the application of a number of such principles. The most important of these which the radio statute sets aside is the principle of acquiring a certain property right by user.
>
> It is a long and established principle of law that if a citizen openly and adversely possesses and uses property for a long period of time without opposition, or without contest, he acquires title by adverse possession. This is known as property by right of user. Congress wrote into the radio law the provision that user should [have] no effect upon the right of the Commission to provide for the use of any wave length by a new and different person if the public interest would thereby be served.[34]

The support of the major commercial broadcasters was so central to this reform that the term *capture* seems understated; the regulatory structure that evolved was from the very beginning concocted by incumbent interests. Senator Dill, who knew the origins of the 1927 Radio Act as its author, noted that "broadcasters themselves suggested the inclusion of the words 'public interest' in the law as a basis for granting licenses . . . by a resolution which the National Association of Broadcasters passed in 1925."[35] From the NAB's lips to government statute.

This approach fortified the rights of large existing stations while appropriating broadcasting rights for entrants—a double bonus. In promoting the 1927 legislation, industry advocates universally stressed "public interest" and incumbents began lobbying against entry even before the Radio Act was passed. As reported after the 1925 Radio Conference, the industry concluded that "there should be a limit upon the total number of broadcasting stations, and this limit can be fixed and maintained only by Federal authority."[36] Some radio interests went much farther: "A spokesman for Westinghouse, which already had four stations on the air, expressed the view that fifteen stations would serve the whole country adequately. For him, the purpose of regulation would presumably be to stop in their tracks the hundreds of new stations about to invade the air."[37]

At the time, 1922, there were more than five hundred stations on the air.

The proposal to reduce all U.S. broadcasting by 97 percent was, of course, but the fantasy of a wishful industry lobbyist. Yet with the advent of the Federal Radio Commission in 1927, the growth of radio stations—otherwise accommodated by the rush of technology and the wild embrace of a receptive public—was halted. The official determination was that less broadcasting competition was demanded, not more.

3

Protection by Subtraction

Taylor Branch has divided government agencies into two categories: "deliver the mail" and "Holy Grail." "Deliver the mail" agencies perform neutral, mechanical, logistical functions; they send out Social Security checks, procure supplies—or deliver the mail. "Holy Grail" agencies, on the other hand, are given the more controversial and difficult role of achieving some grand, moral, civilizing goal. The Federal Radio Commission came into being primarily to "deliver the mail"—to act as a traffic cop of the airwaves. But both the FRC and FCC had a vague Holy Grail clause written into their charters: the requirement that they uphold the "public interest, convenience and necessity."

—Erwin G. Krasnow, Lawrence D. Longley, and Herbert A. Terry

AN INDEPENDENT AGENCY, the Federal Radio Commission was composed of five members, nominated by the president and confirmed by the Senate. These appointees would determine which organizations would enjoy the right to broadcast, on what frequencies, at what physical locations and power levels, and under what rules. The previous scheme had relied on first-in-time common-law property traditions, "pioneering rights," as Chancellor Francis Wilson described them in his *Oak Leaves* opinion, or "right of user," in Senator Dill's framing. The new scheme relied on three votes. In the process, a "right" was transformed into a "privilege." A license was a special grant determined by politically appointed individuals, whose preferences the supplicants receiving those grants were required to understand.

The short-lived FRC provided a natural experiment on the impact of regulation. The industry launched in 1920, and on July 1, 1926, there were 536 radio stations broadcasting in America.[1] Then came Secretary Hoover's announcement, on July 9, 1926, that the federal government would no longer enforce wireless rights. On January 15, 1927, *Radio Broadcast* magazine reported the following:

> 181 new stations were operating;
> 148 stations were being built;
> 280 stations were being planned;
> 150 stations had increased power;
> 70 stations had requested higher power;
> 104 stations had changed wavelength.[2]

By February 1927, the number of operating stations had officially risen to 733.[3] Despite great progress in technology and the rapid spread of AM radios, with $170 million invested in receiver sets in 1925 (up 215 percent over 1923),[4] the market was widely perceived as overcrowded, with listeners experiencing too-frequent static interference. In addition, broadcast stations were in a state of legal limbo.

The FRC's initial action was to order broadcast stations to return to their authorized frequencies.[5] Then, on April 5, 1927, it dealt with the first substantive reform ideas, the most important of which was to expand the "broadcasting frequency band." AM frequencies extended from 550 to 1500 KHz; the proposal was that the top end be raised to 2000 KHz. This would have made room for about half again as many stations, sufficient to accommodate each one then on the air. The FRC rejected this plan because of "the manifest inconvenience to the listening public which would result."[6]

The "manifest inconvenience" was asserted to be that consumers would be forced to buy new receiver sets, but that was clearly not the case. The existing band would continue to host broadcasts, and existing receivers would continue to tune them in, with reception presumably much improved. Augmenting the AM band would additionally create the option for customers to buy upgraded sets that played audio on both the old and new frequencies. The rejection of band enhancement did not protect listeners, but rather deprived them of a wider array of radio programs. A second proposal would have reduced the frequency separation of assigned slots and thus have created more channels to host more competitors, no new radios necessary. This, too, was rejected.

Major broadcasting incumbents supported the argument of user inconvenience, denouncing "short-sighted would-be broadcasters and selfish set manufacturers."[7] The "would-be broadcasters" were out of luck. FRC hearings on the policy approach garnered "little criticism" and included "the opinion of one Department of Commerce official that 'the success of radio broadcasting lay in doing away with small and unimportant stations.' "[8]

The political spectrum had made a grand appearance. Less than two months into its existence, the FRC had done the bidding of influential corporate interests, exacerbating scarcity and attributing the outcomes (inevitably) to technical reasons.

Still, broadcasting companies feared "such dangerous propositions as the pressure to extend the broadcast band; . . . the fatuous claims of the more recently licensed stations to a place in the ether; and the uneconomic proposals to split time on the air rather than eliminate stations wholesale."[9] They need not have worried. Proposals for additional spectrum, time-sharing, and reduced separation between assigned channels went nowhere. As *Radio Broadcast* magazine noted in June 1927, "Broadening of the band was disposed of with a finality which leaves little hope for a revival of that pernicious proposition. . . . The commissioners were convinced that less stations was the only answer [*sic*]."[10]

Congress had little problem with this approach, but was disgruntled over the geographic distribution of radio stations. The Radio Act required that there be "fair, efficient and equitable radio service" for each state. Senators and representatives from the South and West thought the Northeast was receiving far more than its share. Congressman E. L. Davis (D-TN) won legislation in March 1928 dividing the United States into five zones and mandating that the FRC assign licenses equally across them.

In response, the FRC issued two proclamations. The first, dated May 25, 1928, was General Order 32. It informed 164 stations that their renewal applications were not persuasive, and that they would have to demonstrate their contribution to the "public interest" at Washington, DC hearings starting on July 9, 1928. Of the 164, some 110 sent representatives; at least 90 percent of these stations secured renewals. Those that could not afford to defend their licenses mostly lost them. The nonrenewals did not result in more stations in other regions.

General Order 40, issued in August 1928, reconfigured broadcasting rights for 94 percent of stations. It established forty clear channels, where high power could be used, and thirty-four regional channels, where stations were restricted to lower power levels. The remaining fifteen AM channels were for local stations to broadcast with still lower power. To fit stations into this grid, the FRC imposed reductions in coverage areas (via power limits) and mandated time-sharing. A few stations received exemptions, mostly those owned by or affiliated with the two national radio networks, NBC and CBS. But the rules were a disaster for more than one hundred smaller stations that were off the air within a year.[11] Stations owned by nonprofit institutions, Robert McChesney noted, were especially hard hit:

> The number of broadcasting stations affiliated with colleges and universities declined from ninety-five in 1927 to less than half that figure in 1930. The number of overall nonprofit broadcasters would decline from over 200 in 1927 to some sixty-five in 1934, almost all of which were marginal in terms of power and impact. By 1934 nonprofit broadcasting accounted for only 2 percent of total U.S. broadcasting time. For most Americans, it did not exist.[12]

The FRC seldom revoked licenses or explicitly denied renewals. Instead it promulgated administrative rules, generally dressed up as engineering edicts. The decision to freeze the AM band, preserving what had been established by the Department of Commerce in 1924, was an example. This ruling reflected the economic view that the "public interest" would not be served by using bandwidth for one thing (radio broadcasting) rather than something else (mostly, idle spectrum).[13] These trade-offs were not determined via market rivalry, where consumers gravitated to the technologies or services they valued most. The selections were state diktat.

It is informative that nonprofit radio broadcasters had gotten off to such a promising start under the common-law property rules, only to be crushed when the "public interest" was formally promoted. McChesney was perplexed that "some still characterized the Radio Act of 1927 as some sort of 'progressive victory' that was 'passed in the best interest of the citizenry.'"[14] No matter the popular perceptions, the audiences drawn by the strongest stations, owned by enterprises such as RCA and Westinghouse, climbed substantially both absolutely (even given the impact of the Depression) and relative to smaller stations. The *Harvard Business Review* summarized the situation in 1935:

> The point seems clear that the Federal Radio Commission has interpreted the concept of public interest so as to favor in actual practice one particular group. While talking in terms of public interest, convenience and necessity the Commission actually chose to further the ends of the commercial broadcasters. They form the substantive content of public interest as interpreted by the Commission.[15]

Program Choices

The metaphorical "traffic cop" coordinates flows without regard to the composition of the vehicles or their cargo. But the new spectrum regulators were hardly agnostic about what should be broadcast.

In its first year the Federal Radio Commission was considering the relative worth of "conflicting claims of grand opera and religious services, of market reports and direct advertising, of jazz orchestras and lectures."[16] By 1929 the FRC was evaluating license renewals based upon the "fairness" of the political material aired by the station.[17] Irresponsible advertisements and vulgar speech were attacked in ways "which would have been clearly unconstitutional if applied to the printed media."[18] Ithiel Pool found it of interest that

> a commission attorney [in 1930] argued that if the commission had jurisdiction to prevent interference in radio reception, it surely had authority to protect the public from "influences of a more dangerous kind. . . . Control for one purpose and not for the other is not in harmony with the avowed intention of Congress to regulate radio communication for the best interests

of the many. It thus becomes imperative for the commission to be guided
by a station's past program record."[19]

The FRC chose to favor stations on the basis of program quality, expressing a strong preference for outlets featuring popular shows with large, diverse audiences. Stations broadcasting a particular point of view were derided as "propaganda stations" and ranked lower. "Numerous non-profit
stations would fall victim to this logic and see their hours reduced and the
time turned over to capitalist broadcasters, often affiliated with one of the
two networks."[20] Political speech was not protected but targeted. Competition was not nurtured but preempted.

And the FRC had no trouble finding stations that, in their eyes, did not
deserve to stay on the air.

Dr. John R. Brinkley, dubbed the "goat gland doctor,"[21] took to the airwaves in Kansas to publicize his special therapy, which involved implanting
the gonads of a "young Ozark goat" into a man's private area to rejuvenate
his virility. Brinkley billed $750 for the service and let the patient bring his
own goat (charging something like a restaurant corking fee). Hundreds
lined up for the surgery; in 1928, his Kansas hospital grossed $150,000.
Thousands more listened to his health tips over the air, where he expanded
his focus to include prostate enlargement and ailments afflicting women.
He gained a vast audience and was nearly elected governor of Kansas in
1930 as a write-in—and would have won had some fifty thousand ballots
not been disqualified for misspellings.

Alas, Dr. Brinkley's application for renewal of his radio license was
denied. Running to Mexico, he broadcast to his U.S. listeners via a high-
powered station just across the border from Del Rio, Texas, until U.S. regulators persuaded the Mexican government to revoke that license, too.

With sharper political overtones, the Reverend Bob Shuler ran afoul of
FRC licensors. A popular Los Angeles radio evangelist who railed against
corrupt local politicians, "Fighting Bob"—a "scrapper for God"—acquired
a large and dedicated following. "Back in the Tennessee Mountains, where
I come from," he thundered, "they would not allow city doctors to strip young
girls seeking restaurant employment." Given that physical exams shocked
and offended him, he had countless objects of disgust in the City of Angels

to choose from. Radio station KCEF, purchased for $25,000 in 1926 by Shuler's Trinity Methodist Church, ultimately reached a reported 600,000 listeners.[22]

"Fighting Bob" railed against the L.A. establishment, charging that "the mayor let a gangster run the city" and had rerouted a public street to connect to the gangster's property. The "chief of police . . . allow[ed] commercialized vice to flourish," while "the police framed the head of the Morals Efficiency Association and killed a woman to cover the frame-up." Shuler once took to the air to declare that should a prominent, unnamed local citizen not give him $100, "I will go on the air next Tuesday night and tell what I know about him." After the church enjoyed an immediate surge in donations, it was revealed that the apparent extortion was an inside joke aimed at a congregant aware of its harmless intent. The church kept the windfall.

When Shuler sought to renew KCEF's radio license in 1930, he filed the application with characteristic gusto, bragging that the station had "thrown the pitiless spotlight of publicity on corrupt officials, and on agencies of immorality, thereby gladly gaining their enmity and open threats to bring pressure to bear to 'get' this station's license." Truer boasts have seldom been made.

The Federal Radio Commission scheduled the renewal for a hearing, the first volley targeting licensees on the edge. Renewals without hearings are cheap; renewals with hearings require counsel who charge by the hour. And a pro forma decision becomes one fraught with risk. In this instance, the FRC conducted a sixteen-day trial in which the parties opposing KCEF's renewal were formally represented by "a former city prosecutor whom Shuler had driven from office."[23]

If Shuler had been a sensational voice on the radio, he also was extremely popular with the general public, and much of what he said turned out to be true. He really had ousted a chief of police and forced a mayor to duck reelection after a murder accusation that Shuler alleged was "corroborated and uncontradicted."[24] Nonetheless, the FRC determined that KCEF had been a "forum for outrageous and unfounded attacks on public officials."[25] Shuler's commentary was, in the Commission's words, "sensational rather than instructive."[26] Renewal denied.

By 1930, the political spectrum's powers were proving just as impressive as Herbert Hoover dreamed they might be. Station licenses were limited in number and out-of-the-mainstream political speech was censored. The quid pro quo was in place.

Public Interest Broadcasters Excluded by Public Interest Regulation

Once the regulators got their groove, it was not just medical mountebanks and crusading moralists who grabbed their attention. Regulators established that a colorful on-air personality was not a survival trait. In 1929, the FRC set forth its view:

> The tastes, needs, and desires of all substantial groups among the listening public should be met, in some fair proportion, by a well-rounded program, in which entertainment, consisting of music of both classical and lighter grades, religion, education and instruction, important public events, discussions of public questions, weather, market reports, and news of interest to all members of the family find a place.[27]

The statement is in fact a radical policy formulation. It proclaims that competitive forces will not be allowed to test alternative formats and that regulators will enforce particular programming menus. The FRC—without any inquiry into the consequences—had determined that radio programs were best delivered not via specialty shops but by department stores. And that those generic outlets must be designed to their liking.

The embrace of generalized radio formats was no random misconception. It supplied a rationalization for the bargain between policy makers and major commercial radio incumbents baked into the Radio Act. The latter did offer well-rounded programming in pursuit of large, profitable audiences. Regulation protected this model in three respects.

First, it secured the broadcasting rights of the stations that could most easily demonstrate they served "all members of the family." Second, the definition of "public interest" to mean a particular programming model created a legal rationale for denying licenses to upstarts intending to offer other formats or competing content. Third, stations tempted to offer novel programs for niche audiences were constrained by the threat of license nonrenewal.

With strong industry support, the FRC sought to extinguish any broadcaster with a cause. The result, over time, was the homogenized "lowest common denominator" programming much like that later described as a "vast wasteland." But being a spectrum allocator means never having to say you're sorry. As Peter Huber writes of the "vast wasteland" complaint: "Almost everyone but broadcasters themselves knows this to be true. What is often forgotten is that this despoliation has occurred under the watchful eye of a federal commission that has had almost absolute control over the airwaves for seven decades."[28]

Regulators achieved mediocrity in radio and, later, TV broadcasting via structural design and regulatory controls. With few rivals, each potential broadcast audience was too large (and valuable to advertisers) to squander on educational shows or point-of-view programs. The latter had limited appeal and attracted requests for free "equal time" and "fairness doctrine" challenges of license renewals. The networks programmed little news, and perspectives were uniform. As the president of CBS News, Richard Salant, was to say: "Our reporters do not cover stories from *their* point of view. They are presenting them from *nobody's* point of view."[29] This curious boast motivated Edward Jay Epstein's classic 1973 study of television, *News from Nowhere*.

We grasp how strange this is by looking at unlicensed media in (less regulated) twenty-first-century America. Bookstores, publishers, magazines, newspapers, websites, social media and blogs often specialize, other times go mass market. Diversity is a feature, not a bug.

But in 1927–1934, the FRC rested confident in the knowledge that its programming vision should be enforced as law. This crushed those upstarts that had gotten a toehold before the FRC was empowered to define and enforce the "public interest." Consider the plight of WCFL, the radio project of union activists, and WEVD, a station dedicated to the politics of socialist leader Eugene V. Debs.

WCFL was launched in Chicago in 1926, when the Chicago Federation of Labor spent $250,000 to acquire an existing station.[30] The organization pledged to broadcast as a "non-profit listener-supported station dedicated to serving the interests of workers and their communities."[31] A creation of the self-described "radical reformer" Edward Nockels, secretary of the CFL, the station had the mission of inserting "labor messages into entertainment shows . . . and [giving] labor unions free air time to communicate with workers."[32]

In 1929, when WCFL asked for a license renewal, the FRC bluntly restricted the station's power and broadcast hours, limiting it to daytime-only operation. The Commission's fundamental concern regarding WCFL was the diminished standing of "propaganda stations (a term which is here used for the sake of convenience and not in a derogatory sense),"[33] a passage that ostentatiously parades the agency's fluency in the bureaucratic arts. There was, the FRC wrote, insufficient space "in the broadcast band for every school of thought, religious, political, social, and economic, each to have its separate broadcasting stations, its mouth-piece in the ether. If franchises are extended to some, it gives them an unfair advantage over others, and results in a corresponding cutting down of general public service stations."[34]

After investing in radio and developing an audience for "its mouth-piece in the ether"—this too was surely not meant in a derogatory sense—WCFL learned that the market interventions its partisans may have advanced in theory had surprising consequences in practice. The opportunity it had won prior to the "public interest" regulatory regime collapsed under the weight of government control. Rejecting WCFL's request for longer hours and higher power, the FRC determined that "there is no place for a station catering to any group. . . . All stations should . . . serve public interest against group or class interest."[35] WCFL's Nockels declared that the FRC had thrown in with the "plutocratic mob" and called on Congress to investigate. "We are sure if this investigation were made, it would show how these special interests privately and underhandedly held their little conference with the Federal Radio Commission and got their swag."[36]

In April 1929, WCFL sought to reverse the new restrictions and be restored to "full-time status" on its 970 KHz channel assignment. The request was denied by the FRC on the remarkable premise that, while WCFL publicly billed itself as a station devoted to labor issues, it actually aired very little on the topic, no more than one hour per day. This critique frontally conflicted with the FRC's stated rationale for reducing WCFL's transmissions, namely that it favored (in McChesney's words) "'general interest' broadcasters with 'no axes to grind.'"[37]

With the flexibility of the "public interest" standard, licensees could be squeezed not only if broadcasts were too pointed, but also if they were too rounded.

The regulatory defeats imposed severe operating losses and forced a reorientation. In May 1932, WCFL's union owners reached an agreement with the National Broadcasting Company. WCFL would enjoy around-the-clock broadcasting rights and higher power via a Chicago AM license owned by NBC. It also gained access to popular NBC programs, began to sell commercial time, and phased out its labor-oriented programming. WCFL became just another radio station.[38] When the station was finally sold, in 1978, to Amway, "it was," according to historian Elizabeth Fones-Wolf, "so devoid of any union or public service programming that few listeners realized that WCFL had a connection to organized labor."[39]

WEVD in New York grew out of a December 1926 decision by the Socialist Party to gain airtime for its views. The following August, after raising the necessary funds from sympathizers, the group purchased WOSM, a failing station in Woodhaven, Queens. It was renamed WEVD to honor Eugene V. Debs, the five-time Socialist Party presidential candidate who had died in October 1926.

The FRC transferred the broadcast license following the purchase, but the Socialist Party soon found its operations under challenge. FRC General Order 32, issued in May 1928, forced WEVD's owners to explain why their license should not be terminated. The party retained its rights, but not without a Commission rebuke. Calling the station "the mouthpiece of the Socialist Party," the FRC ordered it to "operate with due regard for the opinions of others."[40]

The station did regard the opinions of others, and rejected most of them. The Socialists had acquired a broadcast station specifically to bring a new perspective to the airwaves: "The purchase of a labor station . . . will guarantee to minority opinion in America its right to be heard without censorship. With radio now privately owned and controlled, a station like [WEVD] is the only cry in the wilderness."[41]

That cry met its "public interest" fate. In late 1929, WEVD was moved to a new frequency, 1300 AM, where it shared time with three other stations and was permitted to broadcast just fifty hours per week.[42] As a nonprofit station that eschewed commercials or corporate sponsorship, WEVD quickly ran out of money. It sought assistance from the *Jewish Daily Forward*. The newspaper pumped hundreds of thousands of dollars into the station and soon owned the license outright.[43] The station remained on the air, but WEVD's editorial policies became those of the relatively centrist *Forward*, crowding out radical voices. The new owners avidly pursued advertising revenues, promoting everything from, as Nathan Godfried writes, "noodles to furniture to headache remedies to Coney Island excursions."[44] In 1936, Socialist Harry Laidler would lament: "I wonder what Eugene V. Debs, if he were alive, would say about a Debs radio which gives any amount of time to those who were advocating the election of candidates of a capitalist party but which gave practically no chance for candidates of the Socialist party to present their message."[45] WEVD was eventually sold in 1981 to Salem Media and became WNYM, airing Christian programs.

"Public interest" regulation—strongly supported by nonprofit stations, left-wing activists, and the American Civil Liberties Union—proved a death trap. The natural experiment supplied by the alternative regimes in place over the seven years before 1927 and the seven years after yields stark results. In the earlier period, priority-in-use rules limited regulatory discretion and let some two hundred nonprofit flowers bloom. These ventures competed for listeners and expanded democratic discourse. Under "public interest" spectrum allocation, the FRC squeezed the life from these intriguing, free-spirited upstarts.

4

Myth Calculation

We thought the phrase ["public interest"] never really meant anything to users of the airwaves and to those who regulate the industry. . . . A lot of games have been played with it, and there have been
a lot of empty promises made to serve the public interest.
            —Lionel Van Deerlin (D-CA), chairman of the House
                   Telecommunications Subcommittee

EVERY GOVERNMENT REFLEXIVELY SEEKS to control emerging media.[1] Between 1912 and 1927, warlords in northern China asserted exclusive authority over all wireless communications, treating radios as military equipment.[2] Japan's military control of airwaves lasted even longer. The U.S.
Navy had similar plans but was opposed in the early 1920s by Secretary of
Commerce Hoover, who used his bureaucratic infighting skills to open up
substantial scope for commercial radio services. In Great Britain, radio
spectrum was placed under the jurisdiction of the Post Office, which established a state monopoly, the British Broadcasting Corporation. The BBC
restricted views challenging policy makers, most infamously banning Winston Churchill in 1938 when he sought to criticize Prime Minister Neville
Chamberlain's "Peace in Our Time" treaty with Nazi Germany.

Later, the BBC censored popular music, leading rock stations in the
1960s to operate offshore as "pirates."[3] South Africa allowed no broadcast
television until 1976. The apartheid government's explicit aim was to sup-

press the "little black box," thus guarding "Afrikaner nationalist ideology."[4] Television was subversive because "South Africa would have to import films showing race mixing," while "advertising would make [nonwhite] South Africans dissatisfied with their lot."[5] West Germany licensed no private broadcast television station (relying exclusively on state-owned media) until 1984.

The governmental urge to control media continues. China, in addition to routinely blocking Internet searches for "subversive" terms, permits only thirty-four foreign films per year to be exhibited.[6] Countries from Saudi Arabia to Vietnam impose similar restrictions. Despite poetic language about the net being beyond the reach of regulators, many governments try to block certain forms of expression, and often succeed.[7]

In the United States, the regulatory reflex encounters a more formidable barrier. As Supreme Court Justice William O. Douglas wrote, "The First Amendment's ban against Congress 'abridging' freedom of speech . . . create[s] a preserve where the views of the individual are made inviolate."[8] The rise of wireless was instantly seen as supplying a new form of free speech. Radio pioneer David Sarnoff made the case in 1924, calling radio "the bar at which great causes will be pleaded for the verdict of public opinion." His article, "Uncensored and Uncontrolled," published in the *Nation*, argued that the emerging electronic media should be afforded "the same legal status as newspapers. . . . Freedom of the press should be made to apply" to radio.[9] It was widely understood "by many people of the 1920s," as Mary Mander put it, "that control of broadcasting had consequences for democracy."[10]

Yet the urge to regulate was primal: "Congress has good reason for this jealousy as to the control of radio," C. C. Dill offered in his authoritative tome, *Radio Law,* published in 1938 just after he left the Senate. "Nobody can even imagine what radio may someday mean to the human family. . . . Nor can anyone now even dream of the possibilities of television."[11] Policy makers were drawn to the light.

The debate over the 1927 Radio Act acknowledged the conflict between licensing and the First Amendment. Congress's solution was to regulate and yet, ostensibly, not regulate. A provision was included to prohibit government censorship. But this was at odds with the central thrust of the law.

As Dill explained: "The provision which forbids the Commission to censor radio programs does not prevent the Commission from determining whether or not a station's programs are in the public interest. The extent of the 'twilight zone' between censorship and the refusal to renew a station license because of the service rendered, is undetermined."[12]

The ultimate resolution was to treat broadcasters as second-class First Amendment citizens. Licensees exercised some rights to free expression, but not those of newspaper or book publishers. David Sarnoff's plea to make radio broadcasters "uncensored and uncontrolled" was rejected for purportedly technical reasons. Radio was characterized as uniquely prone to chaos. As summarized in the 1969 Supreme Court opinion in *Red Lion Broadcasting Co. v. FCC*, which upheld the constitutionality of the Fairness Doctrine, "broadcast frequencies constituted a scarce resource whose use could be regulated and rationalized only by the Government."[13] In this view, the state was not compromising free speech but creating it.

The logic was wrong. Rembrandts are scarce, but prices work well, as a visitor to Sotheby's might attest, in determining ownership. Rights to use radio frequencies can also be defined and auctioned, as the FCC and more than thirty regulatory authorities worldwide have recently shown via "spectrum auctions," counting their billions in dollars, Euros, British pounds, Swiss francs, Guatemalan quetzals, or Nigerian naira. Prominent jurists, from the famously liberal David Bazelon[14] to the famously conservative Robert Bork,[15] dealt devastating criticisms to the *Red Lion* ruling. Constitutional law expert Laurence Tribe wrote that it is premised on "a profound fallacy about spectrum scarcity."[16]

Despite the intellectual error, the "technical reasons" approach permitted policy makers to sidestep the Constitution. The Radio Act not only authorized a "traffic cop for the airwaves" but gave the government considerable power over what was communicated. Regulators eagerly crafted content controls with uplifting—if Orwellian—language. The Radio Act contained a rule obligating stations to provide "equal time" in covering political candidates; this helped protect incumbent officeholders and had the perverse effect of blocking broadcasts of presidential debates. The Fairness Doctrine, which began to develop in the late 1920s and then blossomed with television in the 1940s, suppressed controversial speech in general. Some legislators wished the Radio Act would go much farther. Inspired by the State of Tennessee's victory in the 1925 Scopes trial, South Carolina Senator Coleman Blease introduced an amendment to ban radio "discourses" on the topic of evolution. "I am willing for the world to know," he thundered, "that on this proposition I am on the side of Jesus Christ."[17] Alas, Congress was not on the side of Blease; his effort yielded only "comic relief."[18]

Yet while Darwin escaped, the power vested in the FRC—passing, word for word, to the creation of the Federal Communications Commission in 1934—would allow policy makers to exclude speech. Unorthodox views became subject to threats of license nonrenewal or victims of station transmission limits. The equal and opposite reaction: bargains between licensees and regulators became endemic. Successful commercial stations gladly exchanged slices of freedom for lucrative favors—each officially determined to further "public interest, convenience, or necessity."

The purpose of a standard is to rule out things that are *sub*standard. By that criterion, the public interest standard does not merit the name. William Mayton astutely observes that whatever a public agency decides is, by definition, in the public interest.[19] As Senator Dill conceded, "It covers just about everything."[20] Yet in testimony to the strategic uses of vacuity, the "standard" has blocked public scrutiny and proved kryptonite to legal challenge. It plays a critical role in spectrum allocation from 1927 to the present.

How the system operated with respect to competition and innovation can be seen with the invention of FM radio.

Edwin Howard Armstrong was born in 1890 and attended Columbia University. Between his junior and senior years, Armstrong invented a "regenerative circuit" to amplify—and greatly improve—AM radio reception.[21] He quickly patented his discovery. Graduating from Columbia in 1913, he enrolled in graduate studies the same year and joined the faculty as an instructor.

In 1917, Armstrong left for the U.S. Army Signal Corps, developing military radios and earning the rank of major.[22] France awarded him a Chevalier de la Légion d'Honneur and permitted him to use the Eiffel Tower for radio experiments.[23] Following the war he returned to Columbia, working without salary to reduce his administrative duties and focus on research. By

4.1. Radio inventor Edwin Howard Armstrong marries Marion McGinnis in 1923 and builds her a boom box—one of the first portable receivers—as a wedding gift.

age twenty-three he had become a millionaire from the sale of patents; his further inventions—including the superheterodyne receiver—would be sold to the Radio Corporation of America. By his early thirties, he was RCA's largest shareholder and a man of means.[24] Married in 1923, he honeymooned in Palm Beach, Florida, where he presented his new bride with the world's first civilian portable radio.

Armstrong plunged his wealth back into research. In 1933, his greatest breakthrough occurred: the creation of frequency modulation, or FM, radio. This technique used wider bands than amplitude modulation (AM) and encoded its information in frequency changes rather than changes in signal strength. Armstrong argued that FM delivered better audio—"high fidelity."

Other experts shared his enthusiasm.[25] According to the historians Christopher Sterling and Michael Keith, when Armstrong unveiled his invention at the Institute of Radio Engineers in New York City in 1935, "most of the listening engineers were astonished by what they heard." Many predicted that FM would eclipse AM, in a "march of science which will obsolete the system now in use."[26] Alas, there was much more to marketplace success than simply building a better mousetrap. There was the FCC.

And the FCC was not impressed; it doubted the new format would work. In January 1936, a Commission report stated that the VHF band (where Armstrong had devised his FM equipment to operate) was virtually worthless for communications stretching beyond ten miles. Armstrong had already demonstrated that FM signals were clearly received more than eighty miles from the transmitter, but he set about building more formidable test facilities to rebut the FCC's technical arguments.[27]

He made some progress. In mid-1936, FM was allocated a single megahertz (near 42 MHz), which would support four experimental broadcast channels.[28] The regulatory headwinds Armstrong encountered suggested that FM was taking a back seat to television, an emerging technology owned by companies with AM radio interests. Television received more than 50 MHz, an allocation that would support eight video channels per market. This, said Armstrong, gave a "virtual monopoly of the frequency bands" to fledgling broadcast TV.[29]

In 1937 Armstrong constructed twenty-five prototype FM receivers to gauge reception from a 50,000-watt FM station he built in Alpine, New Jersey, using $60,000 of his own money. John Shepard III of the Yankee Network, a New England radio chain, conducted further experiments. Armstrong and his collaborators were excited by the results. The signals traveled much farther, were received with greater clarity, and were far less sensitive to interference than AM broadcasts. And despite their greater range, stations did not drown one another out; receivers automatically tuned into the clearest signal.

Armed with impressive field data, Armstrong returned to the Commission. He was surprised when his request to use an unoccupied VHF frequency was greeted with hostility. The Radio Corporation of America, the Columbia Broadcasting System, and American Telephone and Telegraph all argued against FM. RCA, which owned the National Broadcasting Company, was then the largest AM broadcaster in the country; CBS was second. Neither saw a need for additional radio stations. AT&T was concerned because FM was ideally suited to provide wireless relay service, and in fact already linked several New England radio stations. These point-to-point transmissions competed with AT&T's business of transporting network radio programs via coaxial cable.

In battling these interests at the FCC, Armstrong grasped a cause beyond better technology. As his biographer explains, he

> saw in the development of FM the opportunity to free the U.S. radio system of oppressive restriction and regulation. An almost unlimited number of FM stations was possible in the shortwaves, thus ending the unnatural restrictions imposed on radio in the crowded longwaves. If FM were freely developed, the number of stations would be limited only by economics and competition rather than by technological restrictions. Small stations and new networks would have a chance to grow, reducing the need for FCC regulation and lessening the domination of the industry by a few corporations.[30]

The inventor considered his new broadcasting platform "as great an invention as the printing press, for it gave radio the opportunity to strike off its shackles."

Despite the odds against him, Armstrong battled to victory in 1940, four years after he first petitioned the FCC. The 42–50 MHz band was set aside for FM broadcasting and divided into forty channels, enough for at least two thousand stations nationwide.[31] By late 1941, sixty-seven stations were on the air or had submitted applications to the FCC, and some 500,000 households had purchased FM receivers. FCC Chairman James Lawrence appears to have been concerned that RCA and NBC were monopolizing the market, and to have seen FM as providing healthy competition.[32]

But at just that instant, U.S. entry into World War II froze civilian electronics production. FM station construction halted, and electronics manufacturers—which had been selling some fifteen hundred FM radios a day by late 1941—ended radio production by mid-1942.[33] Major Armstrong rejoined the Army Signal Corps. FM relay units were soon deployed by U.S. forces, including General George S. Patton's Third Army, whose furious march through France in 1944 was far too speedy for wired communications.

As the war came to a close, Armstrong's hopes soared. The few stations that had been authorized in 1941 had continued broadcasting, and listeners loved what they heard. Yankee Broadcasting, for instance, covered most of New England with a sound quality that clearly surpassed that of the big AM network affiliates. When wartime production controls were lifted, industry analysts forecast sales of some five million FM radios per year. But this prospect of success led FM's opponents in Washington to regroup.[34]

NBC and CBS petitioned the FCC in 1944 with a bold proposal: toss every FM station off its assigned frequency and relocate the entire industry up the dial. All existing equipment—transmitters owned by stations, receivers owned by listeners—would become obsolete. Proponents claimed the frequency switch would help FM stations avoid "ionospheric interference," a threat alleged to emanate from sunspots.

FM broadcasters and equipment makers protested that ionospheric interference existed solely in the minds of their rivals. Armstrong submitted voluminous data to the FCC. So did the U.S. government's own spectrum expert, D. H. Dellinger of the National Bureau of Standards, who had studied the matter extensively. His analysis showed that "fears about interference problems with the existing FM band were unfounded."[35] On the other hand, purveyors of the sunspot theory suffered from a financial conflict of interest, as their coalition was led by television interests seeking to use the existing FM band. But the FCC accepted their sunspot argument and embraced their policy suggestion.[36]

Nearly all of the expert technical testimony—and there was a lot of it—went against the FCC's position. Congress had created the Radio Technical Planning Board (RTPB) in 1944 specifically to retain leading radio frequency scientists to advise on postwar allocations. The FM panel of the RTPB voted 19–4 against moving FM, saying there existed "no technical evidence to indicate that . . . presently assigned . . . spectrum would be improved by any shift in the present allocation."[37]

Yet the FCC insisted, again basing its policy on "technical reasons." In this, the agency cherry-picked its data and rejected scientific consensus. It focused on predicted events: sunspots *might* become more severe in future years and cause more interference. This speculation was impossible to disprove. The Commission then ignored obvious economic evidence. That the FM business interests unanimously viewed the reallocation as disastrous was not relevant. Yet these were precisely the parties that had the most to lose from sunspots if they were actually a serious threat to FM broadcasting.

Finally, regulators employed the strategy of "too early, too late." When interests and regulators are lining up a policy initiative, they brush off criticism on the grounds that nothing has been formulated and it is too early to object. Then, when the policy is crafted, regulatory backers promptly announce that it is too late to change, and that critics should have spoken up while the policy was being put together. The FM decision was a perfect example. In announcing concern over FM frequencies in May 1945, the Commission put off a determination until later in the summer, when solar interference could be better observed. Then, at the end of June, the FCC shocked FM broadcasters by voting unanimously to uproot the FM band.[38]

The allocation for FM radio at 42–50 MHz granted in 1940 was given to television broadcasting; FM was placed at 88–108 MHz. If the relocation costs had been nil, the new band might have been preferred: it contained more space. But the relocation costs were enormous. Existing transmitters and receivers instantly became worthless. Early adopters who had purchased radio sets often costing one hundred to five hundred dollars (1940s dollars) found themselves owning junk. The word went out: buy new FM radios at your own risk.

If there had been new FM radios to buy. Armstrong spent two years developing receivers for the higher band. When stations began broadcasting on the new frequencies, the public was extremely reluctant to buy the new sets, which seemed to be a speculative investment.

The FCC also reduced station power. Armstrong's commercial-free station in Alpine, New Jersey, was cut back from 50,000 watts to just 1,200. The stated rationale was to promote local broadcasting, a recurring theme in FCC public interest regulation. To squeeze in more local assignments, each station's coverage area had to be reduced. The practical effect was to make the FM stations less viable. With smaller audiences, they had a harder time competing for advertising dollars against established AM stations.

As Armstrong's biographer wrote, "The series of body blows that FM radio received right after the war, in a series of rulings manipulated through the FCC by the big radio interests, were almost incredible in their force and deviousness."[39] In 1948 Senate hearings, Armstrong testified: "The effort has been to mold the allocation of FM into a form where it will become a network subsidiary, unable to take the leading role which its technical merits would give it if left unhampered by regulation."[40]

4.2. A 1941 Stromberg Carlson console radio receiver. The FM dial—42–50 MHz—was made obsolete in a 1945 FCC spectrum allocation reversal that destroyed the fledgling technology. Steve Geary/The Radio Museum.

By then, the threat to AM radio was over. Not until the FCC approved stereo broadcasting for FM in 1960—twenty-six years after Armstrong first showed that this was technically feasible—did FM rise from the dead. Audiophiles flocked to FM "high fidelity," and the masses followed. FM stations multiplied in number and audience share, and in 1979, FM surpassed AM in listeners.[41] By 1985 there were as many FM stations as AM, and by 1995 substantially more (nearly seven thousand FM stations to fewer than five thousand for AM).[42]

Armstrong did not live to see his technology triumph. On January 31, 1954, "despairing that his innovation was a market failure"[43] and locked in an acrimonious lawsuit over patent royalties owed him by RCA,[44] he wrote a letter to his wife, dressed neatly, and walked straight out the window of his fancy thirteenth-floor apartment at New York City's River House. "It seemed incredible to him," wrote Lawrence Lessing, "that in this country, by means of restrictive regulations and slippery measures, a superior scientific advancement could be overwhelmed by the shoddy and the expedient."[45]

Perhaps the greatest irony is that Edwin Armstrong, a prolific inventor, leading scientist, and famous Columbia University professor, preferred to go by the title Major. A great patriot, not only did he serve in both world wars, he donated his key radio technologies to his country, waiving intellectual property rights. His contributions to science and state were such that he would be honored by a U.S. Medal of Merit in 1945, and his picture would adorn a U.S. postage stamp in 1983. Nonetheless, his greatest innovation was stymied and his entrepreneurial spirit crushed on spectrum allocation determinations made by the U.S. government according to "public interest, convenience and necessity."

5

Eureka-nomics

The [FCC] has ignored the fact that tremendous wealth attaches to the most desirable licenses. . . . Instead of adopting regulations that would reflect the actual value of these licenses, the Commission has buried its head deeper into the regulation books and considered more obligations for these special stewards who, in turn, are usually willing to comply with whatever the Commission asks, as long as the cost of compliance is slight.
—Mark Fowler and Daniel Brenner

THE POLITICAL SPECTRUM PROVED fertile ground for the creation of stories about the social obligations of wireless licensees, even as the substance of these tales was largely nil, "arguments without arguments."[1] With gravitational regularity, deep irony has been observed: to ameliorate a purportedly unique form of "physical scarcity," policies have restricted competition far more tightly than the asserted natural constraints. A lexicon arose to explain why a departure from regular rules of order was required, why competitive market forces were not themselves up to the task at hand.

The claim for special status is not so special. To procure protectionist legal props, interest groups far and wide assert that their marketplace is inimitably trapped in a vortex of chaos. "Just about every industry thinks it's 'special,'" notes U.C. Santa Barbara economist Ted Frech. "You should hear hospital administrators talk about how special they are, so that economics doesn't apply."[2]

As with the regulatory violence dealt WEVD's dissenting views or Major Armstrong's FM technology innovation, however, the economic consequences of exclusionary policies do assert themselves, even if under cover of "public interest." Yet they were too little noted. Spectrum misallocation became a "silent crisis."[3] Individuals financially independent of the political bargains have been unable to invest the effort required to examine the basis for regulation or scrutinize the impact of its implementation. Until a curious British intellectual began to ponder these topics in the 1950s.

An "Interesting Error"

Ronald Harry Coase, born in England in 1910, wore leg braces as a child and was relegated to nonacademic courses—literally, basket weaving—until he was ten years old. As he later explained, British schools of the era treated those with physical handicaps as they did those considered "mental defectives."[4] Somehow, Coase overcame both the disability and the tarnish and found himself on the faculty of the London School of Economics by 1935.[5] There he studied public utilities, including the British Broadcasting Corporation. In 1951, he moved to the United States, taking a teaching position at the University of Buffalo. He decided to take a deep dive into the "political economy of radio and television."[6] During research at Stanford's Center for Advanced Study in the Behavioral Sciences in 1958–1959, he investigated the specifics of radio spectrum allocation. By 1959 Coase—now on the faculty of the University of Virginia—was ready to challenge orthodoxy.

Justice Felix Frankfurter had framed thinking on spectrum policy in his 1943 *NBC* opinion. "Radio inherently is not available to all," reasoned Frankfurter. "That is its unique characteristic, and that is why, unlike other modes of expression, it is subject to governmental regulation."[7] Justice Frank Murphy added that "owing to its physical characteristics radio . . . must be regulated and rationed by the government." Physics begged for administrative control: "Otherwise there would be chaos, and radio's usefulness would be largely destroyed."[8]

In a landmark 1959 essay in the *Journal of Law and Economics,* Coase probed this consensus and found not clarity but confusion. What seemed distinct about radio spectrum was in fact true of a wide array of social coordination problems.

> It is a commonplace of economics that almost all resources . . . are limited in amount and scarce, in that people would like to use more than exists. . . . It is true that some mechanism has to be employed to decide who, out of the many claimants, should be allowed to use the scarce resource. But the way this is usually done in the American economic system is to employ the price mechanism, and this allocates resources to users without the need for government regulation.[9]

Why wasn't radio spectrum treated the same way? Coase concluded that policy makers were operating under a "misunderstanding." He quoted former U.S. President and later Chief Justice William Howard Taft, who conceded, "I have always dodged this radio question. . . . Interpreting the law on this subject is something like interpreting the law of the occult. It seems like dealing with something supernatural."[10] Coase saw that the politicization of radio spectrum, under the 1927 Radio Act, allowed influential interest groups to tilt markets in their favor, but assumed the result owed to random error. He also accepted the Supreme Court's fanciful history of radio "chaos," ignoring the key role of "aerial property rights" and market competition in the development of the broadcasting industry.

Still, his exposition of the underlying economics was brilliant: he exposed the defense of radio regulation as entirely circular. The broadcasting market was said to fail, but the state had suppressed the process by which coordination occurs. Instead of letting spectrum rights be owned and assigned through competitive bids, government had assumed the role of central planner. Coase saw a different solution, consistent with the nature of airwaves, technological innovation, free speech, and the creation of consumer welfare. He proposed replacing administrative control with "the pricing mechanism."

This idea had been floated. In 1951, University of Chicago law student Leo Herzel had published an article proposing that the government auction frequency rights, allowing competing licensees to determine services and technologies.[11] The suggestion drew a sizzling rebuttal from professor Dallas Smythe, who attacked Herzel for a "doctrinaire" approach that ignored the "unique nature of the broadcast business." Smythe charged Herzel with propagating abstract theories from "the realm in which it is merely the fashion of economists to amuse themselves."[12]

The dismissal was predictable. Those dialed into spectrum allocation issues were vested in the prevailing paradigm; experts reliably rallied to its defense. Other eyes (even William Howard Taft's) glazed over. As the *Federal Communications Bar Journal* commented of the Herzel-Smythe exchange, the "whole discussion will be over the heads of most readers."[13] But Coase grasped the nature of the conflict and its far-reaching implications for both public policy and economic analysis.

Were there truly no limits imposed on radio use—"open access"—the result could well be chaos. Yet how limits are best configured is not obvious, nor is it a strictly "technical" matter. A radio station could be given an exclusive nationwide right to broadcast on a channel reserved for its use, and so reduce interference for listeners tuning into its programs. But protecting that activity would simultaneously interfere with other opportunities— say, for additional stations to broadcast and for listeners to enjoy a greater range of audio choice. The immediate question is whether the gains from one approach exceed the gains from another. The higher-level question concerns how to most effectively decide that basic decision, weighing the trade-offs.

The task was not, as conventional wisdom offered, to eliminate interference. With conflicting demands there would be costs; interference was a two-way street. And there existed myriad possibilities. Perhaps AM radio signals are more efficiently sent via cable, leaving the AM band for other services. Or perhaps they should be sent via higher frequencies, or by FM technology (available circa 1934), or by satellite (where hundreds of radio channels have been beamed to U.S. subscribers since 2001), or via digital rather than analog formats, or by using more (or less) bandwidth, or by getting customers to buy more expensive radio receivers with better antennas. Or by just forgetting about broadcasting and instead distributing records or tapes—or streaming audio, piggybacking on broadband networks.

Coase did not, of course, predict SiriusXM, Pandora, Spotify, iHeart Radio, or Apple iTunes in 1959, but he did imagine a galaxy of expanding possibilities. He saw that the preferred solutions would drastically change with technology and market innovation. How would society discover and select among them? The comparative advantage of regulators had been assumed. But a comparative assessment was necessary.

Competition regularly outperformed state monopoly elsewhere; perhaps it could here. The glib counterattack on Herzel and the categorical defenses of administrative control sidestepped any realistic analysis. Dallas Smythe had contended, for instance, that because law enforcement was of great social importance, radio spectrum should be administratively set aside for the exclusive use of police departments. "Surely it is not seriously intended," he scolded Herzel, that public safety organizations "should compete with dollar bids . . . for channel allocations."[14] Herzel's rejoinder returned fire: "It certainly is seriously suggested. Such users compete for all kinds of other equipment, or they don't get it."[15] By budgeting and buying resources, costs are revealed. This transparency thus assists vital government functions. If police departments were given raw materials and then charged with making their own squad cars, law enforcement would be worse off. Funds would be squandered and vehicles would be poor—exactly the outcome when police departments are given spectrum and told to construct their own networks even when better wireless services, at lower cost, are available in the marketplace.[16] In fact—at the very moment—police communications were handicapped by ill-suited spectrum choices made in Washington. As a 1968 presidential commission was to report: "Police and other public safety radio services . . . may be unable to obtain vital spectrum resources, while those allocated to other categories go unused in the same area (e.g., frequencies reserved for forestry services were only recently made available to the New York City Police Department)."[17]

Yet the spectrum policy issue could not be decided on theory alone. "On reading Herzel's article," Coase wrote in 1993, "I did not immediately jump to the conclusion that market pricing would be superior to regulation by the FCC." While it was clear that "the Federal Communications Commission conducted its affairs in an extremely imperfect way," could markets do better?

A clue was supplied by the experts who had protested too loudly. "The question of whether pricing should be used . . . was . . . clinched for me . . . by Dallas Smythe, who had been chief economist at the Federal Communications Commission. His objections were so incredibly feeble . . . that I concluded that, if this was the best that could be brought against them, Leo Herzel was clearly right."[18] Accepting the traditional history of failure in broadcasting, Coase ascribed the outcome to "the real cause of the trouble [which] was that no property rights were created in scarce frequencies." That would invite trouble under many circumstances. And the remedy was clear: assign spectrum rights to owners and unleash the forces of competition.

> Certainly, it is not clear why we should have to rely on the Federal Communications Commission rather than the ordinary pricing mechanism to decide whether a particular frequency should be used by the police, or for a radiotelephone, or for a taxi service, or for an oil company for geophysical exploration, or by a motion-picture company to keep in touch with its film stars, or for a broadcasting station. Indeed, the multiplicity of these varied uses would suggest that the advantages to be derived from relying on the pricing mechanism would be especially great in this case.[19]

Coase had uncovered a deeper truth. The concept of market failure could not be considered in absolute terms. Its existence had to rest on the relative success of government intervention. That simple insight was radical. When Coase submitted his article to the *Journal of Law and Economics,* he was told he had made a fundamental error: he had ignored the effect of "externalities"—impacts on third parties who (neither buyers nor sellers) have no say in a particular activity yet are affected by it. Wireless emissions were thought to inherently create such spillovers. Broadcasters would have an incentive to blast away, reaching large audiences while preempting others who might like to use airwaves differently. To say that private property in spectrum could salvage this mess seemed to ignore basic theory. "Chicago economists," wrote George Stigler, later to win a Nobel Prize, "could not understand how so fine an economist as Coase could make so obvious a mistake."[20]

Coase assured the editors that he had correctly analyzed the situation and that "even if my argument was an error, it was a very interesting error."[21] Don't laugh—it worked. The paper was published and, the following year, an editor—Aaron Director, an economist at the University of Chicago Law

School—hosted a seminar for Coase to explain himself. To a skeptical audience, Coase asserted that externality problems—including radio interference—might be more efficiently handled by market bargaining than by regulation. Economic theorists (and spectrum policy makers) had believed that unregulated actors would be confused by mixed signals, prices that did not take account of all costs or benefits. Policies (taxes, subsidies, or direct regulation) were seen to correct the situation, yielding better information and proper incentives.

Coase ingeniously showed that the same trade-offs that challenged market actors could also sabotage efficient outcomes designed by regulators. That had been incorrectly omitted from the analysis. The effectiveness of social coordination by law had been assumed. But it was the friction—summarized as "transaction costs" in private sector dealings—that caused the problems. If such costs were assumed to be zero for private as well as public actions, there would exist no market failure to fix; spontaneous bargaining would eliminate externalities and resources would flow to their highest valued uses.

George Stigler described the reaction: "We strongly objected to this heresy. Milton Friedman did most of the talking, as usual. He also did much of the thinking, as usual. In the course of two hours of argument the vote went from twenty against and one for Coase to twenty-one for Coase. What an exhilarating event! I lamented afterward that we had not the clairvoyance to tape it." This constituted the one and only moment that Stigler was to witness a "sudden Archimedian revelation"—a *Eureka!*—in his long and distinguished academic career.[22]

Coase did not argue that free markets operated perfectly. Just the reverse: both public and private actions confronted coordination problems. His objection was to the deus ex machina approach asymmetrically applied. Regulatory process was seen as freely available and inherently reliable, inevitably solving dilemmas that stymied others. It was the view, as H. L. Mencken colorfully put forth in a broader context, that the "government is in essence not a mere organization of ordinary men like the Ku Klux Klan, the U.S. Steel Corporation or Columbia University, but a transcendental organism composed of aloof and impersonal powers, wholly devoid of self interest."[23]

5.1. Ronald Coase, 1910–2013. An English-born economist who came to the United States in the 1950s, Coase questioned conventional wisdom about radio spectrum, uprooted traditional theory, and won the 1991 Nobel Prize in economics. Weidenbaum Center on the Economy, Government and Public Policy/Washington University in St. Louis.

Coase's 1960 "The Problem of Social Cost"[24] extended what he had discovered in his 1959 article on radio spectrum. It showed that pollution problems were a by-product of beneficial activities, but that polluters often escaped having to pay for environmental damage as they would for other inputs. The resources were not markedly different; the legal rules were. The economic result was that what was dear looked cheap, and was squandered.

The implication was that many inefficiencies could be addressed by improving ownership rights and enabling gains from trade. Rather than dictating how broadcasts should be conducted, the state defines and sells frequency rights, allowing buyers to competitively determine what wireless services to offer and how to deliver them. Instead of mandating how smokestacks are built, government could limit air pollution levels and permit markets to exchange the emission rights.[25] The latter rewards firms and technology creators for advancing pollution abatement, while the former pays bureaucrats to devise rules that may or may not work—and leaves their salaries and fringe benefits unchanged whether their plans succeed or fail.

This approach brought a "disruptive clarity" to the management of all natural resources.[26] Coase's 1960 article ultimately became the most widely cited study in the social sciences,[27] and its author was awarded a Nobel Prize in economic science in 1991. Yet "had it not been that these Chicago economists thought that I had made a mistake in the article on the Federal Communications Commission," Coase observed, "it is probable that 'The Problem of Social Cost' would never have been written."[28]

Imagining Markets

Having rearranged basic economic theory, it was perhaps child's play for Coase to deconstruct traditional thinking on spectrum allocation. The myth was that wireless services were too important to be left to market forces. But this was an illusion: clashes over the use of a limited resource were a recurring economic reality. The policy problem was how best to channel that rivalry.

Competing spectrum owners were key. "The reduction of interference," wrote Coase, "may require costly improvements in equipment, and operators on one frequency could hardly be expected to incur such costs for the benefit of others if the rights of those operating on adjacent frequencies have not been determined." New rules could improve matters: "The institution of private property plus the pricing system would resolve these conflicts."[29]

Regulators maintain order with clumsy tools, poor incentives, muted customer feedback, and other people's money. As a result, they often misconstrue trade-offs and focus on the wrong tasks. "It is sometimes implied that the aim of regulation in the radio industry should be to minimize interference," wrote Coase. "But this would be wrong. The aim should be to maximize output." In fact, virtually every wireless communication imposes some level of interference on other services.

Drive the country at night listening to AM or FM radio. As you move between cities, one station's signal will fade while another broadcast gradually displaces it on the same or an adjacent channel. During the transition, your receiver struggles to distinguish between them. The squeals of older analog radios provide a more impressive soundtrack for this science project, but even newer digital FM radios exhibit sparks and gaps. The conflict between the stations struggling to reach your ears could be remedied by eliminating one of them, but is that the best solution? Too often, the administrative solution is to do exactly that—ensuring strong support from the stations that remain while garnering little if any kickback from listeners whose modest skin in the game does not justify hiring a lobbying consultancy to fight for the stakes involved in tuning in to an additional heavy metal—or classic country—station while driving the Interstate.

Coase suggested that spectrum markets allocate bandwidth. Communications experts greeted his idea with howls of laughter. Invited to testify before the FCC in 1959, Professor Coase was asked an opening question: "Are you spoofing us? Is this all a big joke?"[30] In 1965, an FCC official explained the hostility: "After the initial shock of rationally considering the use of the pricing mechanism in frequency allocations, the virtually unanimous view of communications specialists would be that the multiplicity of users both national and international, . . . the interference characteristics of radio with signals at relatively low energy levels interfering at diverse points many hundreds of miles away, . . . and the hundreds of thousands of licensees involved in addition to the many millions of consumers make the pricing mechanism unworkable for frequency allocation."[31]

The RAND Corporation, a prominent think tank, was more interested. Coase and economists William Meckling and Jora Minasian were asked to develop the idea of spectrum markets in a 1962 monograph. But the organization soon found that it had stirred a hornet's nest. One academic referee, on reading a draft of the paper, wrote to RAND:

> Time has somehow left the authors behind. On the domestic scene, they ignore the social, cultural, and political values which have come to inhere in mass communications, in particular, broadcasting, as well as fifty years of administrative law developments. On the international level, it would appear that it has been kept from them that everywhere but in the United States, communications are almost totally a state function and monopoly. . . . I know of no country on the face of the globe—except for a few corrupt Latin American dictatorships—where the sale of the spectrum could be seriously proposed.[32]

That was the nice part. In the cover letter accompanying the review, the reviewer (who was affiliated with the think tank) "raised other types of considerations which bear on RAND's 'public relations' in Government quarters and in Congress." Publishing the tract could have dire consequences.

These would exist even if the [report] did a fine job of showing clearly what is wrong with the present system and gave a balanced evaluation of the advantages and costs of several alternative solutions. In that case, our vulnerability would be far less to the fire and counterfire of CBS, FCC, Justice, and most of all—Congress. But as the [report] is presently designed, I am afraid that to issue it . . . is asking for trouble in the Washington–Big Business maelstrom because we haven't in the first place measured up to the intellectual requirements of the problem selected for the study.[33]

RAND, having commissioned the study, refused to publish it.

Scientific Design and Social Progress

Conventional wisdom could scarcely have been more wrong. While the "price system" has not displaced the 1927 Radio Act, far-reaching experiments in radio spectrum liberalization have provided Coasian proof of concept, as I detail below. We have learned that markets can work, spectacularly well. And that Latin American dictators would be the last policy makers to sell spectrum—public auctions are not the way to channel special favors to your brother-in-law. Once spectrum policy samizdat, the 1962 RAND paper was published in 1995 and is now proudly offered as a free download.[34]

Thankfully, despite the best efforts of RAND and its reviewer to bury the report, Coase's pioneering economic analysis has gifted us insights into how spectrum can be usefully harnessed for wireless communications. Science, of course, plays a leading role.

Radio spectrum is an "invisible resource"[35] defined in frequencies. Waves of electrical energy travel through this space, and their patterns convey information. The more frequently the waves cycle, the shorter their length. A measure is supplied by this variation. A Hertz, the standard unit of measurement of radio waves (named for the nineteenth-century German physicist Heinrich Hertz), denotes one cycle per second. A kilohertz (KHz) is one thousand cycles per second, a megahertz (MHz) one million, a gigahertz (GHz) one billion.

Guglielmo Marconi knew about radio signals and sought to harness them for communications. In 1895, he did. The idea that electronic messages could be sent through open space, cutting out the physical connection be-

5.2. Scientist and entrepreneur Guglielmo Marconi ponders radio, 1909.

tween transmitter and receiver, was so astounding that the breakthrough was named for what it was not—*wireless*.

But Marconi thought that only one radio transmitter could operate in any geographical area.[36] That stifling constraint was relaxed in 1900, when he was awarded patent no. 7777 for a "tuned circuit" enabling frequency division.[37] This innovation sparked a quest for bandwidth that continues to rage. Advances have occurred along the extensive margin, utilizing higher frequencies, and the intensive margin, packing more traffic into a given band.[38]

Extensive Margin. In the early 1920s, radio use was confined to a band from 3 KHz to 3,000 KHz, dubbed "medium waves." The AM radio dial, 500–1500 KHz, generated the most interest. But by 1930, when the Federal Radio Commission claimed jurisdiction over some 60 MHz, the communications spectrum had expanded twenty-fold. By 1936, the new FCC was regulating access to more than 300 MHz.[39] Fast forward to 1990, when useful transmissions were routine up to about 16 GHz (16,000 MHz).[40] By the dot.com boom of the late 1990s, networks using frequencies from 28–40 GHz were attracting billions of dollars in investment capital. Today, radio communications stretch all the way to 300 GHz.[41] While the great

Table 5.1. Spectrum Bands

Band	Location	Wireless Services (Partial)	Approx. First Use
Medium frequencies	300 KHz–3 MHz	AM radio	1920s
High frequencies	3 MHz–30 MHz	Shortwave radio	1930s
Very high frequencies	30 MHz–300 MHz	FM radio, broadcast TV	1940s
Ultra high frequencies	300 MHz–3 GHz	Broadcast TV, mobile phones (1G, 2G, 3G, 4G), cordless phones, wi-fi, Bluetooth, satellite radio	1950s
Super High Frequencies	3 GHz–30 GHz	Fixed microwave links, wi-fi, cordless phones, satellite TV, "wireless fiber," 5G	1950s–1990s
Extremely high frequencies	30 GHz–300 GHz	Short-range wireless data links, remote sensing, radio astronomy, 5G	1990s

Sources: Thomas W. Hazlett, *Optimal Abolition of FCC Spectrum Allocation*, 22 JOURNAL OF ECONOMIC PERSPECTIVES 103 (Winter 2008), 107; 4G Americas, *5G Spectrum Recommendations*, White Paper (Aug. 2015).

majority of productive wireless applications access frequencies of less than 60 GHz, and while frequencies under about 6 GHz are still considered "beachfront property," technological progress has opened vast new vistas. The relentless climb through frequency space is reflected in spectrum taxonomy, ascending from medium to high to very high to ultra high to super high to extremely high frequencies. Scientists may be running out of modifiers, but they were alert to have adopted logarithmic scales early on. There is now some 100,000 times more usable spectrum than there was when Congress declared it a scarce resource in 1927.

Intensive Margin. Increasing traffic capacity within a given band expands the intensive margin. The precellular mobile telephone transmissions that began in the United States in 1946 consumed 120 KHz of bandwidth per circuit (each circuit hosting one conversation). By 1950 they used only 60 KHz; 50 KHz in the mid-1950s; and 25 KHz by the mid-1960s. Better radios and smaller channels enabled more information to flow.

Progress accelerated with first-generation (1G) analog cellular networks (with U.S. licenses distributed between 1983 and 1989). These produced huge gains in capacity by splitting up markets into sectors or "cells." Radios powered down, transmitting just to the nearest base station, which connected the mobile user with the "public switched network" linked to everyone else.[42] This allowed each cell to reuse the allocated bandwidth. Only a few dozen subscribers could simultaneously make phone calls in St. Louis, Missouri, in the 1970s, as residents vied for a small number of area-wide circuits. Cellular, however, enabled thousands to chat all at once using similar bandwidth.[43]

By the 1990s, U.S. 1G networks were being upgraded to 2G. These new networks abandoned analog for digital standards, delivering higher-quality voice while adding data services. Driven by Moore's Law (processing power, for a computer chip of a given cost, doubles about every eighteen months), networks became more capacious and functional.

Let the upgrades roll. In just over a decade—from 2G systems deploying digital data services in 2000 (GPRS, or General Packet Radio Service) through 3G systems (CDMA2000 and WCDMA) deployed in the early to mid-2000s, to 4G LTE (long-term evolution) systems deployed starting in 2010—standard speeds for mobile data applications rose from 9.6 kilobits per second (kbps) to more than 10 megabits per second (mpbs), a thousand-fold increase.[44]

For a given network, bandwidth bolsters speed and capacity. In 2012, *PC* magazine tested AT&T and T-Mobile wireless download speeds in thirty major U.S. markets. "All other things being equal," its reviewer noted, "less spectrum means slower speeds and more congestion." Under the subheading "Why Spectrum Matters," the magazine added, "If you want better wireless Internet, you need to choose a carrier who knows the answer."[45]

The first cellular telephone call, wild and untethered, was made in October 1973 from the corner of Sixth Avenue and 54th Street in New York City. Marty Cooper, a radio frequency engineer and vice president of Motorola, led the team that was about to drop an invisible dime. He elected to dial the person least likely to celebrate: Joel Engel, an AT&T engineer working on a competing project at Bell Labs in New Jersey.

> "'Joel, this is Marty Cooper. . . . I'm calling you from a cellular phone, but a real cellular phone, a hand-held portable cellular phone.' And there was silence on the end of the line," Cooper remembered.[46]

The annoying taunt bent history. Or did it? From Marconi to Cooper, there had been steep progress, which continues. In one century our ability to use the airwaves for communications has increased about a trillionfold. Cooper attributes these astounding gains to advances in four areas:

- Frequency division
- Modulation techniques
- Spatial division
- Increase in magnitude of the usable radio frequency spectrum

The last pushes out the extensive margin, the first three the intensive margin. Cooper estimates that "of the million times improvement in the last 45 years, roughly 25X is attributable to the extensive margin, 5X to frequency division; 5X to modulation advances [and] 1600X to spatial division."[47]

This numerical endorsement of cellular technology—largely accounting for the stunning gains in spatial efficiencies—testifies to the value of quiet. Whereas high-powered television stations blast energy over large areas from one central location, cellular systems produce millions of tiny "broadcasts" across tightly packed slivers of space. Capacity is theoretically limitless: just add cells. Alas, this "densification" is not free. A leading scientist in the precellular 1950s explained the basic reality: "We . . . have found more efficient ways to use spectrum. . . . However, all of those uses generally involve a decision between a problem of how much money you want to spend and what you want to get accomplished. Almost any method of making use of the spectrum more efficiently requires much more elaborate apparatus at both the transmitter and the receiver."[48]

Today the air is charged by the buzz over 5G. Samsung's trials on the streets of New York deliver more than a gigabit per second over a distance of "up to two kilometers." This tops 4G LTE (long-term evolution, the leading fourth-generation standard) by well over ten to one. Samsung is experimenting with spacious 28 GHz frequencies, vast and uncrowded. An engineer will say that this relative lack of use is because signals of that frequency travel light, fade quickly, and are flummoxed by fog, fowl, or foliage. The economist replies that some ingenious engineer will soon make these airwaves pay.

Both are right. "Samsung says it has greatly mitigated these problems by sending data over any of 64 antennas, dynamically shaping how the signal is divided up, and even controlling the direction in which it is sent, making changes in tens of nanoseconds in response to changing conditions." In short, the network will better target each beam—spatial division to the nth power.[49]

Going forward, mobile networks (actually, the networks are fixed—it's the users who are mobile) will rely on radical localization. Technology suppliers will put base stations in your living room.[50] Your house becomes a cell, and your family enjoys all of a carrier's spectrum allocation to itself. Carriers are mixing and matching networks, shifting data traffic to fixed home, office, or Starbucks via wi-fi connections, and stretching wide-area bandwidth (via cellular licenses) farther by using smaller cells. So networks subdivide—splitting *macrocells* (covering a several-mile radius) into *microcells* (one to two kilometers) and *picocells* (one hundred meters) and *femtocells* (ten meters).

But more cells may not be the best answer. Perhaps, more bandwidth—or different frequencies, new modulation techniques, tweaked business models, or wired connections to every seat in the theater or football stadium—will keep wireless customers just as happy. Or maybe the spectrum should be diverted, at least in part, for more important uses like emergency services. Or pizza delivery. There are countless options.

The wireless entrepreneur searches for those that work best, comparing input costs (what resources are worth to customers elsewhere) with output price (what value they create in a given wireless deployment), and so calculates profit. That is her quest. It is also her risk, motivating her to seek and exploit every little efficiency. When value is added and profits flow, the funds generate a bounty—to enjoy and to share with partners, bankers, investors, and suppliers who help make the happy result happen. This elicits cooperation from strategic players who might otherwise not pitch in.

But in wireless markets this change agent can be severely handicapped. Spectrum is a key variable in the efficiency equation, yet the Spectrum Store keeps dreadfully short hours and stocks laughably sparse inventories. In a different world, the marketplace would allow entrepreneurs to sample frequencies, gauge their performance, and see their prices. Airwaves used one way would be available to accommodate more pressing needs. With competing entrepreneurs seeking to buy low and sell high, underutilized bands would turn into busier and more productive frequencies. But the task is subtle, incorporating guesses as to future consumer behavior, technology trends, and evolving market rivalry. These baffle the best of experts. The words of Arthur Schopenhauer apply: "Talent hits a mark no one else can hit. Genius hits the mark others cannot see."

Degree of difficulty is here exacerbated by nondivine intervention: the Spectrum Store is not your friendly neighborhood lumberyard. Under traditional allocation rules, spectrum is put to use via specific directives issued by regulators. These can be fashioned such that conflicts between radio users are limited. "Tight regulatory control over the use of spectrum [enables] the regulator to ensure that excessive interference does not occur," as a recent textbook by British spectrum experts Martin Cave, Chris Doyle, and William Webb put it.[51]

But define "excessive" given that "Interference occurs in all radio systems."[52] There is no doubt some important interest in protecting existing operations, but there is a pull the other way to welcome new technologies and innovative applications. These involve a clash of interests. The optimal balancing point is invisible. It is determined by how firms and individuals think about things that have yet to happen. As Coase, Meckling, and Minasian elucidated in their once controversial study,

> There are various combinations of resources—transmission power, antenna height and directivity, frequency of transmission, method of propagation, etc.—that can be utilized to achieve a given level of (received) power at a point distant from the point of transmission. The range of alternative combinations is determined by technology—the state of the arts—and is an engineering problem. The "proper" combination actually to use to achieve

a given goal is, however, an economic problem and is not (properly) soluble solely in terms of engineering data.[53]

This passage exudes masterful clarity. And the authors took their crisp insight farther, illuminating processes that might most reliably discover the information needed to realize society's best outcomes. In this, the economists did not have to reinvent the wheel. They found markets provided, in myriad other contexts, the incentives and counterpressures to overcome stasis. With these competitive forces, far more could be made of the airwaves. While advances in wireless may look smooth and relentless when viewed from high altitude—and Cooper's Law an inexorable fact of nature—they actually come in fits and starts. Until a decades-long battle to make airwaves available for cellular phone service was won, for instance, experiments were stymied, deployments were thwarted, and ignorance about future possibilities reigned. "Who's going to develop a promising technology," asks former FCC Chief Engineer Michael Marcus, "when the only thing an investor obtains is the right to hire a lawyer to lobby the FCC for permission to deploy?"[54]

Coase grasped the importance of the Spectrum Store. The government allocation system "lacks the precise monetary measure of benefit and cost provided by the market," and regulators "cannot . . . be in possession of the relevant information possessed by the managers of every business which uses or might use radio frequencies, to say nothing of the preferences of consumers for the various goods and services in the production of which radio frequencies could be used." He quoted FCC member Robert E. Lee, exasperated over a 1950s controversy: "I am finding it increasingly difficult to explain why a steel company in a large community, desperate for additional frequency space, cannot use a frequency assigned, let us say, to the forest service in an area where there are not trees."[55]

With markets, that problem is easily resolved. Coase saw the forest.

PART TWO

SILENCE OF THE ENTRANTS

One of the great mistakes is to judge policies and programs by their
intentions rather than their results.

—Milton Friedman

THE RADIO ACT OF 1927, later incorporated in its entirety into the 1934
Communications Act, claimed to bring order to the airwaves. It did that, in
a manner, and much more. In preempting the development of markets in
radio spectrum, it instituted a system of "public interest" regulation. Over
the ensuing decades, a string of communications suppliers would try to
enter existing or newly imagined service spaces. Many of these initiatives
were driven by the development of new technologies; in other cases, entrepreneurs sought to satisfy demands unmet by existing firms. The resulting
struggles graphically demonstrate key dynamics of the political spectrum.

6

The Death of DuMont

Concentrated power is not rendered harmless by the good intentions
of those who create it.

— Milton Friedman

IN THE 1970S, VIEWERS COULD turn on their television sets and find their
choices limited to NBC, ABC, and CBS, with—occasionally—PBS or an
unaffiliated station popping up. New York, the largest media market in the
country, boasted three independents. The Brookings Institution scholar and
former FCC economist Robert Crandall posed the mystery aptly in 1978:

> As viewers of American television flip across their dials . . . they are silently
> mocked by seventy-five or more channel demarcations where their sets
> do not respond. Instead of many choices of program, they must reconcile
> themselves to three, four, or perhaps five.
> Sometimes they must wonder why. . . . The real reason has little if any-
> thing to do with electronic phenomena—with either a shortage of channels
> or, as some would have it, the inherent inferiority of UHF. Rather, the limita-
> tion exists because of the FCC's desire to make sure that viewers are offered
> a big dollop of edification with each swallow of entertainment no matter
> how edifying the edification or how entertaining the entertainment.[1]

Between 1941 and 1953, the FCC had allocated a huge swath of spectrum
for television broadcasting, setting aside eighty-one channels consuming

486 MHz of prime frequencies (6 MHz per channel).[2] In issuing broadcasting licenses to fill this vast space, the FCC was not content to be a mere traffic cop but instead sought to carefully plan the structure and form of television service. The key tool was its ability to severely restrict the number of competing stations, making the artificially scarce licenses extremely valuable. The FCC assigned them without monetary payment, but licensees had to pledge to satisfy a vaguely defined public interest standard.

This plan dictated technology and erected insurmountable barriers to entry. As bad as the monopoly power it created, the lock-in of inefficient airwave use was worse. As one set of experts commented decades later: "The effect of this policy has been to create a system of powerful vested interests, which continue to stand in the path of reform and change."[3]

The Federal Communications Commission began allocating VHF radio spectrum for television in 1939, and in 1945 it established a plan for channels in the top 140 markets.[4] But three years later, prompted by the rush of postwar applications, it froze assignments after having issued 108 VHF TV station licenses and no UHF licenses. It would issue no new licenses for another four years.

The spectrum allocation decision facing the FCC was profound in its implications: the agency could define its task as bringing television *service* to all parts of the country—or television *stations*. The life or death of networks depended on the difference. In 1952, the FCC announced its choice, issuing a TV allocation table in its heralded Sixth Report and Order.[5]

DuMont, the weakest of the four TV networks then operating via the 108 licenses issued before the freeze, had aggressively pushed the FCC to support competition by emphasizing the scope of television service rather than the number of local stations. It submitted detailed proposals with sophisticated analytics. To survive, it needed a plan in which virtually all American households could tune in to four or more stations. Without such an opportunity for nationwide coverage it would be unable to match its rivals' economies of scale. With smaller audiences, it would have to charge less for advertising, reducing its revenues and diminishing what it could afford to pay for program production. Audience share would decline still further, leading to a financial death spiral.

The alternative position, favored by the larger networks, particularly CBS, would "provide each community with at least one television broadcast service."[6] The localists knew that increasing the number of cities with three TV stations would force the FCC to leave fewer markets with four or more.

Consider two rival plans for the northeastern United States.

Plan One: The Commission assigns licenses to six of the twelve VHF channels (2–13), locating them in New Haven, Connecticut, and allowing them sufficient power to reach homes from Boston to Philadelphia. The stations all transmit from one spot in New Haven so that the every-other-channel allocation is sufficient to eliminate almost all cross-station interference even on inexpensive TV tuners. This approach allows everyone living in the Boston–New York–Philadelphia corridor to receive five commercial stations and one noncommercial station, with distinct programs on each. But the stations are regional, not local.[7]

Plan Two: The Commission ensures that there are TV stations in Philadelphia, Trenton, Albany, New York City, New Haven, Hartford, Manchester, Providence, and Boston. Putting stations in more markets means that, to mitigate interference between them, the FCC imposes lower power limits, restricting how far signals can travel, and leaves many local channels idle—"taboo" in Commission terminology. If Channel 4 is on the air in Boston, it is probably blacked out in Providence, Hartford, Albany, and Manchester. These taboo channels shrink the TV dial, but more cities have stations.

DuMont pleaded for Plan One. But the FCC chose Door No. 2.

> If it were not for the FCC's TV allocation plan, which created low power, local stations, we could all have access to a great many more channels. . . . The essence of the DuMont Plan was to have fewer cities with TV stations, but to have each station cover a large geographical area, spanning a number of cities. Such a plan would permit the creation of new networks and increase the number of choices available to each viewer . . . increas[ing] diversity of programming.[8]

Give television pioneer Allen B. DuMont credit. He was a technological innovator of the first order, and he had the derring-do to create imaginative content. DuMont launched Jackie Gleason's *The Honeymooners,* as well as the science fiction of *Captain Video and His Video Rangers,* written by, among

others, Arthur C. Clarke.[9] And he had the sense to strenuously object to the
FCC plan, which claimed to distribute licenses so that

- 100 percent of U.S. households could receive at least one TV
  station;
- 98.6 percent could receive at least two stations;
- 90.0 percent could receive at least three stations;
- 81.4 percent could receive at least four stations;
- 71.9 percent could receive at least five stations;
- 55.0 percent could receive at least six stations.[10]

A casual reading might not reveal how this allocation would crush network competition. Yes, a third national network appears to suffer a coverage disadvantage of only about 10 percent compared with the top two, and
the fourth network a disadvantage of less than 20 percent. (This assumes
that the top network is the first to acquire an affiliate in each market; the
second-largest network then acquires the next affiliate; and so on.) In fact,
the competitive disparities were far larger due to huge differences between
VHF (channels 2–13) and UHF (14–83). To deliver a video broadcast over
the same geographic space as a VHF transmission, a UHF station would
have required ten times the power, illegal under FCC rules. Even if allowed,
the UHF tuners in TV sets were largely incapable of the clarity of even a
cheap VHF tuner.[11]

The "UHF handicap" was severe. But the FCC looked past it on the glib
assumption that UHF technology would quickly improve and the disparity
would disappear. That projection was convenient but wrong. In 1959 the
FCC concluded that the allocation plan had been a "dismal failure." In
1961, it found that "commercial UHF stations cannot compete with VHF
stations on anything like an equal basis."[12]

Withdraw that faulty assumption, and the FCC's calculations for the
1952 TV allocation table reveal the competitive carnage to come. An inventory of "technically comparable" coverage areas shows that there was just
one market in the United States where viewers could receive as many as
seven TV stations, and that this market accounted for less than 1.6 million
households—6.0 percent of the nationwide total. Fourteen markets offered

Table 6.1. Quality-Adjusted Broadcast TV Coverage in 1952 FCC Plan

Stations per Market	Number of Markets	Cumulative Markets	TV Households	% of All TV HHs	Cumulative TV HHs	Cumulative % TV HHs
7	1	1	1,578,216	6.0	1,578,216	6.0
6	1	2	3,950,412	15.0	5,528,628	21.0
5	1	3	38,034	0.1	5,566,662	21.1
4	14	17	4,299,762	16.2	9,866,424	37.3
3	36	53	7,666,316	29.0	17,532,740	66.3
2	57	110	5,913,824	22.3	23,446,564	88.6
1	52	162	3,027,171	11.4	26,473,735	100.0

Source: Thomas L. Schuessler, *Structural Barriers to the Entry of Additional Television Networks: The Federal Communications Commission's Spectrum Management Policies,* 54 SOUTHERN CALIFORNIA LAW REVIEW 875 (1981), 938 (Table 14). Coverage calculations include both VHF and UHF stations, using "technically comparable" contours per the FCC's *Sixth Report & Order* (1952).

reception of four or more stations, accounting for about 37 percent of the U.S. population. Nearly two-thirds of the U.S. population was stuck with three or fewer TV channels.

Coverage areas were radically reduced from the initial FCC estimates when adjusted for real-world conditions—that is, the UHF handicap. It was not the case that 81.4 percent of households could receive at least four stations; only 37.3 percent could. It became apparent that a third network would struggle, while the fourth would languish. DuMont saw the threat and pushed a rival plan under which each of four broadcast networks could gain coverage in 95 percent of U.S. homes.

In the DuMont plan, moreover, with the lone exception of Baltimore, every one of the top one hundred markets avoided "intermixture." This meant that each market was either exclusively VHF or exclusively UHF. VHF stations would compete with other V's; UHF stations with other U's. This would better allow the UHF stations to successfully vie for audience share, as it would encourage viewers in UHF-served areas to invest in UHF tuners and antennas. In contrast, the plan adopted by the FCC featured twenty-one intermixed markets, pitting outmatched UHF stations against

Table 6.2. Broadcast TV Coverage in 1952 DuMont Plan

Stations per Market	Number of Markets	Cumulative Markets	TV Households	% of All TV HHs	Cumulative TV HHs	Cumulative % TV HHs
7	6	6	9,332,818	35.5	9,332,818	35.3
6	1	7	346,016	1.3	9,678,834	36.6
5	13	20	3,968,651	15.0	13,647,485	51.6
4	113	133	11,382,908	43.0	25,030,393	94.6
3	8	141	597,783	2.2	25,628,176	96.8
2	15	156	845,559	3.2	26,473,735	100.0
1	0	156	0	0	26,473,735	100.0

Source: Schuessler (1981), 928 (Table 8).

VHF stations, spelling doom for the former and dominance for the latter. As well as being better for DuMont, the alternative TV allocation plan was spectacularly better for viewer choice than the FCC's.

"The DuMont project," a 1958 congressional report noted, "was exemplary for its breadth of understanding of the problem and for its professional quality."[13] It rejected unwarranted assumptions and proposed licenses that would support the competitive networks that had arisen. FCC regulators were less impressed. Throwing doubts about UHF to the wind, they imposed their own allocation plan. Allen DuMont's gloomy prediction soon came true: in September 1955 the DuMont Network was shuttered. "It is widely recognized that the collective set of spectrum allocation policies at the FCC . . . posed an impenetrable barrier to the entry of a fourth national television network."[14] None would emerge until 1986, when the Fox Television Network launched.[15]

It is apparent what NBC, CBS, and ABC got from this deal. But "localism," the public interest justification, was not the result. In 1952, the Commission's *Sixth Report* called for some 2,002 TV stations, 1,433 (72 percent) of them UHF. Yet a decade later, there were just 102 UHF stations on the air—a decline from the high-water mark in 1954, when 121 operated. Very few of the small-market UHF stations survived to broadcast at all.

6.1. An artistic rendition of the DuMont Television Network logo. The broadcaster, which launched in 1946, went dark in 1955—murder by spectrum allocation. U.S. TV licenses were assigned such that no more than three commercial stations could reach most households.

Some regulators may have genuinely believed that the FCC plan would serve the goal of localism. But a Brookings Institution study, published in 1973, suggested otherwise: "Statistical analysis reveals that the FCC has abandoned its stated policy objectives in granting licenses. Local ownership, news and public affairs programming, and local origination [of programs] all *detract* from the likely success of an application" for a TV license.[16]

If the Commission did not actually promote localism, what did it do? Northwestern University law and economics professor Matthew Spitzer commented that "the FCC's primary goal appears to have been to place at least one over-the-air service in every large community. This corresponds neatly to placing broadcasting stations in as many congressional districts as possible."[17] The government's preference for wide distribution of federal largesse is well known. An old Washington joke has it that the optimal weapons system is a tank that may or may not fire but is manufactured in all 435 congressional districts.

In a comprehensive 1981 study, Thomas Schuessler of the University of Arizona Law School concluded that "the three existing networks will continue to operate in the relatively competition-free haven bestowed on them by the Federal Communications Commission."[18] In fact, Schuessler's excellent historical analysis notwithstanding, the triopolistic haven was about to end. The deregulation of cable television would unleash the competitive

Table 6.3. TV Station Allocations in 1952 and TV Stations Broadcasting in 1962

| | 1952 Allocation in FCC's Sixth Report & Order | | |
	No. of UHF Stations	No. of VHF Stations	Total No. of Stations
Commercial	1,271	498	1,769
Educational	162	71	233
Total	1,433	569	2,002
	On the air in 1962 (% of assignments)		
Commercial	83 (6.5)	458 (92.0)	541 (30.6)
Educational	19 (11.7)	43 (60.6)	62 (26.6)
Total	102 (7.1)	501 (88.0)	603 (30.1)

Source: Douglas Webbink, *The Impact of UHF Promotion: The All-Channel Television Receiver Law*, 34 LAW AND CONTEMPORARY PROBLEMS 535 (Summer 1969), 546.

assault that hidebound regulators and nervous TV station owners had long feared. But why did it take until the 1980s? For one, because the FCC and Congress had declared years earlier that they had solved the competitive problem in television with the All Channel Receiver Act of 1962.[19]

Years later, Newton Minow was still claiming it. Seeing how cable television and satellite-delivered video programming had rocked the "vast wasteland," Minow claimed in 1995 that his 1961 speech had "accomplished two things: it provoked the wrath of the broadcasters . . . and it helped wake up the American people to the fact that television had fallen far short of its creators' goals. . . . Thus the commission worked to promote cable and other new technologies, to increase the number of educational stations, and, in 1962, to retrieve them from the twilight of UHF by requiring all set manufacturers to make their sets capable of receiving UHF signals."[20]

We all forget history.

The All-Channel Receiver Act (which the FCC proposed to Congress and lobbied for) mandated that, starting in 1964, all TV sets sold in the U.S. include tuners that covered both the VHF and UHF TV bands. Yet UHF broadcasts remained inferior when viewed on standard television receivers.

Table 6.4. Maximum Broadcast TV Coverage for a Fourth Network, 1952–1979

Network Rank	1952 Coverage (% U.S. TV Households)	1965 Coverage (% U.S. TV Households)	1979 Coverage (% U.S. TV Households)
4th	37.3	28.4	35.8
5th	21.1	12.5	15.7
6th	21.0	11.1	15.4

Source: Schuessler (1981), 987 (Table 33).

Even when nearly all TV sets in use contained all-channel tuners, UHF stations languished. A 1978 U.S. Senate report announced: "The intent of the All-Channel Receiver Act of 1962 has not been realized. UHF television broadcasting remains sorely disadvantaged within the national television system."[21]

In response, the FCC formed a new task force, which in 1980 reported that "neither changes in broadcast power nor in receiver performance could be expected to greatly reduce the UHF reception disadvantage."[22] The Commission created, and until 2016 observed, a "UHF discount." When determining how many broadcast TV licenses a company owned, for purposes of limiting ownership concentration in a policy crafted in 2003, it counted a UHF station as just one-half of a VHF station.[23]

Why did the FCC call for the All-Channel Receiver Act? In the 1950s, the UHF band had proven an economic killing field. A 1959 FCC report noted that nearly one-half of the 369 UHF construction permits the Commission had issued had been given back to the government, and that 90 of the 165 UHF stations that had gone on the air had subsequently failed.[24] This carnage was a political problem for regulators. While protecting the Big Three commercial networks in the 1952 TV allocation table, the FCC had always pointed to the availability of UHF channels as fertile ground for competitive growth. The space it had reserved for thousands of UHF outlets formed the crux of the agency's argument that it had not simply done the bidding of the major networks.

But with few UHF stations on the air and fewer still offering popular programming, customers were refusing to buy TV sets that included, at a higher price, UHF tuners—so manufacturers stopped making them. The fraction of televisions equipped to receive UHF signals fell from around 20 percent in 1953 to just 6 percent in 1961.[25] Consumers resisted paying for an extra function that did not really function.

Plans were floated to fix the failure. One idea was to homogenize the broadcast bands, going all-VHF or all-UHF. Another, to expand VHF from twelve to fifty channels, was discussed in a 1959 FCC report. While conceding that this was the preferred approach, the Commission was stymied. There was no practical way to reallocate the additional VHF bandwidth from other services, including fire and police departments. The idea was dropped.[26]

The reform that received the most consideration was "de-intermixture." The FCC would go market by market, establishing uniformity in each. Some (large) markets would be allowed six or seven VHF stations and no UHF. Others would be allocated six or seven UHF stations and no VHF. This would limit transition costs and permit the development of as many as seven national networks, a substantial improvement for viewer choice.

The three existing networks fiercely opposed de-intermixture, as did their VHF affiliates. A compromise was hatched: de-intermixture was dropped, and instead, Congress would mandate UHF dials on TVs. FCC member Robert E. Lee made the bargain explicit: "Congress in effect made a deal with the Commission—drop de-intermixture, and we get the all-channel television bill."[27]

But requiring UHF tuners did not achieve much because the tuners did not do much. VHF stations still came in with far more clarity, over longer distances. Douglas Webbink found that the number of UHF stations did increase following the all-channel mandate, but that the entrants were mainly educational stations operating with direct subsidies.[28]

The vast majority of UHF TV licenses allocated by the FCC—some 1,543—remained unclaimed.[29] UHF broadcasters were, as a group, losing money. In 1966, the FCC implicitly conceded the trend when it dramatically reduced the number of available commercial TV licenses from 1,271 to 590.[30] A 1970 article in the *Harvard Law Review* concluded that the mandate had proven "inadequate. . . . UHF stations which have survived remain anemic in finances and derivative in programming."[31] In 1972, the economist Rolla Park wrote that "UHF broadcasting remains a marginal proposition. . . . Only about one-third of all allocations for commercial UHF stations in the top 100 markets are in use."[32] Former FCC General Counsel Henry Geller wrote three years later that the Commission had left UHF TV "crippled and struggling."[33]

Nonetheless, the "all-channel" rescue effort was expensive. The FCC had claimed that after the mandate was implemented, there had been no discernible TV set price increase.[34] Webbink, however, estimated that the mandate cost consumers an extra $85 million to $110 million annually, at least $500 million between 1964 and 1968. His conclusion: "The All-Channel Receiver Law was a mistake."[35]

Looked at globally, broadcast television was a socially transformative technology. In important dimensions, federal spectrum regulators had never had a more influential role to play. But on nearly every margin the "public interest" determinations made to allocate radio spectrum for broadcast TV undermined competitive forces and innovative ideas that might have delivered much more robust consumer choice, free expression, and economic growth. The daring DuMont Network died for the FCC's sins. Other enterprising creations were deterred altogether. And the exclusions would last generations, as the regime set aside copious bandwidth that would become locked into an underperforming and obsolete architectural design, blocking vast emerging opportunities. That resulted in a spectrum policy crisis that entrepreneurs and regulators would eventually confront. They are dealing with it this very day.

7

"Thank God for C-SPAN!"

The government has almost always acted to restrict and restrain competition and output in television markets in order to protect the economic interests of members of the industry. . . . In this respect the FCC has behaved no differently than various now defunct federal regulatory agencies such as the Civil Aeronautics Board and the Interstate Commerce Commission.

—Bruce M. Owen

MAHANOY CITY, PENNSYLVANIA, is nestled in the Appalachian Mountains, eighty-six miles from Philadelphia. Broadcast TV signals reach it, but not perfectly and not everywhere. John Walson worked in the community as a lineman for Pennsylvania Power and Light in the years following World War II. On the side, he ran a small consumer electronics store. In 1947, when the first black-and-white TV sets arrived with their 12½-inch screens, they carried price tags of $450 to $575—two months' wages for the average U.S. worker. To demonstrate the items, Walson would load customers and sets into his pickup truck, scale nearby New Boston Mountain, and plug into an antenna attached to a seventy-foot pole.

Then the brainstorm: cut out the truck. Get PP&L to run a line from the hilltop to his shop, and transmit the Philadelphia broadcast signals to TV screens in his store. He did. But then customers asked to buy not only Walson's showroom TV sets but also his showroom reception. So he strung more lines. His business went from retailing electronics to building and operating a communications network. Residential customers paid two dollars a month for antenna service. By mid-1948, Service Electric had 727 subscribers. An industry was born.[1]

Cable television was "spectrum in a tube." It featured the same basic technology as broadcasting, except that (to reprise Einstein's fabled analogy) it restored the cat. The network created its own bandwidth, taking TV programs where they did not otherwise reach. "CATV"—community antenna television—systems popped up throughout the 1950s, serving two million households by 1960.

Then, in the early 1960s, something crazy happened. Cable TV systems began to serve not just the TV-deprived neighborhoods beyond the reach of off-air reception, but markets where off-air television was well established. To protect their turf against a growing competitor, incumbent broadcasters began demanding that the FCC take action. The broadcasters had two lines of attack. One was that cable was stealing broadcast signals, retransmitting them for a fee, and not paying for the content. That legal issue concerned copyright, and the U.S. Supreme Court twice denied the broadcasters' claims. In the *Fortnightly* (1968) and *Teleprompter* (1974) decisions, it found that CATV systems simply extended signals and did not appropriate them, and that cable operators were no more infringing on broadcasters' copyrights than the manufacturers of television sets or antennas were. In any event, protection of intellectual property was not the FCC's job. In the 1976 Copyright Act, Congress amended the law to give some protection to broadcasters' content ownership rights.

The second anticable angle concerned the broadcasters' obligations under the public interest standard. Were new competitors to reduce their profits, broadcast TV licensees would be less able to fulfill the promises made to supply news, information, and educational programming. This, the argument went, would diminish the public interest.

The Commission had refused calls to regulate cable TV systems in the 1950s. But when cable began competing with local broadcasters, the FCC changed its mind.[2] In the broadcasters' first major victory, *Carter Mountain Transmission Co. v. FCC*,[3] the Commission refused to issue a microwave license to a radio operator who wished to transport broadcast TV signals for delivery to a cable TV operator in Wyoming. A Wyoming TV broadcast station had filed the complaint in order to alert the Commission to the danger of allowing existing stations to be challenged by cable upstarts.

A stream of new rulemakings followed *Carter Mountain*. The premise was that cable TV was simply a niche product and that it ought not to be allowed to disrupt the television market. As the FCC wrote in 1965, it realized the "valuable contribution of CATV in bringing new . . . service to many places," even as "CATV service should be supplementary to and not cripple local TV broadcast service."[4] No demonstrable harm had yet come to broadcast TV, but the Commission feared that the public interest would suffer if it did. "The essential purpose of our policy is to take hold of the future—to insure a situation where we or the Congress, if it chooses, can make the fundamental decision in the public interest on the basis of adequate knowledge."[5]

The FCC claimed its position was a defense not of the powerful and highly profitable commercial TV networks, but of fledgling UHF operators too weak to resist market forces. The "unique" all-channel TV tuner mandate in 1962 had established that "Congress and the American public have staked a great deal on the development of UHF."[6] If unregulated cable TV systems brought new competitive content into a market, they could "siphon" audiences. If a cable system "should destroy a broadcasting operation, the CATV would not thereafter 'render the required service,'"[7] including the local, educational, news, and public affairs programs vital to the health of democracy. New rules were thus needed "to prevent or ameliorate adverse impact of CATV competition."[8]

It is important to understand the FCC's tortured reasoning. While the agency had set aside enough spectrum to accommodate eighty-one broadcasting channels per market, it had imposed technical constraints and licensing choices that resulted in only a handful of strong broadcasts in each city and just three commercial networks nationwide. UHF channels had little hope of challenging the Big Three, and as we have seen, the assignments rarely resulted in success. The few stations that survived attracted minuscule audiences (they were sometimes called "UFO stations," a comment on how often they were seen).

Nonetheless, UHF figured prominently in FCC "public interest" determinations. It gave regulators the perfect poster child. When cable TV looked as if it might intensify video competition by winning over VHF audiences, the FCC leapt to take hold of the future, citing the potential damage the new rivalry might bring not to lucrative VHF stations but to the floundering UHF. The danger was imminent and the threat was to be nipped in the bud. "We cannot sit back," the Commission wrote in 1966, "and let CATV move signals about as it wishes, and then if the answer some years from now is that CATV can and does undermine the development of UHF, simply say, 'Oh, well, so sorry that we didn't look into the matter.'"[9]

Incumbent TV station profits had morphed into a proxy for the "public interest." The benefits of pro-consumer competition were acknowledged— "CATV . . . makes possible the provision of a variety of program choices, particularly the three full network services, to many persons"—but instantly dismissed. Cable TV was "a service which, technically, cannot be made available to many people and which, practically, will not be available to many others." The prediction fared poorly when tested in actual markets, but it provided justification enough for the Commission to then craft rules to "ameliorate the risk that the burgeoning CATV industry" posed for "television broadcast service, both existing and potential."[10]

Fledgling cable TV operators gamely fought back. They retained Herbert Arkin, statistics professor at City College of New York, to show that cable television did not affect TV station revenues and was unlikely to do so.[11] It was a perverse argument that accepted the protectionist premise of FCC policy, but that was how the game was played. The National Association of Broadcasters countered by investing $25,000 to commission a study led by the respected MIT economist Franklin Fisher. This analysis, sufficiently rigorous that it was published in the venerable *Quarterly Journal of Economics*,[12] predicted that CATV systems would offer additional channels and fragment television viewership. Finding that higher audiences were positively correlated with higher station revenues, Fisher and his colleagues concluded that unimpeded competition would reduce UHF station revenues in future years. The Commission found the analysis compelling, and "the bulk of the [FCC's stated] evidence came from 'the Fisher Report.'"[13]

But Fisher's main prediction—that UHF stations are "likely to be discouraged from entry by [even] a relatively small increase in CATV penetration"—

turned out to be false. The key omitted variable in his study was signal clarity. His econometric analysis did not account for the positive effect of CATV—retransmitting local broadcast TV signals—in eliminating the "UHF handicap." Excluding this impact led Fisher to report that existing TV broadcasters would be harmed by competitive entry and that "the impetus of CATV is toward a reduction in local television service."[14]

Others reached different conclusions. The RAND Corporation economist Rolla Park, in a 1970 study, found that "even with 100 percent cable penetration, the balance works out in my model so that cable helps UHF independents."[15] Park's research, though influential at the FCC, came too late to stop the agency from erecting barriers that blocked cable TV growth for well over a decade. In this, it followed a well-established pattern. As Ithiel de Sola Pool put it, "In all countries, cable television's advocates have had to combat established broadcasters for the right to create cable systems."[16] That the FCC had no authorization (in the 1934 Communications Act or elsewhere) to regulate cable TV systems was but a modest impediment. It asserted that it was excluding new media merely to protect its authority over old media. In 1972 the U.S. Supreme Court signed off, ruling that cable regulation was "reasonably ancillary" to the FCC's statutory purpose to foster broadcasting, a service that could be "placed in jeopardy by the unregulated growth of CATV."[17]

The circularity of this argument bears note. Broadcasting was regulated because spectrum was a "physically scarce" resource limited by nature. But when "spectrum in a tube" promised (or threatened) to relax that scarcity, the government was allowed to extend its powers to preserve the very limitations that justified regulation in the first place. In Pool's words, the FCC put its "thumb on cable."[18]

From 1965 through 1970, the FCC aggressively discouraged the upstart rival by imposing federal franchises and "anti-siphoning rules." The agency imposed arcane, detailed mandates, "assum[ing] the characteristics of a central planner,"[19] banning CATV systems from featuring:

- movies that had been in theatrical release more than two years before the cablecast;

- sporting events that had been telecast in the community on a non-subscription basis during the previous two years; and
- series programming of any type.

The "traffic cop" for the airwaves was now prohibiting episodic TV programs over wires. *Mad Men, Breaking Bad, Duck Dynasty,* and *Game of Thrones* would have been illegal in the 1960s. With these restrictions, cable remained a small, mostly rural service that, according to scholars Stan Besen and Robert Crandall, was "held hostage by television broadcasters to the Commission's hope for the development of UHF."[20] Alas, the policy did not even help its intended beneficiaries.[21]

This anticompetitive equilibrium held too long, but not forever. Despite the broadcasters' best efforts, the idea of a "wired nation" excited media activists: "In 1970," wrote the UCLA law professor Monroe Price, "dreams of a new communications system were upon us, and vast implications were being forecast."[22] The promise of leapfrogging the scarcity imposed by the FCC generated waves of hope and hype. "The new technology," according to Price, "was supposed to enrich bankrupt cities, bring ballet to the balleto-mane, challenge the power and influence of AT&T, make our streets secure, disalienate youth, and deliver the paper in the morning." New networks would help the poor find jobs, expand adult literacy, and "improve information on dental care."[23] New York City Mayor John Lindsay was exuberant about cable television franchises, "urban oil wells under our city streets."[24]

Broadcasters were also investing in cable, both to hedge their bets and to exploit its potential as a channel for more creative video content. Programmers were, at the same moment, eager to secure their rights. In this mix of excitement and fear, the various parts of the television business came together to cut a deal, soon rubber-stamped by the Commission. This was standard "public interest" operating procedure. Odd bedfellows—captains of business and wide-eyed social dreamers—joined hands. Despite the Commission's best efforts, new technologies were emerging and economic interests were realigning. In what became known as the Consensus Agreement, "the licensee's vise-like hold on its assigned frequency under the doctrine of broadcaster responsibility"[25] began to loosen.

The order issued by the FCC in 1972 retained the patina of micromanagement. Besen and Crandall commented, "The 1972 cable rules can only
be described as baroque."[26] But an important corner had been turned: some
small obligations of cable operators had been peeled off. Further reforms
came in a 1974 order. In 1977, the federal courts overturned the FCC's ability to regulate premium programming such as HBO, launched in 1972.
The deregulation wave of the Ford and Carter years swept reforms farther
along. By the end of 1980, federal cable TV franchises had been abolished
and content restrictions mostly eliminated. "In less than twenty years,"
Besen and Crandall wrote in 1981, "the FCC has argued that no regulation
of this new industry is required, then regulated the industry . . . to preclude
the development in any but the smallest television markets, and, finally,
appears to be moving to almost total deregulation."[27] All that remained was
for Congress to preempt rate regulations imposed by local governments.
That came via the Cable Communications Policy Act of 1984.

The ancien régime was crushed. Flowers sprouted almost instantly. By
1979, national cable programming services were being established, showing what the future of television might offer.

Allowing "spectrum in a tube" to compete with spectrum allocated by the
FCC permitted underlying economic forces to assert themselves. There was
huge pent-up demand for additional video choice and vast untapped capacity to produce programs viewers wanted to watch. Television was transformed between 1978, when cable reached few households and offered
virtually no original programming, and 1988, when a majority of U.S.
homes subscribed and dozens of premium and basic cable channels were
available. Some of the entrants were so phenomenally successful, in fact,
that by this time they had achieved monopoly status of their own.

In 1989, a typical U.S. cable TV system cost about $600 per subscriber
to build, but could be sold for $2,400, market-based evidence of excep-
tional profits. This outcome was aided by the unholy alliance struck by cable
operators and municipal governments, in which exclusive franchises were
awarded in exchange for political support. Taxation by regulation—subsidies
for popular projects—were the public part of the deal. The private side included payoffs for City Hall staffers, campaign contributions, and shady
stock distribution programs in which cable company shares went to prom-

Table 7.1. Cable Television Program Networks, 1979

Network	Genre
Nickelodeon	Children
ESPN	Sports
C-SPAN (House of Representatives)	Public affairs
The Movie Channel	Movies
Black Entertainment Television	Minority
Galavision (Spanish)	Minority
Modern Satellite Network	Varied
SPN	Varied
PTL	Religious
Trinity	Religious

Source: Stanley Besen & Robert W. Crandall, *The Deregulation of Cable Television*, 44 Law & Contemporary Problems 81 (Winter 1981), 109.

inent local residents with political pull. Those who were given equity interests had an incentive to pressure city council members to award their company a monopoly franchise. (Awards were legally "nonexclusive," but municipalities generally issued only one per market.)[28] While this practice was dubbed "rent a citizen," it was not a lease but a purchase.[29]

The television market grew more competitive with the launch of nationwide satellite rivals DirecTV in 1994 and EchoStar in 1996. Cable operators, initially in denial, running TV ads claiming that satellite TV did not work, were forced to react. In 1999 they doubled their annual capital outlays to nearly $11 billion, then increased them again in 2001, to more than $16 billion.[30] These funds were spent to digitize networks and greatly expand channel capacity, moving to counter "the Death Star," industry slang for their rival now beaming competitive product from space. But the upgrades did much more: they enabled two-way data services via "cable modem." Cable operators, having been deregulated and needing no additional federal, state, or local approvals, exploited the opportunity by migrating to Internet services. Competing against slower dial-up connections and a dead-end telephone technology called ISDN (integrated services digital network), their broadband offerings began to enlist substantial market

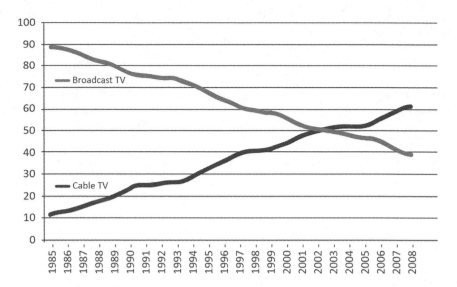

7.1. Prime time viewing shares for broadcast TV versus basic cable, 1985–2009. During the period, cable eclipses broadcasting. The data are from Nielsen TV Ratings (2010). Broadcast TV includes network, independent, and public stations. Cable TV includes basic and premium channels.

share. This, in turn, forced phone carriers to upgrade their product, developing (at long last) DSL (digital subscriber line) services and, in the mid-2000s, fiber-optic connections like Verizon's FiOS. The U.S. broadband race was on, taking the mass-market Internet to a new level.

But first, it was plain old cable that won consumers. Viewer ratings and household subscribership tell the story. In 1985, broadcast TV accounted for almost 90 percent of all prime time TV viewing; in 1994, 74 percent; in 2004, just under 50 percent. By 1988, more than half of all American homes subscribed to cable; by 2000, 85 percent did. In 2013, subscription television service—cable, satellite, telco TV (phone carriers being permitted to offer video courtesy of the 1996 Telecommunications Act)—had, despite the FCC's best guess, become the dominant media. Homes that relied solely on broadcast TV made up just 7 percent of all U.S. TV households.[31]

As the FCC backed away from managing the market, the constraints of scarcity loosened. This expansion of the conduit reduced the opportunity cost of transmitting any one TV program to potential viewers. Instead of losing a huge audience should a given show appeal to only a small slice of the population, the use of a TV channel became cheap. Risky endeavors became affordable; experiments aimed at niche audiences became the new normal. Video producers were no longer confined to least-common-denominator content.

The Big Three had dominated television by offering perhaps one hundred hours of original programming per week, less during summer reruns. With the liberation of cable, viewers gained access to thousands of hours per week. The flood produced broad diversity in genre and quality. Bruce Springsteen sang that there were "fifty-seven channels and nothin' on,"[32] but fifty-seven gave the viewer a far better shot than three. A richer mix brought more bad, by some estimates, but unquestionably more creative content appealing to targeted audiences: *The Larry Sanders Show* (HBO), *Queer Eye for the Straight Guy* (Bravo), *Blue's Clues* (Nickelodeon), *Chappelle's Show* (Comedy Central), *The Walking Dead* (AMC), *Through the Wormhole* (Science), *Bridezilla* (WE), *Amazing Earth* (Discovery), *Mythbusters* (Discovery), *I Shouldn't Be Alive* (Animal Planet), *Dexter* (Showtime), and *The Wire* (HBO)—none of which could have aired under the old regime. By 2010, cable networks were dominating television's Emmy Awards.[33] In 2013, the TV series chosen by the Writers' Guild as the best written of all time was HBO's *The Sopranos*.[34]

Investment dollars poured into Hollywood and its wannabes worldwide. Between 1983 and 2011, outlays for cable shows multiplied year on year, starting from less than a quarter billion annually to more than $20 billion. The torrent of original video programming has ignited keen interest. In 2002, audience ratings for cable nets, as a group, shot past those for broadcast television channels, gaining well over a two-to-one advantage by 2013.[35] Of course, by then the next generation of television had arrived, with broadband networks—themselves a product largely of cable TV deregulation—delivering video direct to customers from websites. In Fall 2015, of the ten new TV shows generating the most social media buzz, two were produced by Netflix and just one by the venerable CBS.[36]

Perhaps nowhere has the change been more profound than in the news and public affairs category. Recall that in 1964, when the FCC adopted strong measures to protect the "public interest" by blocking cable from

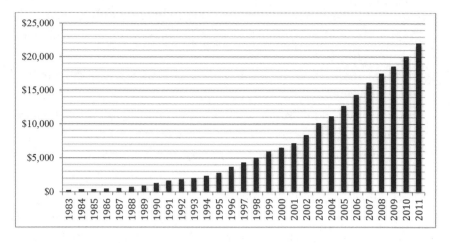

7.2. Programming outlays, in millions of dollars, for basic cable TV networks, 1983–2011. Given the opportunity to compete with broadcasting for viewers, cable TV invested heavily in new and diverse content. Data from SNL Kagan, *Economics of Basic Cable Networks* (1998), 5; (2012), 9–10.

competing with broadcasters, the Big Three networks each ran just fifteen minutes of nightly news.[37] It was widely assumed that save for sensational events like the Army-McCarthy hearings or the Kennedy assassination, news could not attract audiences. Regulators nonetheless believed it offered large societal payoffs in terms of an informed citizenry and was therefore worthy of subsidies. News and public affairs coverage was a loss leader forced upon the broadcaster in exchange for a lucrative license.

That mandate formed the central tenet in the FCC's policy of protecting TV stations from cable. "News, information, and public affairs programs are the heart of broadcasting in the public interest," wrote Newton Minow in 1961. A chapter in his 1964 book *Equal Time* was titled "News, News, Never Enough." It was the regulatory mission.

But it was not the regulatory outcome.

With deregulation, one of the first content services launched (in 1980) was Ted Turner's Cable News Network (CNN), followed (in 1982) by a second CNN channel, Headline News (to become HLN). The twin services proved popular and profitable. Diverse and controversial opinions were aired, with shows like *Evans and Novak, Crossfire, Capital Gang,* and *Both Sides with Jesse Jackson.* In the new multichannel environment, CNN found it could profitably devote full-time news coverage to breaking issues of public importance. Its international import was underscored in 1991, when it televised first the Gulf War and then the collapse of the USSR. The pivotal moment came when the image of Russian President Boris Yeltsin confronting a Soviet tank commander on the streets of Moscow was seen around the world via CNN International. In 1992, U.S. businessman Ross Perot announced his independent run for president on CNN's *Larry King Live,* bringing a public forum to American politics as never before. Perot advanced electronic democracy again by returning to King's show a year later to debate Vice President Al Gore on the North American Free Trade Agreement. When the Markle Foundation sought to award a grant for serious and substantial TV reporting of the 1996 U.S. presidential campaign, it attempted to endow the Public Broadcasting System. But PBS was too hidebound. Instead, Markle—a nonprofit—bestowed the funds upon a for-profit: CNN.

Once CNN demonstrated that news could pay for itself, broadcast networks dramatically increased their output: early morning, daytime, and all-night news programs were expanded; "news magazines" appeared throughout prime time schedules. And green shoots popped up, left, right, and center on the cable TV menu.

In 1996, a conservative brand appeared in the form of Rupert Murdoch's Fox News Channel. By 2002, Fox News had surpassed CNN in viewer ratings. MSNBC, a joint venture of General Electric and Microsoft, also joined the fray in 1996. It faked left with Chris Matthews and Phil Donahue, then went right with Laura Ingraham, Tucker Carlson, and Alan Keyes. It took until 2005, in the wake of Hurricane Katrina, before MSNBC found its inner progressive. President George W. Bush was pummeled nightly by the channel's prime time lead, Keith Olbermann, and viewer shares rallied. The new competition swamped the segment's first mover. *New York* magazine reported in 2010 that "The rise of Fox News on the right and MSNBC's follow-up pincer movement on the left have trapped and isolated CNN inside its brand."[38]

The diverse speech that began to flow with cable TV has burst the floodgates with the emergence of the mass-market Internet. Every website is a channel, every blog a micro-niche. Many are appalled by the content. Ideologues roam free; wingnuts wax poetic; polemicists shout loudly. Online shaming and cyber bullying have emerged as poignant social issues. "The Internet allows us to see what other people actually think," notes futurist Clay Shirky. "This has turned out to be a huge disappointment." Yet this transparency is a feature, not a bug. "There's no way," Shirky writes, "to get Cronkite-like consensus without someone like Cronkite, and there's no way to get someone like Cronkite in a world with an Internet; there will be no more men like him, because there will be no more jobs like his."[39]

Supreme Court Justice William O. Douglas noted long ago that "free speech . . . may indeed best serve its high purpose when it induces a condition of unrest . . . or even stirs people to anger. Speech . . . may have profound unsettling effects as it presses for acceptance of an idea."[40] Justice Oliver Wendell Holmes had been more concise: "Every idea is an incitement."[41]

The marketplace of ideas emerged in far fuller force with the deregulation of cable. Once artificial scarcity was stripped away, Americans were treated to not just the ideological rivalry of CNN, MSNBC, and Fox News, but of diverse news and information sources such as Bloomberg TV, CNBC, and Fox Business, of BBC America, Al Jazeera America, Vice, and Trinity Broadcasting, Discovery and National Geographic, Fusion, Adult Swim, and IFC. Comedy Central, a cable channel owned by Viacom, launched savage social satire with *South Park* and popular political commentary with *The Daily Show* and *The Colbert Report*. The programs were seen to be "changing politics"[42]—and they excited "South Park Conservatives"[43] even as they "revitalized the Democratic Party."[44]

News from *somewhere* came to life when the FCC stepped back. Open markets achieved what the TV allocation table promised but never delivered. That wasteland is now lush and fertile.

Even before Cable News Network, there was a gambit called C-SPAN, the Cable-Satellite Public Affairs Network, conceived by Brian Lamb circa 1977. Lamb was then a reporter for *Cablevision* magazine, having worked as a press liaison for the U.S. Navy during the Johnson administration and then for the newly formed Office of Telecommunications Policy, headed by Tom Whitehead at the Nixon White House. As a policy maker he became acquainted with the barriers erected to stop cable TV, and the reforms under way that could relieve the blockage. He saw a new video distribution plat-form rolling out with abundant channels but little programming, and won-dered what might fill the void.

The idea Lamb came up with, gavel-to-gavel coverage of the U.S. House of Representatives, had one thing going for it: there was no competition for the content. And talk was cheap and plentiful in Congress. But Lamb imag-ined what others could not. He formed a nonprofit corporation, procured backing from cable TV operators, and created a charter with an ongoing funding mechanism that ensured editorial independence.

C-SPAN went live in 1979. It obtained permission to place cameras in the House, later in the Senate, and put the government on television. When congressional sessions were dormant, committee hearings were often available—if not, there were think tank seminars, political speeches, book festivals, campaign events, White House press briefings and candi-date debates. There were also Iowa caucuses and New Hampshire rallies, marches on Washington and state of the state speeches in California, tours of presidential libraries, and commemorative services at Civil War battle-fields. Events would be aired whole, not cut or edited, and live whenever possible. In short, C-SPAN offered the educational public affairs program-ming that TV stations licensed according to "public interest" would do any-thing to avoid.

Lamb had the field to himself. Emerging cable operators, frantic for distinctive, high-quality, cheap programming to call their own, were happy to sign up. The network accepted no advertising. C-SPAN's sole means of support was the license fees paid by cable, satellite, and (later) telco TV operators that carried its program. In 1996, that fee would be set at 6¢ per subscriber per month and generate about $28 million a year—tiny com-pared with the $700 million CNN was spending on programming. But with its minimalist production style, noncelebrity anchors, and elastic supply of no-cost public affairs content, that amount was enough to put C-SPAN on the dial—followed by C-SPAN2 and, in some areas, C-SPAN3.

The network's model was serious, politically balanced, and hands-off. It would let speakers—politicians, scholars, authors, regulators, lobbyists, and citizens calling in on the telephone—express themselves without interrup-tion. It presented a wide cross-section of opinion on controversial issues, letting subscribers hear the arguments, evaluate their proponents, and make up their own minds. Amazingly, it worked. Loyal viewers, appreciative of the democracy-advancing product, commonly participate in the network's many call-in programs by first blurting out: *Thank God for C-SPAN!* C-SPAN, wrote David Corn in the *Nation,* became the "one undeniable advance in the land of politics-and-the-media." It "opened up Congress and Washington."[45] C-SPAN features political coverage that Michelle Malkin has called "scrupulously independent and nonpartisan."[46] On Lamb's retirement in 2012, *USA Today* agreed:

> In an era when cable TV personalities are known for their loud voices, strong opinions and big egos, C-SPAN founder Brian Lamb is a refreshing exception. Scrupulously non-partisan, he never gives his opinions or even his name on the air, nor do any of C-SPAN's regular hosts. It is, in all, a valuable reminder of older values in an era dominated by bombast, hyperbole and political spin.[47]

Consider C-SPAN's role in expanding the public square for election coverage. For decades following the 1927 Radio Act, which required radio licensees to grant equal time to news coverage of candidates for public office, campaign debates were rarely broadcast. The equal-time rule required that all candidates, including those from myriad minor parties, be included. Radio and television stations found it uneconomical to devote airtime to political discourse. *Equal* time, with its chilling effect, became *no* time.

When cable's unregulated "spectrum in a tube" arrived, C-SPAN quickly supplied the solution that had eluded regulators. As a cable channel, it had abundant time to fill, and Lamb's business model shielded programmers from commercial pressures. (When asked in 1996 how many people watch C-SPAN, Lamb answered that he and his staff "never, ever, ever look at that.")[48] The network altered the political landscape, bringing the Iowa caucuses to national audiences, televising speeches in New Hampshire and rallies in South Carolina. It aired primary debates and a massive amount of footage on national, state, and local campaigns.

In the 1999–2000 presidential primary campaign, neither major party had an incumbent running for reelection, which put both parties' nominations up for grabs. But the broadcast networks hosted just two debates, both

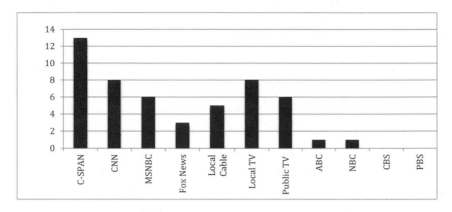

7.3. Televised presidential debates, 1999–2000 (through the New Hampshire primary). Cable TV, initially blocked by regulators to protect public affairs programs on broadcast TV, came to supply the overwhelming share of such content to U.S. voters. Data from Thomas W. Hazlett, "Digitizing Must Carry Under *Turner Broadcasting v. FCC* (1997)," 8 *Supreme Court Economic Review* 141 (2000).

for Democrats, one on NBC's *Meet the Press,* the other on ABC's *Nightline.* Local stations hosted a few more in Iowa or New Hampshire. But cable channels televised far more, and it was C-SPAN that brought the most debates— thirteen—into voters' homes.

Pundits marvel. David Corn calls C-SPAN "perhaps the closest-to-perfect Washington (and media) institution there is."[49] The American Historical Association gave the network an award, noting that "many Americans— including a fair share of the American Historical Association's membership—rightfully value C-SPAN as an achievement of historic significance." The FCC itself states, "The channel has earned a reputation for fairness and neutrality."[50] Such "fairness and neutrality" is won via independence from the FCC. "I'm a huge believer," Lamb said in 2011, "that I don't want government anywhere near me."[51] Still he is occasionally flummoxed:

> Our network [will carry] a panel discussion among so-called media experts, and they will sit there and decry the state of current-affairs [programming] as if we don't even exist. They will appear on this network, in front of our microphones, saying, "What this country needs is an opportunity for people to be heard. What it needs is an opportunity for long-form discussion. What it needs is an opportunity for more voices out there to get a chance to speak

their minds." And I'm sitting there watching, saying, "What's wrong with my brain—am I missing something here?"[52]

There is much conundrum to go around. "Cable television made possible an unlimited transmission of video signals," Thomas Sowell wrote in *Knowledge and Decisions*. "The whole structure of the industry—networks, affiliates, advertising patterns—could have been undermined or destroyed by the new technological possibilities."[53] As in other industries, regulators sought not to welcome disruptive innovation but to suppress it. The "public interest" objectives employed as ad hoc rationales for incumbent protection were achieved only when the regulatory barriers collapsed.

Thank God for C-SPAN? Yes, be reverent if you may. But the prayer might well include a homily on the blessings of deregulation. God was pulling for democracy all along. We got C-SPAN when rules said to counter the vast wasteland passed on to Regulator Heaven.

8

Lost in Space

Interest groups . . . have proved remarkably successful at hijacking government.

—John Micklethwait and Adrian Wooldridge

IN 1993 I RECEIVED A PHONE CALL from an attorney representing an FCC applicant for a new service it hoped to provide: satellite radio. The spectrum allocation issue was being decided, and the lawyer had heard that I might be knowledgeable about the nature of radio competition. Cutting quickly to the nub, he told me: "We want an economic study showing that satellite radio won't take any revenues from terrestrial radio broadcast stations."

I replied that I understood the task, but had a preliminary question: "How are you going to keep this study from your investors? The point of their exercise, it seems to me, is to take audience share, and then revenues, from your terrestrial radio rivals."

The lawyer laughed. "Well, you know how the game is played."

Using sound business judgment, the party hired another economist. But the FCC continued—and continues, decades later—to demand that satellite digital audio radio services (DARS) not compete with traditional radio stations. This is why SiriusXM—the merged combination of the two firms that eventually won licenses for the service—can offer two hundred or more channels of programming but must broadcast all of them *nationwide*. No customizing by market or region is permitted. Any local content must be aired nationally.

This policy was adopted to protect "localism." In fact, it kills it.

In 1990, a firm called CD Radio petitioned the Federal Communications Commission to allocate spectrum for satellite digital audio radio service. Three other firms also notified the FCC of their interest in supplying satellite radio service, asking for 12.5 MHz per license, or 50 MHz total. Given the FCC's "inclination towards cautious deregulation,"[1] a formal rulemaking was convened.[2] All four applicants were required to submit detailed information regarding their technology and business plans. Each firm expected to deliver approximately thirty channels of digital audio programming, but they differed in their revenue models, describing both subscription and advertising-only services.

NASA and other U.S. organizations objected on the grounds that satellite radio would interfere with vital communication links. Opposition from radio broadcasters was more intense. AM and FM station owners, acting through their trade group, the National Association of Broadcasters, said the new service would cripple local radio. Cities and towns would suffer, losing community-oriented programs aired by terrestrial broadcasters. National satellite entrants would steal audience share, hurting existing stations, and thus harm the "public interest."

Of course, the new competition would deliver value to listeners, too. The FCC summarized its dilemma:

> Satellite DARS . . . could expand and complement the audio programming choices now available to listeners. By offering a nationally based service, satellite DARS providers could target niche audiences that have not been served by traditional local radio. . . .
>
> It is also apparent that satellite DARS, to some extent, will compete with terrestrial radio. Proposed satellite DARS systems will provide 30 or more channels of national digital audio programming. . . . Some of these DARS channels may provide programming that is similar to what is available on local stations. . . . We request comment . . . regarding advertising revenues that may be lost due to competition from satellite DARS.[3]

In other words, competition is good, but competition is bad. The bad got more attention. Existing radio broadcasters told the FCC that satellite would decimate their industry. "Struggling stations" might be "pushed over the financial precipice" as satellite radio services would "immeasurably injure terrestrial radio stations by siphoning off listeners with their thirty or more channels of new programming."[4] The new listening choices would produce a "diversion of audiences and fragmentation of advertising," undermining "the ability of traditional radio stations everywhere to provide quality local programming and community services."[5]

Despite the threat to life as we know it, the Commission eventually did allow satellite radio to move forward. In 1997 it auctioned SDARS licenses —not the four requested, but two, with the spectrum allocation accordingly reduced from 50 MHz to 25 MHz. High bids totaled about $173 million; the winners were XM (formerly American Mobile Radio Corporation) and Sirius (formerly CD Radio).[6] Satellites were launched in 1999 and 2000, and the first subscribers received service in 2001. The new operators supplied diverse program menus featuring far more than thirty channels. In 2004, "shock jock" Howard Stern announced that he would jump from broadcast to Sirius. By 2006, when Oprah Winfrey launched her own channel for XM, the service had some fourteen million subscribers, each paying about $13 a month.

But the broadcasters' efforts to deter satellite had not gone unrewarded. Not only had the launch of their rivals been delayed by several years, but handicaps were imposed as well. In particular, the FCC imposed a key restriction: satellite was not permitted to air local-only programs. Because roughly 80 percent of the radio industry's advertising revenue came from local rather than national advertising, this offered important financial protection. Yet even with this restriction, by 2004 the Commission was bombarded with broadcaster filings showing how dire the competitive situation had become, as satellite carriers had "devoted substantial bandwidth," according to the NAB, "to compete directly with local broadcasters with local content."[7]

The complaint stemmed from the use, by XM and Sirius, of terrestrial repeaters. These devices, stationed on the ground, receive satellite radio feeds from space and then broadcast them locally. They improve reception, particularly in areas where tall buildings, tunnels on roadways, or leafy trees tend to interfere with the satellite signal. But these repeaters could also insert customized content just in Houston or Denver or Boston. What was cheap and technically easy, however, was legally *verboten*. The satellite licenses required that all programs be broadcast nationwide.

A paradox was that the terrestrial radio industry was itself nationalizing content. By the 1990s, independent local stations were becoming rare, bought up by large national chains like Clear Channel. Programs became coast-to-coast shows, from talk radio's Rush Limbaugh and Dr. Laura Schlessinger, to turnkey music formats. Satellite radio is an efficient way to distribute such homogenized content (indeed, satellites beam the shows to AM and FM stations, which then retransmit them), and allowing these streams to flow directly to consumers would naturally pressure existing broadcasters to differentiate their local products.

That prospect was resisted by the NAB, which blasted SDARS repeaters as "a crutch for a technology that is not up to the task of providing the seamless, mobile coverage promised by proponents." In 2002 an NAB official demanded that the FCC "put a halt to this ruse of a terrestrial repeater network." The trade press headlined scandal: "NAB Accuses XM of Local Programming Plot."[8]

The FCC blinked. While its rules did not actually ban "local programming," which is a form of free speech, satellite licenses mandated that all broadcasts had to be uniform. Program menus in Boston had to be identical to those in Denver. So eager were customers to receive local news, weather, and traffic updates, however, that XM and Sirius began producing market-specific reports and broadcasting them nationwide. Houston weather and traffic would be aired in Denver and Boston and everywhere else. This soaked up valuable bandwidth, a pernicious impact that constituted victory for terrestrial broadcast lobbyists.

Still, the incumbents were far from satisfied. They demanded a total ban on local news and information. "With the addition of local traffic and weather," the NAB told the FCC in 2004, "satellite radio is no longer an exclusively national service, and its impact on terrestrial broadcasting is growing."[9]

Having been asked to save localism by prohibiting local programming, the FCC was now attacked for only partly complying. Broadcasters were outraged. Eddie Fritts, president of the NAB, wrote a letter to Congress calling the effort "part of a longstanding pattern of deception by the satellite radio industry" and accusing satellite radio of undermining the public interest. "When the FCC licensed satellite radio, it intended a national radio service that would supplement, not detract from, the important services of free, local radio. . . . The satellite radio industry has no intention of abiding by the terms of their licensure."[10]

Members of Congress obediently introduced H.R. 998, sponsored by Representatives Chip Pickering (R-MS) and Gene Green (D-TX), a bill to preclude XM and Sirius from "inserting local content on ground-based repeaters." The legislation became moot: the FCC interpreted the two companies' existing license terms as saying what the NAB said they did. Fritts was jubilant. "While a national satellite radio service may indeed offer worthwhile niche programming to a limited segment of the population," he generously conceded, "no company should be permitted to disregard the terms of their licenses."[11]

In February 2007 XM and Sirius, both facing financial challenges, announced that they would merge. The NAB urged regulators to reject the combination as a "merger to monopoly." Of course, had it really created a monopoly, the rival broadcasters would have been delighted: the post-merger firm would raise prices, reducing its subscriber base and leaving more audience share for terrestrial stations. The NAB's opposition was a pretty good sign that the transaction, far from restricting competition, would strengthen it.[12]

The FCC, acting under its authority to approve license transfers, held up the deal for seventeen months, until July 2008. During that time, the merging parties' financial situation deteriorated further; by early 2009, the "merger to monopoly" was facing bankruptcy, a fate averted when shareholders effectively ceded ownership to major bondholders, led by cable TV mogul John Malone.

Despite its regulatory burdens, SiriusXM survives. In 2016, the firm served more than thirty million paying subscribers[13] and offered more than two hundred audio channels. Satellite subscribers had access to at least thirteen "news/public radio" channels, from Fox News to C-SPAN to NPR Now, and another four "politics/issues" channels offering commentary from nonpartisan, conservative, African-American, and progressive perspectives.

Table 8.1. News and Public Affairs Programming on SiriusXM Radio (Feb. 2015)

Channel	Service	Genre	Description
112	CNBC	news/public radio	CNBC simulcast
113	Fox Business	news/public radio	Fox Business simulcast
114	FOX News Channel	news/public radio	Fox News simulcast
115	CNN	news/public radio	CNN simulcast
116	Headline News	news/public radio	HLN simulcast
117	MSNBC	news/public radio	MSNBC simulcast
118	BBC World Service	news/public radio	world news
119	Bloomberg	news/public radio	business news
120	C-SPAN Radio	news/public radio	C-SPAN Live simulcast
121	Insight	news/public radio	"inspiring talk with a sense of humor"
122	NPR	news/public radio	NPR news & conversation
123	PRX	news/public radio	independent public radio
169	Radio One	news/public radio	"Canada's no. 1 public radio news source"
124	POTUS	politics/issues	nonpartisan political talk
125	Patriot	politics/issues	conservative talk
126	Urban View	politics/issues	African-American talk
127	Progress	politics/issues	progressive talk
153	Christina	Latino	advice & current affairs
155	CNN Español	Latino	all-news Spanish language
156	Radio Fórmula	Latino	"news and more from Mexico"
129	Catholic Channel	religious	"talk for saints and sinners"
131	FamilyTalk	religious	Christian talk
80	Rural Radio	family & health	agriculture & western lifestyle
81	Doctor Radio	family & health	"real doctors helping real people"
167	Canada Talks	Canadian	Canadian current affairs & talk
170	ICI Première	Canadian	Radio Canada news & information

Source: SiriusXM website.

There were eleven Spanish-language channels and eight specializing in Canadian content. The service provided far more informational programming than is broadcast terrestrially in any local radio market—and also offered six comedy channels.

Automobiles now roll off the assembly line with dashboard buttons for Pandora, Spotify, AM/FM/HD Radio, Sirius-XM, and USB ports to connect smartphones and MP3 players—a dizzying array of conduits available to pump Pink, Luke Bryan, Bach, *EconTalk,* or *Car Talk* through the vehicle's speakers. Yet amid this expanding array, one service is technically banned from airing local news. This relic of the 1990 argument against satellite radio entry still stands, an attempt to limit free speech and rig markets a generation later.

9

Baptists, Bootleggers, and LPFM

What starts out here as a mass movement ends up as a racket, a cult, or a corporation.

—Eric Hoffer

PROHIBITION OF ALCOHOLIC BEVERAGES in the United States was both a moral crusade and a protectionist racket. It benefited from a coalition. Religious opponents of Demon Rum, many of them Baptists, sought the ban on spiritual grounds. Mobsters, for whom smuggled liquor was a profit center, wished to remove legal rivals from the market. Policy makers, pressured from both directions, found Prohibition an offer they could not refuse. The strange bedfellows procured the Eighteenth Amendment to the U.S. Constitution in 1919, which then required the Twenty-First Amendment to undo it, in 1933. The power generated when Baptists and bootleggers work for a common goal became a paradigm in political economy.[1]

Low-power FM (LPFM) radio rules do not, technically speaking, prohibit microbroadcasting. But they so painfully overregulate it as to prevent such stations from effectively competing in the marketplace. The ongoing trauma of LPFM illustrates how diverse and even antagonistic interest groups can align to forge policies that serve their mutual interests while thwarting those of the public.

AM and FM radio broadcasters, as industry incumbents, have opposed

LPFM, knocking heads with media activists who crusade for new licenses in a campaign of the "have-nots" (community radio) against the "haves" (corporate radio). But the activists, even while championing the cause of LPFM, have sought and obtained debilitating restrictions. Their aim, to keep new entrants pure of commercial taint, precisely matches the agenda of the powerful broadcasters they profess to despise.

The 20 MHz allotted to FM radio (recall the 1945 FCC decision that led to Edwin Armstrong's demise) is dotted with "white spaces" that could host tens of thousands of low-power community stations. Yet the rules have permitted only a few hundred to emerge—typically, only those with the support of established educational or religious institutions. Independent community groups, the touted beneficiaries of the FCC's LPFM rules, have found it nearly impossible to run stations.

Low-cost radio transmitters became available in the 1980s. Alert community groups seized the moment, constructing cheap FM stations and broadcasting across small neighborhoods. Upstarts sprang to life in the many dead spots on the dial. By the 1990s, Radio Free Berkeley, Radio NaGo in Brooklyn, and Steal this Radio in the East Village had become locally famous.[2]

They were also unlicensed by the Federal Communications Commission, and thus "pirates." The Commission put some effort into shutting them down.[3] According to the FCC's FM allocation, there were no licenses available for low-powered FM radio. Minimum station wattage was then 6,000, which extended signals about twenty-six miles. The most powerful FM stations, pumping at 100,000 watts, could send signals hundreds of miles.[4] But there was a healthy supply of dreamers and technophiles. In 1998, the FCC received some thirteen thousand inquiries asking how to obtain a low-power license.[5]

One citizen did more than inquire. In February 1998, the agency received a formal petition requesting that it define a new class of service for low-power FM radio that would allow entrepreneurs and community organizations to do legally what the pirates were doing. A second petition arrived a month later. The FCC, with the backing of Chairman William Kennard, responded with uncharacteristic speed and purpose. In January 1999 it proposed to allow new stations using just 1,000 watts of power. Assuming a hundred-foot antenna, that would yield a broadcast radius of about 8.8 miles, enough to serve a small city or a good part of a large city.

The proposal produced great excitement. Citizens' groups saw the new service as a useful counter to the consolidation of commercial radio stations following the 1996 Telecommunications Act. Some thirty-five hundred parties filed comments with the FCC. As consumer activist Ralph Nader described the situation:

> Ever wonder why radio generally has become so canned, flat and insipid, bereft of local news, stuffed with commercials, mercantile values and the same old, tired junk? . . . Two conglomerates own over 400 radio stations each, all over the country.
>
> . . . At last, the Federal Communications Commission (FCC) may come to the rescue. Right now, the FCC is considering whether to set up noncommercial low-power FM (LPFM) radio stations. . . . Imagine the new voices that could flourish on these micro-stations—service and advocacy groups, universities, community and civic organizations, ethnic groups, arts organizations, seniors groups and others.[6]

When the Commission issued its LPFM rules in January 2000, the action was hailed as a game changer. Chairman Kennard proclaimed: "In creating a low power FM radio service, the FCC has thrown open the doors of opportunity to the smaller, community oriented broadcaster, and will give hundreds—if not thousands—of new voices access to the nation's airwaves."[7] Both the *Wall Street Journal* and the *New York Times* ran largely celebratory news stories.[8]

And then a brawl broke out. The broadcasting incumbents' lobby, led by the National Association of Broadcasters (NAB) and National Public Radio (NPR), deluged Congress with apocalyptic warnings. They opposed LPFM on the grounds that the 13,307 full-power stations on the air (5,009 AM, 8,298 FM) were all that were needed. There was no space on the dial for any more—*technical reasons* dictated that the status quo be protected. "The FM band is already severely overcrowded," said Dick Ferguson, a vice president of Cox Radio and an NAB official. "We do not know how you could add more stations."[9] The NAB warned that low-power stations would "likely cause devastating interference to existing broadcasters and will challenge the

F.C.C. as guardian of the spectrum,"[10] and distributed a CD to House and Senate members containing an audio recording of noise—nerve-wracking, static-laden squeals and clutter. The accompanying letter explained that cacophony would be the consequence of LPFM. But the recording was not of LPFM broadcasts. It had been created for the NAB in a sound studio.

FCC staff officials took the unusual step of denouncing the ploy. The asserted marketplace evidence, wrote FCC chief technologist Dale Hatfield and Mass Media Bureau head Roy Stewart, was a "disinformation effort . . . that purports to demonstrate the type of interference . . . that NAB claims will occur from new low-power FM radio stations."[11] The Commission issued a "factsheet" entitled *Low Power FM Radio Service: Allegations and Facts,*[12] which refuted the interference claims point by point.

That the FCC publicly opposed the incumbents inspired great hope among LPFM partisans. The press sensed that an important story was developing. "When it became apparent that the usually plodding FCC was on a fast track to license low power stations, radio stations already on the air became nervous," wrote Frank Ahrens in the *Washington Post*. "Existing broadcasters have fought the low power proposal with everything they've got."[13] The story was headlined: "Political Static May Block Low-Power FM; FCC, Congress Battle over Radio Plan."

The NAB/NPR lobby put forward a bill to trim the number of LPFM assignments; it passed both houses and was signed into law by President Bill Clinton in December 2000. LPFM activists screamed that 75 percent of potential station assignments had been extinguished. (The actual reduction was later calculated as 43 percent, to 1,300 slots.)[14] In the eyes of the press, the regulators had bravely forged new competition only to be rebuffed by reactionary forces in Congress.

This was not entirely accurate. As Bruno Viani and I showed in a 2005 paper, there were approximately 100,000 low-power radio slots "technically" available in the FM band alone. (Of course, AM and other bands could have been used as well.) Given the interference rules adopted by the Commission, vast new numbers of licenses might have been authorized. But no more than a few hundred materialized.

The key impediments are found in the FCC's January 2000 rules. While they reportedly provoked the ire of NAB, NPR, and Congress, they in fact constituted a huge setback for LPFM. The requested 1,000-watt limit had become a paltry 100 watts, shrinking reception distance to 3.5 miles or less and slashing area coverage by about 85 percent. And the FCC laid on a deep thicket of restrictive rules:

- Only nonprofit organizations could apply for licenses. This radically limited the competitive opportunity, as commercial newspapers as well as radio and TV stations were excluded.
- Initially, only one station could be owned by each organization. Multiple station ownership, up to a limit of ten, was permitted only after a licensee had operated a single station for three years. Meanwhile, the FCC allowed full-power stations—larger by orders of magnitude—to aggregate in the hundreds, capturing the advantages of economies of scale.
- Advertising was prohibited. This blocked revenue sources prized by existing commercial stations and reduced the chances that LPFM could be financially viable. Were local stores, for example, permitted to buy five- or ten-dollar radio spots, reaching neighborhood audiences to announce sales, a new media niche might have emerged.
- Pirates walked the plank. Individuals who had engaged in unlicensed radio broadcasting were banned, eliminating the best pool of talent for the LPFM stations—especially devastating in that they had to be nonprofit operations.

Most of these restrictions were advocated not by LPFM's enemies but by its friends, who were committed to the notion that the new service would be community-oriented, not commercial. But by the time they finished laying out these regulations, the new stations were virtually nonexistent. Those that clung to life needed established, well-funded institutions—churches or schools—to support their overhead. The new, independent radio voices were never created.

The showdown between big broadcasting incumbents and low-power upstarts was much more a "phony war" than a battle royal. The much-publicized NAB/NPR attack against the FCC's allocation plan was mostly about public affairs executives' and policy activists' efforts to get attention.

Congress, in its legislation, focused entirely on the FCC's separation rules (how much frequency space to leave between stations) and ignored the severe power limits and high-cost, anticommercial, antiadvertising rules imposed on LPFM. These were strongly favored by LPFM's champions, who fought for the service, they said, to mimic the "all-volunteer, non-commercial, anti-profit pirate radio station[s]" illegally broadcasting before the LPFM service.[15] But the regime they put forward strangled the upstarts.

FCC Chairman Kennard had stressed many appropriate points, warning that the industry should not "use interference concerns as a smokescreen for . . . a fear of increased competition."[16] But he was unable to steer his own agency away from that smokescreen, or prevent it from quashing financial viability. By the time his supposedly pro-LPFM policy went into effect, the service was dead. Rodger Skinner, the retired broadcasting engineer whose 1998 petition had triggered the LPFM proceeding, and who had hoped to apply for a license to operate a local station, was disgusted. It was a "huge blow," he told me, "when the FCC limited low power FM to non-commercial use only." This left fledgling entrepreneurs "with no way to support a real radio station." He believed that while many activists and pirates wanted low-power licenses just to set up "play radio," he was seeking "a full-fledged new broadcasting service." LPFM "died," says Skinner, the day in 2000 when the FCC rules were issued.[17]

In the event, there was no event. By 2004, some 350 low-power FM stations were reported to be on the air; by 2008, the number had risen to 809. Almost all were located outside the nation's 269 radio markets. Churches and schools, which could cross-subsidize the stations, operated the great majority. In mid-2000, one prominent LPFM advocate noted that the greatest "threats" to LPFM came from "the broadcasters lobby and NPR." But the next "biggest threat comes from fundamentalist Christian radio operators." Not that there was anything wrong with Christian fundamentalists. "All advocates of low power radio are advocates of freedom of expression." Still, "It is deeply ironic that progressives have fought so hard for this service, but right wing churches seem poised to get most of the frequencies. It is all the more ironic because these churches have put little or no effort into establishing the service, and do not seem interested in persuading their Senators to support it—they spend all their efforts on getting as many stations as they can. Prometheus Radio Project seeks to counter this trend by getting as many legitimate, local progressive organizations and publicly minded institutions to apply for stations as possible."[18]

Alas, the churches were organized and relatively flush, while the community activists were operating on ideological fumes. But in the end it hardly mattered. Under the rules, it was impossible for an LPFM station to reach any discernible audience. No public interest was advanced. Idle spectrum continued to be wasted. And the leading organization lobbying for LPFM, the Prometheus Radio Project, was clear about its mission. "We want to make sure that microradio is not for business," said a spokesman.[19] Mission accomplished! The lobbyists at NAB and NPR really do owe them a drink.

Within the political spectrum, LPFM is officially a success story. Congress has championed the cause,[20] media activists tout the service as a landmark democratic victory over corporate interests, and the Commission has publicly supported the technology. In 2007, the FCC declared that LFPM had "flourished for the most part."[21] When, in 2013, additional slots were made available for the first time in more than a decade, LPFM advocates took a victory lap.[22]

As of December 31, 2014, the Commission reported 10,727 full-power FM radio stations and 942 low-power stations.[23] Assuming (extremely conservatively) that the FM stations all broadcast at the lowest power level for Class A stations, 6,000 watts, and LPFM stations all broadcast at the highest levels allowed, 100 watts, the disparity in total wattage exceeded 683 to one. Given that there are tens of thousands of potential low-power slots in the FM band, the competitive alternative to incumbent stations provided by LPFM is effectively nil—despite many years of government "support."

Finally, recall the assertion by NAB and NPR that the FM band is filled to the brim. The same two organizations have since embraced HD radio. After the FCC authorized HD in 2002,[24] some twenty-four thousand new channel slots for full-power radio broadcasts magically appeared in the FM spectrum allocation, three for every FM station then operating. The new signals are broadcast on frequencies within the 200 KHz channels where the stations have broadcast analog signals, and interference is not an issue. As each existing station both suffers the costs and captures the benefits of new operations, they are careful to add services that expand rather than contract the pie. Within their own spaces, frequencies they can monetize, the broadcasters came to believe that the FM band was *under*utilized.

Not that this digital enhancement of product choice for consumers was their first option. HD radio was FM's response to the threat of satellite radio: as an NAB publication put it, the digital upgrade was "critically important for terrestrial stations in view of the launch of two satellite distributed digital audio radio services in 2001."[25] This is what competition does: it makes incumbent businesses uncomfortable enough to innovate. When policy rules welcome the future, technological advance and consumer choice can flower. New spectrum spaces will be discovered where once there was only impossible congestion. All of which prompts a philosophical question: If you were all alone in the forest, and there was radio interference that no one heard, would competitive entry still be contrary to the "public interest"?

PART THREE

ADVENTURES IN CONTENT REGULATION

Unfortunately, the U.S. political system is not especially effective at checking foolish ideas before they influence policy.
—John Mearsheimer

BARRIERS TO ENTRY DO NOT constitute the sum of the mischief done by errant regulatory rules in the political spectrum. The Fairness Doctrine, formalized in 1949 and then abolished in 1987, cast a dark shadow—a "chilling effect"—over public discussion of controversial issues on radio and television. Even when reviewed by the U.S. Supreme Court, this counterproductive, even nefarious, effort went unchecked. "Must-carry" rules, forcing cable TV operators to include all local TV signals in packages sold to subscribers, mirror the Fairness Doctrine in counter-productiveness and in passing muster with a supine Court. In this instance, however, a brilliant and passionate dissent by Justice Sandra Day O'Connor provides some hint that adult supervision is at least theoretically possible. More recent battles over "indecency" in broadcast media serve to illustrate the futility of such attempts to regulate speech—even as they remain surprisingly popular across the ideological spectrum.

Orwell's Revenge: The Fairness Doctrine

The Fairness Doctrine . . . puts the head of the camel inside the tent and enables administration after administration to toy with TV or radio in order to serve its sordid or its benevolent ends. . . . It was precisely the mistrust of the evanescent, narrow, factional views of those in power and the belief that no one has a patent on the "truth" that underlay the First Amendment.
                    —William O. Douglas

THE MOST IMPORTANT SUPREME COURT verdict dealing with spectrum allo-cation since the 1943 *NBC* case resulted when a tiny, daytime-only radio station in Red Lion, Pennsylvania, got its day in court. In 1969, the Court ruled[1] that U.S. regulators could constitutionally order radio and television licensees to broadcast certain programs under rules that aimed to "insure a fair and balanced presentation of all public issues."[2] These regulations were organized under a policy dubbed the Fairness Doctrine.

The keen legal interest in the *Red Lion* verdict grew even keener five years later, when the Court held that no newspaper could be subject to Fair-ness Doctrine–type regulation.[3] For print media the First Amendment's freedom-of-the-press clause made the judgment of editors—more precisely, the publishers who employed them—inviolate. That the *Miami Herald* was an influential medium of public opinion in south Florida was irrelevant. So was the fact that the paper's editors had savaged Pat Tornillo, a candidate for the Florida State Senate, in a shrill, one-sided editorial. A radio station owner—a wireless publisher—was not protected from government's judgments about fairness, but a newspaper owner was.

The anomaly stirred much debate in law reviews in the 1970s and 1980s. Even beyond that civil liberties tussle, however, *Red Lion* is a wonderfully illuminating regulatory episode. As a test case for how well the U.S. legal system could monitor the monitors, it gave a troubling answer: not well. "Nowhere in the massive legal record," wrote Fred Friendly in 1975, "with all its citations and historic precedents, is there evidence of the political origins and motivations of this case." The Supreme Court, and jurists on two appellate courts, "were unaware . . . at the time of the *Red Lion* decision" that the conflict at the center of the case was triggered by the efforts of "men near the seat of national power . . . to utilize the Fairness Doctrine . . . to inhibit and keep off the air what they considered to be dangerous and noxious views."[4]

As noted earlier, regulators began imposing "fairness" obligations on broadcast licensees almost as soon as they had power. In 1928, the Federal Radio Commission renewed the license for Socialist Party station WEVD in New York with the stern warning that it must "operate with due regard for the opinions of others."[5] The following year, the FRC refused an application by the Chicago Federation of Labor to increase the power and hours of WCFL because the station was run "for the exclusive benefit of organized labor."[6] In taking a swipe at "propaganda stations," the Commission signaled its problem with controversial opinions.

A new political atmosphere formed during the Great Depression. U.S. newspapers overwhelmingly opposed President Franklin Roosevelt; some 95 percent of publishers were anti–New Deal.[7] This may have posed little impediment to Roosevelt's agenda, given his sweeping electoral victories in 1932 and 1936 and huge congressional majorities. But following the 1937 court-packing fiasco and his setbacks in the 1938 elections, Roosevelt took an active interest in FCC regulation of radio.

He persuaded a first-rate government lawyer, Lawrence Fly, to take the helm at the Commission; Fly agreed on the condition that he be permitted to hire a cadre of top-notch young attorneys from Harvard Law School. He and his team then crafted a "cross ownership ban," making it illegal for a publisher to own both a newspaper and a radio station in the same market. This provoked a ferocious backlash from Congress, where many members strongly supported—and were supported by—radio stations, including those owned by newspapers.

Representative Eugene Cox (D-GA), who chaired the Select Committee Investigating the Federal Communications Commission, railed against the FCC "Gestapo."[8] Cox worked with congressional leadership on a plan to eliminate the FCC's budget appropriation. Fly's FCC struck back, exposing kickbacks to Cox, a lawyer, who had reportedly taken $2,500 to represent WALB, a radio station in Albany, Georgia, before the FCC.[9] Fly flagged the matter, and scheduled the station's pending license renewal for a hearing—a hostile action. A frothing Congressman Cox blasted the FCC as "the nastiest nest of rats to be found in the country."[10] In furious backroom infighting, a young New Dealer emerged as a key defender of the agency. Texas Congressman Lyndon B. Johnson intervened with House Speaker Sam Rayburn, also of Texas, to support the Commission and quash the budget cuts.

This would generate political chits worth their weight in gold. While the hapless Congressman Cox went rogue and was publicly humiliated for perhaps a few thousand dollars in crony cash, Johnson forged a bond with commission members and staffers that he turned into agency juice, quietly winning highly sought broadcasting rights denied to others.[11] Radio and then television broadcasting in Austin, Texas, became the Johnson family fiefdom via mercantile grant.

While the wages of Lyndon Johnson's lobbying efforts were extraordinarily high, the nature of the inside transaction was pro forma. Any call from a member of Congress inquiring how a particular application was being treated customarily led the FCC to handle it with special care. LBJ biographer Robert Caro describes young Congressman Johnson's bravado: once, while entertaining commissioners at his home, he telephoned a mid-tier FCC employee to check on the progress of a certain radio license application. Lady Bird Johnson later visited an FCC commissioner in his office to follow up on the matter. These gestures clarified the application's importance to the "public interest."[12] Such personal touches, Caro writes, were "described as 'boudoir episodes'—off-the-record contacts with interested persons regarding individual licensing proceedings pending before the Commission."[13]

While his FCC licenses were formally issued in his wife's name,[14] LBJ died a very wealthy media magnate.[15] It is an incidental irony that perhaps the FCC's greatest moment of service to genuine public interest, standing up to a corrupt and arrogant member of Congress, enabled another congressman to procure lucrative favors in exchange for his patronage.

While the FCC survived Congressman Cox's ill-motivated crusade, opposition to an outright ban on newspaper ownership of radio stations still ran high in Congress. The Commission sharpened its focus. If it could not force anti–New Deal publishers to sell their stations, perhaps it could mute their editorializing. It did so via the "Mayflower Doctrine," establishing that information delivered via radio, including news, must be presented "fairly" and "objectively."

The policy was targeted at the unabashedly right-wing Yankee Network, owner of three radio stations in New England, including WAAB in Boston. The network ran commentary from the likes of Father Charles Coughlin, an outspokenly anti-Semitic cleric suspected of pro-Nazi sympathies, whose ideology had drifted from the radical left to the extreme right. By the late 1930s he had found his voice as a staunch critic of "Franklin Double-crossing Roosevelt."[16] In 1939, when WAAB's license came up for renewal, Mayflower Broadcasting filed a competing application. The challenge was dismissed for factual misrepresentations, but Fly's FCC took the opportunity to send a warning to the Yankee Network by scheduling a renewal hearing. While the Commission elected not to withdraw the station's license, it issued an ominous nonendorsement:

> Radio can serve as an instrument of democracy only when devoted to the communication of information and the exchange of ideas fairly and objectively presented. Indeed, as one licensed to operate in the public domain the licensee has assumed the obligation of presenting all sides of important public questions, fairly, objectively and without bias. The public interest— not the private—is paramount.[17]

Yankee got the message and promised to stop editorializing. Enshrined as policy, the Mayflower Doctrine implicitly forbade all other broadcast licensees to editorialize. Anti–New Deal station owners were silenced. Mission accomplished.

The lull lasted until the Commission reversed course in its 1949 report, Editorializing by Broadcast Licensees,[18] which formally gave birth to the Fairness Doctrine. This imposed a two-pronged obligation. First, all radio and TV licensees were required to present "vitally important controversial issues of interest in the community served by the broadcaster." Second, they were required to "provide a reasonable opportunity for the presentation of contrasting viewpoints on such issues."[19]

Setting aside the serious First Amendment issue of whether the government should abridge the broadcaster's editorial freedom, "fairness" is an elusive quality to define. "Fair and balanced" is in the eye of the beholder; confirmation bias is endemic (and likely hardwired into the human brain).[20] Edward Jay Epstein exposed this basic challenge: "The Commission states in its 'Fairness primer' that it is not 'the Commission's intention to make time available to Communists or to the Communist viewpoints'—a notion which brings into question the Commission's concept of 'fairness.' "[21]

In 1963 the government then issued the Cullman Doctrine, which clarified that radio or TV stations were required to provide not just equal time to those responding to a previously expressed viewpoint, but free equal time.[22] This set the stage for the classic test of regulated broadcast "fairness."

WGCB, a radio station in Red Lion, Pennsylvania, transmitted at 1440 kilocycles on the AM dial, using just 1,000 watts of power. It was owned by the Reverend John M. Norris, a devout Christian and ardent conservative. Beginning at 1:12 P.M. on November 25, 1964, WGCB aired a fifteen-minute commentary by the Reverend Billy James Hargis that, for about two minutes, lambasted writer Fred Cook and his best-selling book *Goldwater: Extremist of the Right*. Hargis was a firebrand right-winger from Tulsa, Oklahoma, whose evangelical ministry had purchased the time slot for $7.50. His critique attacked Cook's judgment ("defending Alger Hiss"), his ethics ("a professional mudslinger"), and his publisher, the *Nation* ("a scurrilous magazine which has championed many Communist causes").[23]

WGCB soon received a letter from Cook requesting airtime to respond. Norris responded by offering to sell Cook a fifteen-minute slot for the same price Hargis had paid. Cook declined and complained to the FCC, asserting his right to respond to the personal attack without being charged. The

Commission agreed and so ordered its licensee. Norris refused to comply and sued the FCC instead. The Supreme Court, by a vote of 8–0 (Justice William Douglas was undergoing an appendectomy), upheld the FCC's position, affirming the constitutionality of the Fairness Doctrine.

Red Lion extended the *NBC* precedent, justifying government licensing of the electronic press based on the idea that spectrum was a unique natural resource. Electronic media were distinct from old-fashioned books, magazines, or newspapers. Paper was not scarce and needed no federal government supervision; radio frequencies were, and did. And that necessary control could be leveraged. The Court reaffirmed its ruling in *NBC* that the Radio Act "does not restrict the Commission merely to supervision of the traffic. It puts upon the commission the burden of determining the composition of that traffic."[24] The FCC was not only permitted but positively required to regulate content.

As it had done in 1943, the *Red Lion* Court was practicing economics without a clue; the scarcity it described applies as much to books as to wireless signals. Today, when the *New York Times* is read on smartphones and tablets via wireless connections, the distinction between "radio" and "paper" media has literally disappeared. But in 1969 the Court not only saw bright lines, it confirmed that the FCC's purview included not only "supervision of the traffic" but the "composition of that traffic."

The Court offered just one constitutional check to limit regulators. The Fairness Doctrine was designed to promote balanced coverage of controversial issues. If such a rule, by imposing free time for responses, effectively taxed such coverage to create a "chilling effect," it would presumably violate the First Amendment. As Justice Byron White commented in the majority opinion:

> It is strenuously argued . . . that if political editorials or personal attacks will trigger an obligation in broadcasters to afford the opportunity for expression to speakers who need not pay for time and whose views are unpalatable to the licensees, then broadcasters will be irresistibly forced to self-censorship and their coverage of controversial public issues will be eliminated or at least rendered wholly ineffective. Such a result would indeed be a serious matter, for should licensees actually eliminate their coverage of controversial issues, the purposes of the doctrine would be stifled.[25]

White dismissed this concern by noting that the Federal Communications Commission had not observed such counter-productive outcomes. In a footnote, he quoted the strong words of CBS President Frank Stanton: "We are determined to continue covering controversial issues as a public service, and exercising our own independent news judgment and enterprise. I, for one, refuse to allow that judgment and enterprise to be affected by official intimidation."[26]

This passage was carefully selected. Elsewhere, Stanton had said pretty much the opposite: "Reprisals no less damaging to the media and no less dangerous to our fundamental freedoms than censorship are readily available to the government—economic, legal, and psychological. . . . Nor is their actual employment necessary to achieve their ends; to have them dangling like swords over the media can do harm even more irreparable than overt action."[27]

However cursory the Supreme Court's examination of "chilling effect" was, the simple reality is that neither the FCC's views on its own regulatory agenda nor public statements by news executives constitute compelling evidence. It is a pity that the Court did not look harder at the facts of *Red Lion*. Drilling down into that case—as former CBS News President Fred Friendly did in a brilliant 1975 book, *The Good Guys, the Bad Guys, and the First Amendment*—turns up a gusher.

In September 1963, the Nuclear Test Ban Treaty with the Soviet Union passed the U.S. Senate by a vote of 80–19, and a crowning achievement of the Kennedy administration came to be law. Despite the lopsided vote, the issue had been controversial and the debate heated. Leading Democrats were highly concerned about the rise of conservative radio commentators who had created ad hoc radio broadcasting networks "by buying time and syndicating programs favoring only one side of an issue," as Senator John Pastore (D-RI) put it in a letter to the FCC.[28]

Following their Test Ban Treaty victory, administration political operatives seized on what they had learned in the battle. A key tactic had been to exploit the Fairness Doctrine, requesting free equal time from any radio station that broadcast a commentary opposing the treaty. According to an internal 1964 memo from a Democratic National Committee lawyer, "The constant flow of letters from the Committee to the stations may have inhibited the stations in their broadcast of more radical and politically partisan programs."[29] Pleased with these results, the White House moved toward a more comprehensive strategy. The first stage, initiated in October 1963, created a team to monitor radio broadcasts. Kenneth O'Donnell, a White House adviser and close friend of President Kennedy, sought out Wendell Phillips, a federal housing official who had worked for the *New York Times* and *Denver Post*. Scanning the airwaves from his home in Bethesda, Maryland, Phillips found "extreme right-wing broadcasting . . . irrationally hostile to the President and his programs."[30]

In January 1964, Phillips became the DNC's director of news and information and began hiring consultants for the labor-intensive task of scanning the radio dial. He created information packets to assist Democratic Party groups and allied organizations in requesting equal time, quickly securing some five hundred rebuttal slots. "The idea," writes Friendly, "was simply to harass radio stations by getting officials and organizations that had been attacked by extremist radio commentators to request reply time, citing the Fairness Doctrine."[31]

Phillips also sought to plant news stories and subsidize books. Enter Fred Cook, a *New York World-Telegram and Sun* reporter whom Phillips knew from his reporting days. Cook had embarked on a biography of Barry Goldwater, senator from Arizona and the eventual Republican presidential candidate, with the working title "Goldwater: Fanatic of the Right." Phillips gave Cook access to the DNC's stock of opposition research. "Phillips and the DNC," Cook later recalled, "provided me with most of my materials from their vast files."[32] The DNC arranged for Grove Press to publish the book, under the title *Goldwater: Extremist of the Right*, and agreed to buy fifty thousand copies. Grove ended up selling forty-four thousand copies to the public and seventy-two thousand to the DNC.[33]

As the Goldwater book was being written, Phillips approached Cook about using some of the DNC's data for an article in the *Nation*. Cook jumped at the chance, contributing "Radio Right: Hate Clubs of the Air" to the magazine's issue of May 25, 1964. The piece argued that radio pundits, including Billy James Hargis, were fomenting a climate of hate that had contributed to the assassination of President Kennedy. It also sounded a call for action: "One recourse for liberal forces would appear to be to demand free time to counter some of the radical Right's free-swinging charges."[34]

Phillips's monitoring campaign was in full swing, with more than a thousand demand letters sent to radio stations. By November 1964, when President Johnson beat Senator Goldwater in a landslide, the effort had produced some 1,678 hours of free airtime. In a confidential report to the DNC, Martin Firestone, a Washington attorney and former FCC staffer, explained:

> The right-wingers operate on a strictly cash basis and it is for this reason that they are carried by so many small stations. Were our efforts to be continued on a year-round basis, we would find that many of these stations would consider the broadcasts of these programs bothersome and burdensome (especially if they are ultimately required to give us free time) and would start dropping the programs from their broadcast schedule.

The campaign continued past the election. One such demand letter made it to Norris, requesting free equal time for Fred Cook to respond to Hargis's attack of November 25, 1964; the subsequent disagreement was litigated in *Red Lion*. While the Court flagged the importance of any "chilling effect" that Fairness Doctrine requirements might have, it entirely missed the fact that the very request that inspired the case had been part of a concerted effort not to engage in robust debate but to suppress it.

As Bill Ruder, a Commerce official in the Kennedy and Johnson administrations, conceded: "Our massive strategy was to use the Fairness Doctrine to challenge and harass right-wing broadcasters and hope that the challenges would be so costly to them that they would be inhibited and decide it was too expensive to continue." Adding to the deception, the campaign operated behind front groups—such as the National Council for Civic Responsibility, formed as a tax-free arm of the moribund Institute for Public Affairs—masking the effort's actual purpose. A decade later, the head of the National Council said to Friendly, "Let's face it, we decided to use the Fairness Doctrine to harass the extreme right. In the light of Watergate, it was wrong."[35]

The Court heard no such evidence, or mea culpa, in deciding *Red Lion*.

Criticism of the Fairness Doctrine grew throughout the 1970s. Compelling analysis came from such civil libertarians as Supreme Court Justice William O. Douglas and David Bazelon, senior judge of the DC Circuit, who explained in a 1975 article why he had switched from supporting the

Doctrine to opposing it.[36] Friendly's book was also influential, as was Ithiel Pool's landmark *Technologies of Freedom,* released in 1983. Whatever the intellectual source, change came on August 4, 1987, when the FCC voted 4–0 to abolish the Fairness Doctrine.

Congress was livid. "I knew those lickspittles would do something like this," said House Commerce Committee Chair John Dingell (D-MI). Senate Commerce Committee Chairman Ernest Hollings (D-SC) found the vote "wrongheaded, misguided and illogical."[37] He sponsored legislation to reinstate the Doctrine, which passed the Senate 59–31 and the House by 302–102, only to be vetoed by President Reagan, a former broadcaster. The deregulation stood.

The abrupt switch created a natural experiment: did the marketplace, with the Fairness Doctrine in place, host more informational programming or less? If the post-deregulation marketplace broadcast more news than the market when regulated by the two-pronged mandate of the Fairness Doctrine, one could reasonably infer a "chilling effect." Luckily, the radio market lends itself to just such a study, as radio station formats identify programming by genre, including news, talk, news/talk and public affairs.[38] By examining the popularity of these formats before and after the abolition of the Fairness Doctrine, we can see whether the rule gave us more news and information programming—the "public interest" thesis—or less, consistent with a "chilling effect."

In a paper published in the *Journal of Legal Studies* in 1997, the economist David Sosa and I investigated these questions.[39] Theoretically, the effects of the Fairness Doctrine were ambiguous. While the first prong (the coverage requirement) may have pushed stations to increase informational programming, the second prong (the requirement that equal time be provided for free) would operate as a tax. Because that tax could be evaded by avoiding controversial topics to begin with—offering music and sports, say, instead of news or opinion—the question was whether this incentive outweighed the other impacts. We examined the distribution of radio station formats before and after 1987 to determine which effect prevailed.

Our findings were stark. From 1975 through 1995 the most popular formats were, initially, music-oriented. But starting in 1988, right after the Fairness Doctrine was abolished, information-based formats took off. In

1987, just 7.1 percent of AM stations were classified as news, talk, news/talk, or public affairs; by 1995, some 27.6 percent were. FM stations saw a similar trend of a distinct magnitude. FM news formats climbed from about 1 percent of stations under the Fairness Doctrine to 4 percent in 1995. The shift to "nonentertainment" specifically kicks in following deregulation. The news/talk format did not even exist in 1975, or 1987—but it exploded in the early 1990s. By 1995 it was the most popular in the information-based category, adopted by some 854 AM stations.

There is a strongly negative correlation between the Fairness Doctrine and the quantity of informational programming. This does not prove that the FD had a chilling effect, but is clearly inconsistent with the reverse hypothesis—that the doctrine increased the amount of news and public affairs programming. Some might be tempted to dismiss the change in radio markets as simply the product of the man the *Wall Street Journal* called the "Godzilla of talk radio," conservative Rush Limbaugh, who launched his syndicated nationwide show in August 1988. But did Limbaugh change markets, or did abolishing the Fairness Doctrine enable stations to run controversial programming like his?

The latter is the more compelling view and has even become conventional wisdom.[40] It is perhaps ironic, then, that conservatives in the 1980s largely supported the law. Right-wing activist Phyllis Schlafly testified in 1987: "It is unacceptable that the First Amendment right to speak on radio or television should be limited only to those who have the money to buy a station." Even though it was President Reagan's FCC appointees who proposed to eliminate it, she found the Fairness Doctrine necessary because of the "outrageous and blatant anti-Reagan bias of the TV network newscasts."[41] Her ideological allies at the National Rifle Association and Accuracy in Media agreed. And when Congress voted to codify the Fairness Doctrine in 1987, the House bill was cosponsored by leading conservatives including Newt Gingrich, Phil Crane, Robert Dornan, and Vin Weber.[42] The Senate version was cosponsored by Missouri Republican John Danforth[43] and supported by conservative firebrand Jesse Helms.[44] Reagan's veto angered many in his own party.

Conservatives did not just support the Doctrine, they used it. In 1977, for instance, the FCC received a "fairness complaint of unprecedented proportion" filed by Ernest Lefever of the American Security Council Education Foundation. His analysis purported to document that CBS News "had for several years been one-sided and 'dovish'" in reporting on the Soviet Union's military strength.[45]

Radio markets (not to mention the cable TV and online content markets that followed) supplied more informational content after the Fairness Doctrine was removed. A breadth of choice emerged. "Mr. Limbaugh, the biggest right-wing talker, draws 13.5m listeners a week," observed the *Economist* in 2007. "National Public Radio, which strives to be fair and balanced but leans to the left, draws 20m. The shock jocks who rule many urban markets are also vaguely leftish."[46] Legions of decidedly nonconservative talkers aired: comedian (now U.S. Senator) Al Franken, actress Janeane Garofalo, Jim Hightower, Bill Press, former New York Governor Mario Cuomo, current California Governor Jerry Brown, Harvard law professor Alan Dershowitz among them. The radio network Air America was established with financial backing from wealthy progressives aiming to counter the conservative din. The radio conglomerate Clear Channel, syndicator of Rush Limbaugh, backed the venture and supplied stations to broadcast its content.

Not all survived. Entry was open and exit was easy. "One reason why Air America found it hard to find talent and listeners," opined the *Economist*, "is that most liberals are perfectly happy with NPR."[47] Conversely, Sirius Patriot and Sirius Progress have no trouble coexisting in the "monopoly" marketplace of satellite radio.

The U.S. Supreme Court posited that evidence of a "chilling effect" might change the justices' minds about constitutional limits to FCC authority. While the Court did not have available to it the statistical analysis that would follow removal of the Fairness Doctrine, it might have consulted one of Frank Stanton's news reporters, Dan Rather, as the FCC did in 1985. Rather provided a powerful first-person account of the Doctrine's impact:

> When I was a young reporter, I worked briefly for . . . newspapers, and I finally settled into a job at a large radio station owned by the *Houston Chronicle*. Almost immediately . . . I became aware of a concern which I had previously barely known existed—the FCC. The journalists at *The Chronicle* did not worry about it; those at the radio station did. Not only the station manager but the newspeople as well were very much aware of this government

presence looking over their shoulders. I can recall newsroom conversations about what the FCC implications of broadcasting a particular report would be. Once a newsperson has to stop and consider what a government agency will think of something he or she wants to put on the air, an invaluable element of freedom has been lost.[48]

The regulatory path forged by the Kennedy and Johnson administrations was taken to new levels under President Nixon.[49] Small radio stations carrying fringe commentators were replaced by a more formidable target of intimidation: the broadcast television networks. The tools the Nixon administration used to attack perceived liberal bias also relied on the Fairness Doctrine.

Richard Nixon embarked on a strategy to present his ideas to the country via nationally televised speeches. During his first eighteen months in office, he made fourteen such speeches, far more than Eisenhower, Kennedy, or Johnson over a comparable period.[50] At the same time, the administration sought to blunt televised criticism, both by discouraging broadcasters from airing Democratic rebuttals to presidential speeches (such rejoinders had become a practice in 1966) and by reining in network commentators like Howard K. Smith of ABC, John Chancellor of NBC, and Eric Sevareid of CBS.

The administration employed a dual strategy. Publicly, Vice President Spiro Agnew was unleashed to flail at the "nattering nabobs of negativism" and to ignite populist anger at elite broadcasters. The zenith of this attack was a high-voltage speech in Des Moines, Iowa, on November 13, 1969: "Is it not fair and relevant to question its concentration in the hands of a tiny, enclosed fraternity of privileged men elected by no one and enjoying a monopoly sanctioned and licensed by the government?. . . The networks have dominated America's airwaves for decades. The people are entitled to a full accounting of their stewardship."[51]

The second, less visible, prong had White House staff, with some cooperation from top FCC officials, monitoring the broadcast networks' content. The goal was to challenge license renewals by asserting violations of the Fairness Doctrine. Threats of these actions were conveyed in a stream of private communications to network executives.

White House aide Charles Colson spearheaded this campaign. With the approval of Chief of Staff H. R. Haldeman and Nixon himself, Colson personally visited the New York headquarters of the three television networks in September 1970 to discuss administration concerns. For the next two and a half years, he called CBS Chairman William Paley or President Frank Stanton about once a month, occasionally arranging meetings in Washington or New York. He also called top ABC and NBC executives, though less often. In a July 1971 White House meeting between Stanton and Colson, according to newsman Daniel Schorr, "Colson chuckled that he could never hope for constant fairness from CBS, but maybe they could agree on an 'occasional fairness doctrine.' Stanton smiled appreciatively and said he wanted Colson to feel free to pick up the phone any time he felt he had reason to complain."[52] In 1972, Colson phoned Stanton to say that the administration was considering a five-point plan of action that included a proposal to license the networks directly[53] and a campaign to make license renewals for television stations more difficult.

In some important instances, these efforts resulted in more cooperative broadcast media. A report on the Watergate break-in by Walter Cronkite in the waning days of the 1972 campaign was pared way back after Colson spoke to Paley.[54] In June 1973, CBS announced it would no longer follow presidential statements with immediate news analysis by network correspondents. It was widely believed, Schorr writes, that CBS had been "silenced, or intimidated, or subverted" by the administration.[55] Paley denied this, stating that his only objective was "better, fairer, more balanced" coverage.[56]

The *Washington Post* may have come to feel the most pressure due to its aggressive Watergate reporting and the vulnerability of the broadcast properties owned by its parent, Post-Newsweek. The White House audio tapes that surfaced during the Watergate inquiry included the transcript of a September 15, 1972, meeting in the Oval Office among Nixon, Haldeman, and White House Counsel John Dean.

> PRESIDENT: The main thing is the *Post* is going to have damnable, damnable problems out of this one [Watergate coverage]. They have a television station . . . and they're going to have to get it renewed.
> HALDEMAN: They've got a radio station, too.

PRESIDENT: Does that come up, too? The point is, when does it come up?

DEAN: I don't know. But the practice of non-licensees filing on top of licensees has certainly gotten more . . . active . . . in this area.

PRESIDENT: And it's going to be Goddamn active here.

DEAN: (Laughter) (Silence)

PRESIDENT: Well, the game has to be played awfully rough.[57]

At the end of 1972, a competing application for the license of Post-Newsweek's Miami TV station, WPLG, was filed by Greater Miami Telecasters. GMT featured several prominent investors, including Charles "Bebe" Rebozo, the wealthy backer and close confidant of President Nixon, and William Pawley, a staunch Nixon supporter who in 1973 paid for magazine ads defending the administration against the Watergate "witch hunt."[58] Given these White House connections, many assume the obvious. In the event, Post-Newsweek won renewal of WPLG's license, but only after fighting costly regulatory battles and seeing its stock price plummet.[59]

The idea that government regulation has been used to chill free speech may be further seen in one impressive data point: a 1970 memo submitted by Colson to Press Secretary Herb Klein (and then routed to Haldeman). The memo included seven points followed by a conclusion, here winnowed down to the first and last:

FOR: HERB KLEIN
FROM: CHUCK COLSON
FYI-EYES ONLY, PLEASE
MEMORANDUM FOR H. R. HALDEMAN
September 25, 1970
The following is a summary of the most pertinent conclusions from my meeting with the three network chief executives.
1. The networks are terribly nervous over the uncertain state of the law, i.e., the recent FCC decisions and the pressures to grant Congress access to TV. They are also apprehensive about us. Although they tried to disguise this, it was obvious. The harder I pressed them (CBS and NBC) the more accommodating, cordial and almost apologetic they became. [CBS President Frank] Stanton for all his bluster is the most insecure of all . . .
Conclusion:
    I had to break every meeting. The networks badly want to have these kinds of discussions which they said they had had with other Administra-

tions but never with ours. They told me any time we had a complaint about slanted coverage for me to call them directly. Paley said that he would like to come down to Washington and spend time with me anytime that I wanted. In short, they are very much afraid of us and are trying hard to prove they are "good guys."

These meetings had a very salutary effect in letting them know that we are determined to protect the President's position, that we know precisely what is going on from the standpoint of both law and policy and that we are not going to permit them to get away with anything that interferes with the President's ability to communicate.

Paley made the point that he was amazed at how many people agree with the Vice-President's criticism of the networks. He also went out of his way to say how much he supports the President, and how popular the President is. When Stanton said twice as many people had seen President Nixon on TV than any other President in a comparable period, Paley said it was because this President is more popular.

The only ornament on [NBC president Julian] Goodman's desk was the Nixon Inaugural Medal. [ABC Vice President James] Hagerty said in [ABC President Leonard] Goldenson's presence that ABC is "with us." This all adds up to the fact that they are damned nervous and scared and we should continue to take a very tough line, face to face, and in other ways.[60]

That the Supreme Court could dismiss any "chilling effect" on the accounts given by regulators, buttressed with a footnote to a speech by a broadcast executive, is itself chilling. Frank Stanton may have given stirring lectures on the importance of an independent press, but when top government officials came calling, he aimed to please. As reported privately by Charles Colson to the Nixon White House, "Paley is in complete control of CBS—Stanton is almost obsequious in Paley's presence."[61] And Paley, as Colson put it, was "trying hard to prove they are the good guys."

11

Must Carry This, Shall Not Carry That

> Disguised in procompetition and information diversity language, must-carry rules were an attempt to extend a regulatory edifice crafted to address the problem of spectrum scarcity into an era of spectrum abundance.
>
> —Herman Galperin, *New Television, Old Politics*

THERE MAY BE NO SMOKING GUN memos that prove the must-carry policy is a failure. Yet these rules have fixed markets, protected broadcasters from competition, and excluded valuable programming sought by viewers and even by the regulators themselves. The policy has proven hostile to consumer choice and the First Amendment, and again the U.S. Supreme Court, despite a stinging dissent by Justice Sandra Day O'Connor, professed not to notice. While the regulations were born in a bygone era now upended by deregulation and eclipsed by dynamic technological forces, they continue in effect to this day.

A must-carry rule requires that local broadcast TV signals be included in any video package a cable company offers to subscribers. Congress extended this right to TV station owners in the 1992 Cable Consumer Protection and Competition Act to (once again) protect broadcasting against its cable rivals. TV stations could elect must-carry, under which cable companies would pay no fee to carry them and broadcasters no fee to be carried. The broadcasters also got preference in channel assignments. (Lower channels were generally more popular.) On the other hand, stations could elect to forgo must-carry and negotiate a price to be paid by cable operators. These "retransmission consent" rights had been sought by broadcasters since the Supreme Court refused to recognize them in cases dating to 1968.

There are at least three interesting twists. First, there is the observed impact when cable TV systems, facing limited channel capacity, were forced to eliminate certain high-valued cable TV networks like C-SPAN2 so as to carry over-the-air channels featuring home shopping. Ironically, the government's stated aim with must-carry was to encourage the "widespread dissemination of information from a multiplicity of sources," and Congress's interest was "in preserving a multiplicity of broadcasters to ensure that all households have access to information and entertainment on an equal footing with those who subscribe to cable."[1] Perverse results ensued, shipping and handling extra.

Second, the U.S. Supreme Court reviewed the legality of the must-carry regime under the First Amendment. In the second of two decisions, rendered in March 1997, Justice Anthony Kennedy, writing for a 5–4 majority in *Turner Broadcasting v. FCC,* declared the rules constitutional. A key supporter of the verdict was the prolific legal scholar Cass Sunstein, who dedicated his book *Democracy and the Problem of Free Speech* to the premise that must-carry rules represented a class of government regulations that could enrich public debate and strengthen democratic institutions.

Third, just as the rules withstood the Court, they have survived politically. Yet the great majority of TV stations have no use for must-carry; their bargaining position is strong enough that they can get cable companies to pay for retransmission.[2] These payments totaled $4.9 billion nationwide in 2014 and were rising rapidly.[3]

The case for imposing the carriage mandate frames cable TV operators as motivated to exclude these TV broadcast stations, which compete with them for advertising dollars. But their strongest rivals have not only been carried on cable voluntarily but paid handsome licensing fees for the privilege. Must-carry is quite beside the point for TV stations that viewers want to watch. It has an impact only where it forces carriage of less-popular programming.[4]

The Beginning of Must-Carry

The FCC first applied a must-carry mandate in 1966. "The basic regulatory justification for the rules was that mandatory carriage was necessary for cable to discharge its role as a 'supplement' to local, over-the-air broadcasting, rather than as a replacement for it."[5] Here the Commission protected broadcast television by ensuring that where cable was able to make inroads, it was forced to take its competitors along for the ride.

The retransmission requirements proved onerous. "By forcing cable operators to fill their channels with signals also transmitted over the air," writes Peter Huber, "the FCC had settled on the one policy most certain to perpetuate scarcity."[6] In 1966, cable systems often featured just five channels and rarely more than twelve; even as late as 1986, more than 10 percent of cable households subscribed to systems with no more than twelve channels.[7] In many markets, more than twelve broadcast channels, sometimes up to twenty, were mandated for carriage.[8]

The FCC mandates meant that many cable systems had little room left for nonbroadcast TV content.

The mandate operated as a tax on cable deployments. If cable operators had been free to create their own channel bundles, they would have included popular local TV stations. Many cable systems, however, would have declined to carry the least-watched stations, preferring to devote their capacity to programs unavailable over the air. This would be particularly true in areas with good broadcast reception; there, a simple and inexpensive "A/B Switch" would have let customers access local programs whenever they wanted. (Such switches were typically included with cable set-top boxes and then became ubiquitous in remote control devices.) With the Commission mandating that all cable systems carry *all* local channels, operators could not substitute more valuable cable programming networks for less valuable local stations.

The federal courts overturned must-carry rules in 1985, finding that the FCC had favored one form of communication over another based on a faulty premise. "The Commission," wrote the DC Circuit Court, "has in no way justified its position that cable television must be a supplement to, rather than an equal of, broadcast television. Such an artificial narrowing of the scope of the regulatory problem is itself arbitrary and capricious."[9]

The abolition of must-carry achieved three things. First, it removed a tax, and in that respect helped the furious expansion of the new technology that was making wired video the preferred distribution platform. Second, it created a natural experiment wherein economists could observe how the change in regulations affected cable systems and their subscribers. Third, it made the broadcasters dyspeptic.

The 1992 Cable Act and Turner Broadcasting

Under pressure from broadcasters, the FCC reinstated the rules. After the courts again ruled them unconstitutional, in 1987,[10] Congress came to the rescue with Sections 4 and 5 of the Cable Television Consumer Protection and Competition Act of 1992.[11] The law was the lone legislation to survive a veto by President George H. W. Bush; when the override passed the House in early October, gleeful Democrats chanted *Four More Months! Four More Months!* In addition to reimposing cable rate regulation, the Cable Act obligated systems with more than twelve channels to set aside up to one-third of their capacity for the retransmission of broadcasts in their local markets.[12] Stations, on the other hand, were given the option to negotiate carriage payments or claim must-carry. There was no doubt whom the statute favored. Bruce Owen summarized:

> The local broadcast station has it made: either it can withhold its signal from the local cable operator unless the cable operator pays for it, or the broadcaster can force the cable operator to carry the signal gratis. . . . It is difficult to imagine a more one-sided arrangement.[13]

In the 1960s, the FCC had imposed must-carry requirements on cable systems with the argument that broadcasting was dominant and cable a "supplement." Now the argument was reversed. Cable was dominant and had to be barred from discriminating against weaker competitors.

Within days of the act's passage cable interests challenged the must-carry provision under the First Amendment. The litigation produced two Supreme Court decisions. The first, issued in 1994 and commonly called *Turner I,*[14] addressed the fundamental question of whether a cable TV system or a programmer (such as Turner Broadcasting, the plaintiff) was protected under the free-speech clause of the First Amendment. *Turner I* established that both cable operators and programmers were protected. Cable TV had higher constitutional standing, in fact, than broadcasters (where "physical scarcity" truncated rights), but not as far-reaching rights as those afforded newspapers.[15] *Turner I* then remanded the case to a lower court to determine whether those rights were violated by the 1992 Cable Act.

That adjudication made it back to the Supreme Court in *Turner II,* a 1997 verdict in which a 5–4 majority found the rules permissible.[16] Justice Kennedy's majority opinion stated: "We have identified a . . . 'Governmental purpose of the highest order' in ensuring public access to 'a multiplicity of information sources.' . . . And it is undisputed the Government has an interest in 'eliminating restraints on fair competition . . . even when the individuals or entities subject to particular regulations are engaged in expressive activity protected by the First Amendment.'"[17] The Court found that Congress acted on its belief that, lacking must-carry, "the economic viability of free local broadcast television and its ability to originate quality local programming will be seriously jeopardized."[18]

For Justices Kennedy, Souter, Stevens, and Rehnquist, the economic incentives of cable TV systems were key. Congress concluded that cable operators might use their market power to exclude local broadcast stations from their channel lineups, thus limiting the broadcasters' audiences and thus their capacity to compete for local advertising. Forcing cable systems to carry broadcast signals free of charge prevented this anticompetitive conduct, protecting broadcaster profits and promoting the viability of their programming. Even the 40 percent of U.S. households then viewing (only) off-air broadcast TV could reap these benefits.[19] Justice Stephen Breyer, who supplied the fifth vote, felt no need to justify must-carry rules on competitive grounds. It was sufficient that the Congress intended "to assure the 'over the air' access to a multiplicity of information sources."[20]

This decision has been attacked in numerous forums, but nowhere more thoughtfully than in Justice O'Connor's dissent (joined by Justices Scalia, Thomas, and Ginsburg). She explained the majority's confusion over the difference in "viewpoint neutral" and "content neutral" regulations. Rules that discriminate among either viewpoints *or* types of content are highly likely to be struck down under a "strict scrutiny" legal standard. The must-carry rules did not dictate particular viewpoints, but, explained O'Connor, "Whether a provision is viewpoint-neutral is irrelevant to the question of whether it is also *content-neutral*."[21] It was clear that Congress was discriminating in favor of certain types of speech because it stated as much:

> [*Turner I*] referred to the "unusually detailed statutory findings" accompanying the Act, in which Congress recognized the importance of preserving sources of local news, public affairs, and educational programming. Nevertheless, the Court minimized the significance of these findings, suggesting that they merely reflected Congress' view of the "intrinsic value" of broadcast programming generally, rather than a congressional preference for programming with local, educational, or informational content.[22]

When Minnesota levied a tax on all newspapers, at a uniform rate applied no matter the viewpoint of the publication, the law was nonetheless struck down.[23] Likewise, when Florida legislated a right of reply for candidates to respond to critical newspaper editorials, the Supreme Court unanimously ruled the law unconstitutional even though the right of reply applied to all candidates and partisan perspectives.[24] "Never before has a content-based regulation," wrote attorneys for a cable operator participating in *Turner II*, "selectively targeting a faction of the private press, been adjudged by this Court to escape strict scrutiny analysis provided it is also deemed content-neutral."[25]

The Court had again stumbled in protecting First Amendment rights. But the Constitution's loss is our gain, insofar as we wish to understand the political spectrum. First, as economic policy, must-carry was a bust. Even before the Court ruled, the Federal Trade Commission produced compelling evidence that the rules decreased consumer welfare.[26] Studying the reaction of cable system operators following the invalidation of must-carry rules in 1985, FTC economist Michael Vita found "that systems that sell advertising actually were less likely to drop broadcast stations than were systems that do not."[27] If cable operators had been dropping stations to restrict TV station rivalry, the problem the must-carry rules were said to address, the stations competing in local advertising markets would have been dropped first. Vita also showed that the must-carry rules, which forced cable systems to feature stations with little viewer interest, crowded out more appealing programming and so hurt subscribers.

Turner II created another natural experiment. With the verdict on March 31, 1997—a "surprise until the last minute," the *New York Times* reported[28] —the reaction from financial markets was dramatic. While the shares of major broadcasters, cable operators, and cable programmers were not much affected, stocks in companies owning broadcast TV stations featuring home shopping and infomercials (including Paxson Communications) had the best one-day returns of the entire year, a 13.8 percent bump.[29] The trade press noticed:

> The Dom Perignon was just arriving at Bud Paxson's office Monday morning when he picked up the call from this reporter. "You're talking to the happiest man in America," he said, anticipating a question about his reaction to the Supreme Court ruling that left must carry in place.... Paxson Communications' stock was up $1.75, to $10.75, last Monday after the must-carry decision was issued....
>
> Home Shopping Network Inc. Chairman Barry Diller was the second-happiest man.... Diller noted that people continually ask him, "What are you doing with all those crummy little UHF television stations?" They are "a little less crummy" after last week's decision, he said.[30]

Wall Street grasped that must-carry rules provided windfalls to home shopping entrepreneurs. Retailers like QVC paid cable and satellite operators to retransmit their programs to viewers (5 percent of total receipts from sales to system subscribers was the norm). Now they had an alternative. By buying small UHF stations and converting them to home shopping formats, these networks could get free access to cable. ValueVision International, another home shopping firm, disclosed in its 1997 Annual Report:

> On November 22, 1996, the Company announced that an agreement had been reached with Paxson for the sale of its television broadcast station, WVVI (TV), Channel 66, which serves the Washington, D.C. market, for approximately $30 million.... As part of the agreement, Paxson will be required to pay an additional $10 million to the Company as a result of the United States Supreme Court upholding the "must carry" provision of the 1992 Cable Act. WVVI (TV) carries the Company's television home shop-

ping programming and was acquired by the Company in March 1994 for $4,850,000.[31]

The Home Shopping Network happily told its shareholders that thanks to Congress, its Home Shopping Club was gaining wider distribution: "HSC has also increased cable carriage of HSN 2 and HSN Spree as a result of the [1992] cable re-regulation law."[32] In an essay in the *Wall Street Journal,* I attempted to explain "How Home Shopping Became King of Cable."[33]

Meanwhile, public affairs programmers cried.

C-SPAN is not a publicly traded company. If it were, its shareholders—according to testimony of its founder and then-CEO, Brian Lamb—would have experienced substantial losses from *Turner II.* While C-SPAN was fairly well established, C-SPAN2 and other program innovations (C-SPAN3 and C-SPAN4 were on the drawing board) found themselves expendable. The standard basic cable package in 1993 delivered about forty channels. With more slots consumed by little-watched local TV stations, fewer were open for cable networks. C-SPAN produced no advertising revenues (it was commercial-free; other basic cable networks split their ad "avails" with the local cable system) and failed to generate the audiences of many other cable networks. Hence it was often cut. More than seven million viewers lost access to C-SPAN programs following Cable Act–mandated implementation of must-carry in 1993. While the network's fans complained to operators and the organization fought hard to regain coverage, only "about half" of the lost carriage slots had been recovered by 1995.[34]

The verdict in *Turner II* found must-carry "assur[ing] the over-the-air public 'access to a multiplicity of information sources,'" in Justice Stephen Breyer's words.[35] But C-SPAN offered precisely the sort of informational programs policy makers said they wanted to promote. This vast library of public affairs television programming was developed nowhere else.

In 1998, Gray Davis, a Democrat, ran against Republican Dan Lungren for governor of California. The two men debated each other multiple times, but these debates could be seen in the Los Angeles area only on C-SPAN.[36] In other words, at the instant that the TV stations with must-carry privileges were airing sitcom reruns, religious services, or home shopping, the network being commonly dropped due to must-carry was creating and distrib-

uting precisely the programs the law was ostensibly intended to promote—
and, on that basis, ruled constitutional.

Moreover, the broadcast stations that did present substantial news and
cover politics were almost always the affiliates of the major networks, not
the small independent stations relying on must-carry. Only 750 of the approximately 1,300 commercial television stations in the United States provided *any* local news; the remaining 500-plus carried none whatsoever.[37]
The stations catapulted by must-carry into cable channel slots were drawn
almost entirely from this subset. Must-carry did not bring news and public
affairs onto the dial, it forced them off.

C-SPAN lobbied hard against must-carry, repeatedly sending company
executives to testify before Congress. Congress's response was to try to
buy off the network by offering to extend must-carry rights to C-SPAN and
thus ostensibly protect its coverage. Yet Lamb saw something more. "I am
totally opposed to having politicians require something to be aired, even if
it is my network," he said. "With favors come paybacks." Instead, broadcast
must-carry was written into law, the law was implemented as written, and
the results were as Lamb had warned: mandated carriage of marginal UHF
stations crowded C-SPAN out of millions of homes. "My attitude is one
of controlled rage," Lamb adds. "The government writes a piece of legislation that sounds good, but it is the public service channel that is being
hardest hit."

Congress could not have missed this; it is a pity the Supreme Court did.
The destruction wrought—"Congress has done severe damage to this network"—was well articulated by Lamb, who used his estimable Washington
presence to press the issue. "Even though I testified that [the Cable Act] was
one of the worst pieces of legislation I have seen in 29 years in Washington,
and that it would hurt us, the members of Congress did not listen."[38]

In coming years the Federal Communications Commission would declare that the TV band was far too spacious; given competing demands, a
good chunk of this spectrum should be transferred to other emerging services, including mobile networks. To enable this, the FCC would propose an
"incentive auction" in which perhaps hundreds of TV stations will sell their
licenses back to the government, releasing bandwidth for reallocation. I discuss this process in more detail below. But the irony noted here is that the

price stations demand to vacate the spectrum is pushed up by must-carry rights, which makes marginal stations more profitable. U.S. taxpayers must thus spend billions of dollars to reclaim the favors extended in 1992 law.

At first glance, the "must-carry" fight may look like another case of obsolete rules persisting in a period of dynamic change. But there is more to the circus than that. Follow the bouncing ball:

- The must-carry rules endow over-the-air broadcasters with special cable carriage rights;
- These rules are enacted, and deemed constitutional, on the grounds that all TV stations deliver special content of prime social importance;
- But now the government says that many of the endowed TV stations are not only not special, they are superfluous, blocking the "public interest";
- The state finds itself powerless to revoke (or fail to renew) licenses;
- The government must therefore buy back its licenses, paying a premium because of the must-carry rights that solidified its now faulty allocation of broadcast TV spectrum.

As five of nine Supreme Court justices misconstrued the impact of must-carry, many scholars rooted them on. Even as C-SPAN was being displaced on basic cable lineups by TV stations hawking cubic zirconia, historian Dean Alger claimed that

> cable system owners used the "must carry" rule as an excuse to eliminate C-SPAN. . . . C-SPAN's respected chief, Brian Lamb, along with a good number of citizens in Citizens for C-SPAN, began a crusade over these C-SPAN ejections. Although Lamb misguidedly tried to assign a large part of the blame for these actions to the 1992 Act, most observers saw cable corporation greed and irresponsibility as the principal factor, given that there are forty or more channels on which to place C-SPAN.[39]

Alger—who has taught at Harvard and appeared as an expert on Bill Moyers's PBS shows—artfully packs so much mischief into so short a space.

First, cable operators needed no "excuse" to eliminate C-SPAN from the channel lineups. Far from looking for reasons to abandon the network, cable operators had *created* the service, contributing capital and endowing its corporate management with editorial independence. On this basis, C-SPAN programming has been distributed not only via its cable system founders but by competitive entities: satellite operators, telephone company video systems, radio stations, and the Internet, to users everywhere.

Second, the idea that "cable corporation greed" was responsible for the loss of carriage raises the question of how that volume of greed came to be adjusted upward at precisely the instant the 1992 Cable Act's must-carry mandate kicked in. Just before that pivotal policy moment, cable "greed" had placed C-SPAN on TV sets in more than fifty million households.

Third, that the "respected" if "misguided" Brian Lamb was not an expert on the matter, one whose experience and intimate knowledge of the event could be brusquely dismissed in favor of what "most observers" are alleged to have thought, is a stunning testimonial to grandiose, faith-based pronouncement.

Fourth, that cable operators had no reason to disrupt services because they had plenty of space—"forty or more channels on which to place C-SPAN"—ignores the reality of opportunity cost. Many systems needed to eliminate existing programs to accommodate the newly mandated. That C-SPAN2 *did* lose carriage on thousands of cable systems was documented. That "greedy and irresponsible" cable operators created a second full-time, public service network just so they could drop it as soon as they found an "excuse"—that is fantasy.

But if Alger's take is easy to dismiss—say, because *most observers* consider it unpersuasive—the eager endorsement of must-carry by no less than Cass Sunstein is not. Sunstein, excited by the prospect that the 1994 *Turner I* ruling presaged a change in U.S. legal doctrine, added an afterword to his 1995 edition of *Democracy and the Problem of Free Speech* to hail it. The Court's finding that must-carry was "content-neutral," he wrote, could mean "that Congress will be permitted to regulate particular technologies in particular ways, so long as the regulation is not transparently a subterfuge . . . to promote particular points of view."[40] Legally, Sunstein nails this point, seeing what the Court majority did not: that the Court was taking a radical step. *Turner I* (and *Turner II*) potentially opened all electronic communications to the public interest standard previously confined to broadcast radio and television.

Sunstein heralded the Court's move as paving the way for policies to patch market failures and advance "a system of democratic deliberation."[41] The latter, he argues, is what James Madison meant to promote: his parsimonious First Amendment stricture that "Congress shall make no law abridging freedom of speech . . . or of the press" needs substantial elaboration in order to achieve the vision. This view is widely disputed. The late Supreme Court Justice William O. Douglas powerfully argued for a "laissez-faire regime," appealing to the plain language of the law. "What kind of First Amendment would best serve our needs," he wrote, "may be an open question. But the old-fashioned First Amendment that we have is the Court's only guideline; and one hard and fast principle which it announces is that Government shall keep its hands off the press."[42]

Sunstein recognizes this alternative position, deeming it the prevailing orthodoxy, which he opposes. "The First Amendment should not be an obstacle,"[43] he writes, to substantial regulation of the press. Supreme Court Justices Douglas and Hugo Black, as well as the American Civil Liberties Union, are thrashed for "First Amendment absolutism,"[44] the idea that the government should be stopped from regulating speech even if it might ultimately secure "Madisonian aspirations" that "place a special premium on attention to public issues and exposure to diverse political views."[45]

The problem, to Sunstein, is that "the system of free expression in America" operates as "a system of unregulated markets." News, information, and entertainment are supplied "on the profit principle," the same one used to produce "cars, brushes, cereal and soap."[46] He echoes a very old theme. The unregulated broadcaster's profit seeking, its reliance on commercial advertising, and its "propaganda" were all arguments employed by Commerce Secretary Herbert Hoover as he sought greater government control. The Radio Act of 1927 was instituted to promote the very purposes—responsible public service to protect and enhance democracy—Sunstein claims as policy innovations today.

Broadcasting, said Herbert Hoover, "is not to be considered as merely a business carried on for private gain . . . [but] is to be considered primarily from the standpoint of public interest."[47] Broadcasters' unregulated operation was superseded by "Madisonian aspirations" in 1927. After decades, even the regulators were chagrined to find their handiwork a "vast wasteland." Something had gone deeply awry.

This impressive record garners only fleeting attention from Sunstein. While he observes that "The fairness doctrine was hardly a terrific success on its own terms, and . . . is poorly adapted for contemporary problems,"[48] he does not pause to consider the implications. Others do. Law professors Thomas Krattenmaker and Lucas A. Powe, Jr., for instance, conclude: "Broadcast regulation has been characterized by the very abuses—favoritism, censorship, political influence—that the First Amendment was designed to prevent in the print media."[49]

Instead, Sunstein explores a new approach to constitutional limits, a quest that gained gusto with *Turner I* and *Turner II* and their rejection of "First Amendment absolutism." This development intrigues Sunstein, and he outlines the competing hypotheses with alacrity:

> What was the purpose of the must-carry rules? This is a complex matter. A skeptic, or perhaps a realist, would say that the rules were simply a product of the political power of the broadcasting industry. . . . On the other hand, some people might think that the must-carry rules were a legitimate effort to protect local broadcasters, whose speech may be valuable precisely because it ensures that viewers will be able to see discussions of local political issues.[50]

Sunstein sees a plausible case for the latter explanation, and he finds the hint of a high-minded vision, extending "discussions of local political issues," sufficient to resolve the "complex matter." But what if the actual impact of the must-carry laws—bumping C-SPAN2 out of millions of homes and blocking the expansion of state and local news and public affairs programming on cable TV—were given its due? That was a question not asked.

Ronald Coase, exasperated with the dramatic gap between regulation's announced public interest goals and perverse real-world impacts, wrote in 1965:

> The policy choice should not be put in terms of government action versus the market in the field of radio and television. I am arguing for sensible

government action. I am arguing for a properly functioning market. . . . Of course, the task of building social institutions is not an easy one. But it is not made easier by syrupy talk about broadcasters acting in the public interest. What is wanted is more economics and less humbug.[51]

Alas, we now have a half-century of economic evidence. Of the FCC's two proudest efforts to use regulation to advance democratic debate, the Fairness Doctrine and Must Carry, the actual outcomes are clear: both compromised free speech.

12

Indecent Exposure

The easy access, low cost and distributed intelligence of modern means of communication are a prime reason for hope. The democratic impulse to regulate evils, as Tocqueville warned, is ironically a reason for worry.

—Ithiel de Sola Pool

THE TELECOMMUNICATIONS ACT OF 1996 was a moment of pure bipartisan bliss. A Republican Congress partnered with a Democratic president to write an ambitious law that would usher in a new era of modern technology. While the measure primarily dealt with rules for local and long distance wireline telephone competition, its sweep was broad.

At the White House signing ceremony, CNN founder Ted Turner impishly flashed a "V" sign with his fingers. He was not celebrating a victory for the "public interest." Turner was taunting cable's broadcast industry rivals about the law's mandate of the V-chip. This device, to be implanted in all TV sets sold in the United States after the year 2000,[1] would identify shows containing sex or violence. Parents would be able to adjust their unit's settings to block programs with too much of either.

The V-chip was a flop. A 2004 Kaiser Foundation poll indicated that no more than 15 percent of households had ever used it.[2] Parents rarely monitor their kids' TV viewing by setting digital controls they need their ten-year-old's help to operate. Instead, they let their children watch shows or channels they believe they can trust. Cable TV's family-oriented networks —from Nickelodeon to Animal Planet to Sprout to Disney to TV Land— specialized in such branding. With their homogeneous content genres, cable menus solved the problem that Washington policy makers implanted a chip to fix. The chips' only discernible impact, other than a small bump in the price of TV sets, was that broadcast TV networks were forced to absorb the costs of rating their programs—while Ted Turner, a cable programming mogul, danced.

Political insiders smirked that the innovation revealed a consensus over content: Democrats were offended by violence, Republicans by sex. That it was called the V-chip and not the S-chip reflected a backroom triumph for the Democratic administration. President Clinton hailed the innovation—still required on that 110-inch flat-screen panel you're saving up for—as a social breakthrough. "We're handing the TV remote control back to America's parents so that they can pass on their values and protect their children."[3]

Yet the remote control was never subject to federal jurisdiction, and the task of weeding out video content inappropriate for children was not a task that the political spectrum was up to. Nonetheless, such doomed efforts became prime time hits for politicians and activists across the partisan divide. While a Democratic administration provided a platform for the Vast Wasteland meme in the 1960s, the idea was appropriated by leading conservatives a generation later. Robert Bork, in his 1996 tome *Slouching Towards Gomorrah,* picked up the cudgels. Citing the damage inflicted by smut that spews externalities, Bork complained (quoting conservative media critic Michael Medved) that "Amish kids . . . know about Madonna." Rejecting both free-speech advocates and the perspective of "free market economists" who have caught "the libertarian virus," Bork was gloomy: "perhaps . . . technology is on the side of anarchy." Given that "unconstrained human nature will seek degeneracy," Bork claimed, "it's enough to make one a Luddite." The leading conservative legal thinker was by now willing to go the extra mile. "Is censorship really as unthinkable as we all seem to assume?"[4]

Actually, censorship has long been applied to pornography, obscene expression that "appeal[s] to the prurient interest in sex . . . in a patently offensive way" while lacking "serious literary, artistic, political, or scientific value."[5] Supreme Court Justice Potter Stewart famously said, in 1964, that it was difficult to explain, but "I know it when I see it."[6] Government has latitude to regulate obscenity in print, movies, books, and magazines or on broadcast television. The policy question at issue in wireless is whether controls may stretch further, over nonobscene speech. George Carlin's "Filthy Words" comedy routine, when played on a Pacifica radio station, listed seven words that one could (according to Carlin) not say on TV. But, with a twist of comedic paradox, the words did air on radio. That challenged lawmakers. The material was "filthy" but not *obscene*. The point of Carlin's routine was not prurient but pointed, ridiculing what he considered Puritanical law and custom.

The FCC was not amused. It sanctioned the noncommercial radio station that broadcast the routine and banned such words during times that children might be in the listening audience. In so doing, it extended its censorial powers to "indecent" programming. In a 1978 verdict, *Pacifica*, the Supreme Court ruled that the FCC's actions were constitutional because broadcasts wafted—uninvited—through space, and were "pervasive." This difference in conduit again resulted in a difference in First Amendment protection, distinguishing wireless from print media, albeit with a new rationale for regulation.[7]

It was not an unmixed blessing for those charged with carrying out the law. Now bureaucrats had to know indecency when they saw it. Things got dicier.

Former FCC official Robert Corn-Revere, who later became a prominent First Amendment lawyer, tells the story of what happened at the Commission during the 1990s when a complaint was lodged against the playing of "Who Are You?," a hit song by British rock group The Who. The tune had aired countless times on U.S. radio stations since 1978. The citizen's petition claimed that the recording featured the lyrical refrain "Who the fuck are you?" Corn-Revere, appointed to an ad hoc committee formed to investigate the matter, was surprised. Using their expertise, FCC staff discovered that the offending term—including the most famous of George Carlin's seven words—actually did appear. Twice. WTF?

The FCC group knew that to issue an order banning music that had been relentlessly broadcast on thousands of stations for many years would flunk the laugh test. Being pilloried in the press or on Capitol Hill was something to avoid. So the staffers pondered the creation of a new "public interest" rule. Someone suggested a "British accent exception," or a "first-time offense" pass. Or a "depends on the context" loophole. Alas, a strategy arose. The complaint was recorded and duly filed. It was next pulled out for consideration five years later. Too much time had elapsed, ruled the FCC. The matter was moot. *Next!*

Despite the lack of an action program, censorship in the abstract has been widely popular. This political reality has been witnessed in FCC confirmation proceedings. In 1989, Andrew C. Barrett, a black Republican who had served on the Illinois Commerce Commission, had been nominated for a commissioner slot by President George H. W. Bush. At his Senate hearings he was asked about his background and ideas. Many of these queries are tantamount to IQ questions: if you're not able to formulate the right answers, you're not smart enough to be on the Commission. Barrett quickly discovered the perils of candor in exchanges with Senators Daniel Inouye (D-HI) and Al Gore (D-TN).

> For more than two hours, Senate Commerce Committee members grilled the nominees, resulting in some tense moments, especially for Barrett, whose responses to several questions concerned some members. . . . When Inouye asked how the nominees would treat "indecent and violent" programming, Barrett said he found both "offensive."
>
> But it was the statement that followed that raised some senators' eyebrows. "It seems to me that what we perceive to be indecent is not on television because there is no market. There is a market for indecency out there . . . otherwise they would not be showing it," said Barrett.
>
> Gore said he was "puzzled" by Barrett's response. "I am not suggesting that I believe it ought to be on television," said the FCC candidate. "I am simply suggesting that if there was not a market out there for it, and if there was not viewership out there, I would suggest it would probably not be there. I find it as horrible as most people do."[8]

When the hearing was over, Andrew Barrett hurriedly sent a letter to Senators Inouye and Gore to explain his horror at the terrible programming on television more fully. Barrett was eventually confirmed, but the extra explanation he was forced to make revealed what type of "regulatory analysis" was most vital. Barrett's egregious gaffe was to tell a simple truth about the demand for popular, if crass, broadcast television content.

Under pressure from right and left, the FCC, starting in the 1970s, threatened stations (with fines and license nonrenewals) airing "shock jocks." Nonetheless, radio got racier. Before long, Carlin's forbidden list was sneaking onto television. Bad words were uttered by pop singer Cher at the 2002 Billboard Music Awards, when she—as described by the late Supreme Court Justice Antonin Scalia—"metaphorically suggested a sexual act as a means of expressing hostility to her critics." Not to be outdone, rock legend Bono let loose with similar metaphorical zeal at the 2004 Golden Globe Awards. The Commission imposed fines, but station owners dodged liability when the Supreme Court found that the FCC had botched enforcement of its "fleeting expletive" exception.[9] When Janet Jackson suffered an epic wardrobe malfunction at a Super Bowl halftime show in 2004, millions of flummoxed sports fans scrambled for the replay button. Alas, the FCC did fine CBS/Viacom for airing the indecent instant, but the sanction was again voided by federal courts.

Bork may have regretted that "without censorship, it has proved impossible to maintain any standards of decency,"[10] yet what were poor regulators to do? In fact, the government harassed Howard Stern's radio broadcasting outlets sufficiently that he fled to SiriusXM radio in 2006, saying, "The Super Bowl did us in."[11]

Stern continued to perform for a large subscription audience, soon receiving $80 million in annual compensation to interview porn stars and edgy actors in recovery. YouTube launched an online video streaming service. These escape hatches may be seen as victories for censorship, leaving broadcasting more pristine. That would be wrong. Between 2002 and 2014, a top-rated prime time broadcast television series, *Two and a Half Men,* forged new lower bounds in bawdiness. The show passed full regulatory muster. While the accidental Super Bowl flash was deemed illegal, scripted topless sunbathing by Miley Cyrus, who boasted that she would "take out her twins," was found unexceptional.

It was not for lack of effort that regulators have failed to rein in the raunchy. Consumers, producers, and the expanding market—not to mention the First Amendment—are each formidable. Luddites some might wish to be, but the isolated technophobe can hardly suppress the appetites of others. And imagine the absurdity of trying. In a 2004 case the FCC fined Viacom $3.5 million for, among other transgressions, airing a segment of sitcom *Rock Me Baby* that featured this dialogue:

> BETH: Of course you are [immature]; that's why you laugh every time the South Carolina Gamecocks play the Oregon State Beavers.
> JIMMY: Hey—do you remember that time the announcer said, "The Game-cocks are deep in Beaver territory?"[12]

A serious campaign to eradicate broadcasts with double entendres might be fun to watch. Particularly if this effort was unleashed on, say, presidential debates. Then try programming it onto a V-chip. Then file it in a drawer.

PART FOUR

SLOUCHING TOWARD FREEDOM

The most important argument favoring the flexible license regime is its straightforward adaptability, the ease with which it can be modi-fied to accommodate unpredicted and unpredictable changes in the nature of spectrum utilization.

—William Baumol and Dorothy Robyn

THROUGH A LONG AND WINDING path, liberalization of radio spectrum rights finally pushed past the regulatory labyrinth in the 1970s and 1980s. Mobile phone networks became ground zero in the regime switch. After bottling up cellular for decades, and then managing to squander, almost comically, tens of billions of dollars in public revenues via license lotteries, regulators stood down. Broad, flexible-use rights were awarded by competi-tive bidding.

By the late 1990s, mobile markets were flush with reform. "Spectrum has been privatized—property rights have been created—without anyone using such indelicate terms," noted a communications law treatise in 1999.[1] Rivalry flourished, prices plummeted, and innovative ecosystems formed. The idea that only state-run allocations could prevent a "cacophony of com-peting voices" was reduced to the status of a Barry Manilow ringtone.

But resistance to liberalization was not dead, and top-down, case-by-case allocations have not faded away. Many offer the verdict that a mix of models is essential for progress. Indeed. But that invites the underlying question: What is the proper balance? More fundamental still, how can we find it and continually adjust it? Reform of the political spectrum is yet a work in progress.

13

The Thirty Years' War

The Communications Act directs the Commission to make a finding
of public interest, convenience and necessity before issuing a license
or allowing an existing license to be transferred. The result of this
process is that the Commission is often required to decide what
is the *best* use of a frequency or which user would *best* meet the
public interest. There is, in other words, a "wise man" theory of
regulation.

—FCC staff paper

POLICIES THAT DO NOT ACHIEVE their announced ends have wandered freely
in the political spectrum. Yet it would be highly misleading to focus today
solely on such failures. Key realms of the regime now operate under radically different rules from those delivered under the diktat of the Radio Act
of 1927. One such change permitted the growth of cellular telephone networks. These services were less controversial than radio or TV broadcasting,
which are mass media industries with a heavy influence on public opinion.
And cellular services, if micromanaged, require licenses that are far more
difficult to design than those for broadcasting. A TV station requires little
more than a transmitter to blast one-way signals. A cellular network is an
intricate web of base stations and mobile handsets hosting two-way traffic
flows between radios via atmospheric pathways that are constantly moving
about. Imposing technologies, architecture, devices, and business models

on such systems by administrative fiat would threaten agency overload while imposing far costlier rigidities than those seen in broadcasting. With fewer demands for political control, these bureaucratic costs were simply not worth the candle to policy makers. Mobile wireless networks were thus permitted to configure themselves largely outside of the Public Interest Standard in a marketplace shaped by competitive economic forces.

The steps that led to the creation of wireless spectrum markets were incremental and often overlooked while fights over older technologies and vestigial schemes of no lasting importance dominated public attention. The V-chip, for example, was featured in both President Clinton's and Vice President Gore's acceptance speeches at the 1996 Democratic National Convention.[1] Neither mentioned mobile services, which had been the subject of bold policy action and were being passionately embraced by millions of customers. As few noticed, the universe of spectrum allocation was profoundly altered, and the services unleashed would be transformative.

When Major Edwin Howard Armstrong sought to deploy his new FM radios in 1935, he was confronted by the "'Wise Man' theory of regulation." FCC bureaucrats read his petition and determined whether the new device would work, whether it would interfere with existing businesses, and whether, ultimately, it was in the "public interest." This did not turn out well for Armstrong.

When tech visionary Steve Jobs sought to deploy Apple's iPhone in 2007, he took his innovation to the marketplace. Wireless carriers were offered iPhones produced to their networks' specs; regulators had delegated dominion over defined frequency spaces to these private parties. The networks bid against each other to secure Apple's radio devices. They were soon deployed, and consumers slept on sidewalks to be at the front of the line to buy them.

Armstrong committed suicide in 1954 without ever having seen his disruptive technology make its mark. Jobs died from cancer in 2011 having lived to see his latest futuristic visions, the iPhone and the iPad, sweep the globe thanks to spectrum rules that allowed markets to welcome innovation.

The basic idea of cellular telephone service was introduced in a July 1945 article not in *Popular Mechanics* or *Science,* but in the down-home *Saturday Evening Post.*[2] The head of the FCC, J. K. Jett, related what he had been told by AT&T: that by divvying up wireless networks into "thousands of zones," the company could create vast new communications capacity.[3] Millions of citizens would soon be using "handie-talkies." Licenses needed to be issued, Jett noted, "But that won't be difficult. . . . The only important restriction is that no permit will be granted to an alien." The revolutionary technology would be formulated by scientists within months.

But permission to deploy it would not. The government would not allocate spectrum to realize the engineers' vision of "cellular radio" until 1982, and licenses authorizing the service would not be fully distributed for an additional seven years. That was one long bureaucratic delay. Yet the effort to scale this perverse obstacle had one sublime result. Many of the key structural impediments in spectrum allocation were confronted and overcome. The rules eventually adopted for cellular unleashed a robust wireless marketplace and established a regulatory system far better suited to the demands of a dynamic economy than those of the ancien régime.

Cellular was preceded by mobile telephone service, MTS, launched in 1946.[4] The mobile equipment was unwieldy and expensive—the transceiver could fill the trunk of a sedan—and networks faced tight capacity constraints. The largest markets had no more than forty-four channels.[5] Even at sky-high prices, there were long waiting lists for subscriptions. As late as 1976, the Bell System's mobile network in New York could host just 545 subscribers.[6]

Cellular networks provided an ingenious way to dramatically expand service by splitting a given market up into cells, putting a base station in each.[7] These stations, often located on towers to improve line-of-sight with mobile phone users, were able to both receive wireless signals and transmit them. The base stations were themselves linked together, generally by wires, and connected to networks delivering Plain Old Telephone Service (POTS). The advantages of this architecture for wireless—the radio frequency (RF) interface—were profound.

Mobile radios could use less power because they needed only to reach the nearest base station, not a mobile phone across town. Not only did this save battery life, but transmissions stayed local, leaving other cells quiet. A connection in one cell would be passed to an adjacent cell and then to the next as the mobile user moved through space. The added capacity came from

13.1. A Motorola precellular carphone, 1948. Geoffrey Fors Collection/
Monterey, California.

reusing frequencies, cell to cell. And cells could be "split," adding network capacity. Whereas an MTS system might support hundreds of phone users in a city, each conversation requiring a channel covering the entire market, a cellular system could create thousands of small cells and support hundreds of thousands of simultaneous conversations.

Although AT&T proposed to begin developing cellular in 1947, the FCC rejected the idea, believing that spectrum could be best used by other services that were not "in the nature of convenience or luxury."[8] This view, that cellular would be a niche service for a tiny user base, persisted well into the 1980s. "Land mobile," the generic category into which cellular fell, was far down on the FCC's list of priorities. In an important 1949 proceeding, it was assigned just 4.7 percent "of the spectrum considered useful for the service (i.e., between 25 and 890 MHz)," as a Harvard study later summarized.[9] Broadcast television was allotted 59.2 percent and government uses 25.0 percent. Television broadcasting had become the FCC's mission, and land mobile was a lark. The promotion of TV broadcasting, RF engineer George Calhoun explained, "would stand astride the road to the realization of the cellular idea for more than twenty years."[10]

The government may have thought it was simply making a judgment that TV was a higher social priority than cellular radio development, but this was a false dichotomy. Americans could have enjoyed all the broadcast TV they would watch in, say, 1960 and had cellular phone service too. Instead, TV was allocated far more bandwidth than it ever used, with vast deserts of vacant TV assignments blocking mobile wireless for more than a generation.

How unused were they? Across the 210 U.S. television markets (ranging in size from New York City to Glendive, Montana), the 81 channels originally allocated to the TV band (channels 2–83, with channel 37 excluded) created some 17,010 slots for stations. From this, the FCC planned to authorize just 2,002 TV stations in 1952. In fact, by 1962, there were 603 stations broadcasting in the U.S. Just 3.5 percent of the channels set aside for television were actually transmitting signals a decade after the creation of the 1952 TV allocation table.[11]

Broadcasters vigorously defended the idle bandwidth. When mobile telephone advocates tried to procure UHF frequencies for their services, the broadcasters deluged the Commission, arguing ferociously and relentlessly that mobile telephone service was an inefficient use of spectrum.[12] Why were broadcasters so determined to protect vacant frequency spaces? Given that commercial TV station licenses were severely limited—enough to support only three national networks—the scores of unused TV channels could have been seen as a threat: What if future policy makers got serious about competition? Shrinking the TV band by slicing off chunks for land mobile services might preempt that regulatory turn, protecting incumbent broadcasters from future competition.

Alas, more subtle logic lurks in the political spectrum. Incumbent TV licensees did not fear the prospect of competitive stations—they believed they held sufficient veto power to prevent such policy outrages—as much as they cherished the option value of unused spectrum. Keeping a vacant spectrum parking lot available for future development was seen to accommodate incumbent broadcasters. As things played out, this thinking proved correct. Unoccupied TV frequencies were allocated to licenses awarded to incumbent broadcasters, without payment, during the transition to digital television some years later.

Since the late 1940s, the FCC had been licensing radio common carriers (RCCs) to supply old-fashioned (precellular) mobile telephone service. Its policy was to license just two mobile operators per market—generally, AT&T and a much smaller competitor. In addition, the Commission distributed private land mobile radio service (PLMRS) licenses authorizing wireless links for strictly internal use by companies not in the communications business. These allowed, for instance, workers on an offshore drilling rig to talk with oil company personnel at the home office, an airline to coordinate baggage operations at an airport, or a freight train to check its track assignments. The Commission allocated these licenses according to line of business, creating twenty-one distinct categories, including "manufacturers, utilities, transportation companies, state and local governments . . . taxis, plumbers, and delivery services."[13]

In 1968, there were sixty-two thousand common carrier phone subscribers, almost equally split between AT&T and, collectively, five hundred tiny RCCs (the average "network" having only about fifty subscribers).[14] PLMRS licenses were allotted far more bandwidth (about 90 percent of the 42 MHz set aside for land mobile) and deployed more phones.[15] Compared with the 326 million U.S. cellular subscriptions that existed by 2012,[16] these low-tech radio services were fleas on an elephant. The RCCs were local, barely profitable, generally family-owned businesses.[17] They intensely opposed cellular, seeing it as a sophisticated new rival that would ravage their small-scale operations.

Despite being small-time players, the RCCs proved adept at litigation.[18] This core competency was a product of regulation: the flea did not have to impede the elephant directly, only by running rings around regulators. The mom-and-pop operators seized a force multiplier: Motorola, then a pioneering wireless technology firm. Both RCCs and Private Land Mobile operators were excellent customers, buying radios that cost thousands of dollars each. Motorola in effect held them hostage—if in a friendly way. Its major rival, AT&T, was excluded from selling land mobile radios by a 1956 anti-trust settlement with the Justice Department. Motorola sought to protect its dominant market position by protecting its customers, seeking to deter cellular.

AT&T's Bell Labs, with its amazing endowment of human capital on the frontiers of knowledge, had conceived and developed cellular technology. But as passionate as its scientists were about mobile phones, AT&T enjoyed lucrative monopoly franchises in fixed-line telephony. It convinced itself that mobile services would not add much to corporate sales.[19] The firm, conflicted, was less aggressive in pushing for the new technology than it might have been.

Given AT&T's half-hearted support, the anticellular interests had their way with regulators for many years. While AT&T formally requested a cellular allocation in 1958, the FCC did not respond until 1968. Then, in a 1970 ruling, the agency agreed to deploy spectrum for the new service. It proposed to move 84 MHz from UHF, giving TV stations using channels 70–83 lower assignments, and cobbled together other idle frequencies to set aside 115 MHz in all for new mobile services. (In this, the FCC revealed that mobile service and TV broadcasting could easily coexist; the impediment had been the allocation scheme.) Of this, it slotted some 75 MHz for cellular use by AT&T—the one operator it authorized.

The issue was far from settled. Until 1982, when the FCC began a seven-year licensing process, cellular technology would be caught in a vortex of legal chaos, battered by rulemakings and reconsiderations and U.S. Court of Appeals verdicts.

As a 1991 study concluded, "We believe that, had the FCC proceeded directly to licensing from its 1970 allocation decision, cellular licenses could have been granted as early as 1972 and systems could have been operational in 1973."[20] The study's authors found that the FCC's spectrum allocation process had caused a ten- to fifteen-year delay in cellular service. That estimate is conservative, given the long delays before 1970 and, with licensing, after 1982. These deployment impediments probably delayed technical progress as well, given the flattening of economic incentives.[21]

It made clear business sense for TV broadcasters and the RCCs to seek to delay cellular service. But Motorola's siding with them is ironic. Motorola would become a leading beneficiary of the new marketplace: by 2006 it was the world's second-largest vendor of cell phones, trailing only Nokia, and selling more than 200 million units per year.[22] The firm, whose cell phone manufacturing division was purchased by Google in 2012 for $12.5 billion, describes its history this way:

> Motorola created the mobile communications industry. We invented most of the protocols and technologies that make mobile communications possible, including the first mobile phone, the first base station, and most everything else in between.[23]

Discounting for corporate chutzpah—cellular was hatched at Bell Labs and trials for the first commercial cellular systems were conducted by AT&T on the Metroliner in 1969—it is true that the "Father of the Cell Phone," Marty Cooper, was a Motorola vice president. He famously placed the first cellular call with a mobile handset in 1973.[24] It might as well have been a pocket-dial. At that moment, Motorola's lawyers were placing phone calls of their own, lobbying FCC bureaucrats to keep cellular networks from being built.

This is the kind of mystery that makes the political spectrum entertaining. According to communications attorney Kenneth Hardman, the FCC's decision to push 75 MHz into cellular telephone service "constituted a dagger pointed at the heart of Motorola and its allied private land mobile interests . . . because of the decision's explicit reordering of the land mobile industry structure." Allowing cellular access to radio spectrum would displace Motorola's little cash cow. Land mobile radio may have been the "backwater of the telecommunications industry,"[25] but it was Motorola's backwater.

Even in moving forward the FCC clung to old prejudices. In particular, the agency deemed cellular a "natural monopoly," so "technically complex [and] expensive" that "competing cellular systems would not be feasible in the same area."[26] Not only was there no scope for competition, there was no chance that any company other than AT&T could possess the skills or infrastructure to provide the service.

Not until two decades after it sparked the interest of Bell Labs' visionaries did the FCC seriously consider a spectrum allocation. Yet regulations still

13.2. "Father of the cell phone" Martin Cooper with an early "brick" model. Martin Cooper.

had to be crafted, and the agency spent another decade configuring "a series of Solomonic compromises under its 'public interest' standard."[27] In brief (a fuller version appears in Table 13.1), the tortuous journey wound as follows:

1970: FCC determines, in a proceeding opened in 1968, that wireline phone companies (affiliated with AT&T) should be given monopoly cellular licenses, and sets aside 75 MHz for the purpose. Radio spectrum is reallocated from the TV band, which is pared back from eighty-one channels (486 MHz) to sixty-seven (402 MHz).

1974: The Commission reaffirms AT&T's monopoly position and rules "that only wireline carriers should be licensed" for cellular. With Motorola arguing that only 19 MHz is needed to operate a system, the Commission deems AT&T's request for 64 MHz excessive, paring back its initial 75 MHz cellular allocation to 40 MHz. In an underestimate of damaging consequence, it finds that "64 MHz of spectrum would handle the foreseeable demand for mobile telephone . . . to the end of this century."[28]

1975: Regulators allow firms other than wireline phone carriers— including small RCCs—to apply for cellular licenses.

1981: The FCC abandons monopoly. Seeing that "regulatory policies and technology had changed dramatically"[29] over the previous decade, it authorizes two firms per market.

1982: The FCC finalizes rules and begins accepting applications for cellular licenses. Its plan calls for 734 franchise areas, two per market. A local phone carrier is to receive one; this would typically go to a Baby Bell. (AT&T was being divested at this time, with the local exchange carriers—"Baby Bells"—split off while AT&T retained its long distance and equipment manufacturing businesses.) A "non wireline" licensee would receive the other. In addition, the FCC sets a channelization plan (splitting up spectrum to carry individual phone calls); bars dispatch service for fleet vehicles (while allowing dispatch for individuals); and requires a specific technical format: analog mobile phone system, or AMPS. Some 20 MHz is allocated to each of the two licenses, or 40 MHz in all. The FCC holds another 20 MHz in reserve; in 1986, 10 MHz of this stash is distributed, giving each license 25 MHz each.[30]

By the time this process ended, a child conceived at the same time as cellular would have been thirty-seven years old, and yet the "new" technology was not yet in the delivery room. However one counts the regulatory lag—ten,[31] eleven,[32] or thirty-seven years—it was impressive. It was a joint effort of powerful TV broadcasters and the small radio common carriers that, in 1949, became the first direct competitors to the Bell system licensed by the FCC. They "waved the banner of small business and entrepreneurship" in blocking cellular, "warn[ing] that their little systems would be crushed by the telephone companies."[33] They were right—but the gains for society, and for the many RCCs that snagged nonwireline cellular licenses, were well worth abandoning an outmoded mobile telephone system that accommodated trivial traffic at outrageous prices.

The FCC delayed so long that the technology appeared in other countries first. Bahrain had a two-cell cellular phone system by 1978, months before the FCC authorized two experimental systems, and more than five years ahead of the first U.S. commercial network activation, in Chicago.

But at the end of this long policy trek, something spectacular emerged. Regulators, stymied by their own system, began to embrace a new model.

Radio in the 1920s and television in the 1950s had been socially disruptive, thrusting cheap, ubiquitous, instantaneous communication into the mass market. Yet it was operationally simple: one tower, one transmission, one-way. That much regulators could handle—not efficiently, and not without wasting precious spectrum and crushing upstarts like FM radio—but they could claim to understand the alternatives before them. The potential entrants who failed to acquire spectrum could do little about their plight, and the Commission, posing with the victors, could proclaim its favored technology a roaring success.

The unfolding of cellular brought new complexities. Regulators had little ability to grasp, let alone balance, the intricate options they faced; to guess how science would develop; to anticipate what products innovators might discover; to understand the business models competitive forces might create; or to predict how consumers would respond to any of this. Nonetheless, the FCC reflexively sought to control each aspect of cellular. It did a lot of stabbing in the dark. Regulators might consider a given approach "best" after the AT&T presentation, only to spin in favor of the rival Motorola network a month later. The agency was at the mercy of vested interests. No one on the planet foresaw what the engineers at Bell Labs or Caltech might cook up, or completely understood it after the fact, including the engineers themselves.

The Commission was not staffed by incompetents or slackers. It was tasked with a problem made insoluble by the mechanisms employed. The wizards in the private sector were not omniscient, either. When, in 1980, AT&T asked the respected management consultant McKinsey and Company to estimate how big a business cellular might become, McKinsey forecast 900,000 cell phones in operation by 2000. This proved to be off by more than two orders of magnitude: actual U.S. mobile phone subscribership in 2000 exceeded 109 million.[34] This prediction has become famous as one of the most dazzling errors in tech history, but McKinsey's reputation seems not to have suffered. It was hired by the FCC in 2010 to help craft America's National Broadband Plan.

Bending to necessity, the FCC loosened up. What was becoming an important new sector of the economy—one that would overwhelm broadcast TV and radio by any social or economic indicator you chose—was not susceptible to governance by eight-year rulemakings. The fast-changing world of microelectronics was delivering creative destruction on a frantic schedule that could only be organized on the fly. The FCC did not fly.

Table 13.1. U.S. Mobile Telephone Milestones, 1945–1989

Date	Event
1945	Cellular concept discussed by FCC Chair (after AT&T briefing) in *Saturday Evening Post*.
1946	AT&T begins (noncellular) commercial mobile telephone service in St. Louis.
1947	AT&T proposes a 150-channel mobile telephone system using 40 MHz.
Dec. 1947	Internal AT&T memo proposes cellular technology as mobile network upgrade. D. H. Ring, "Mobile Telephony—Wide Area Coverage," Bell Laboratories Technical Memorandum (Dec. 11, 1947).
1949	AT&T proposes a UHF mobile telephone system, but the FCC instead allocates vast spectrum for television broadcasts. FCC also allocates a small increment of bandwidth for noncellular mobile phone service, distributing licenses to both AT&T and to new radio common carriers (RCCs), forming city-by-city duopolies.
Jan. 1953	Bell's Kenneth Bullington publishes "Frequency Economy in Mobile Radio Bands" in the *Bell System Technical Journal*. Paper includes the cellular concept.
1954	Science writer Victor Cohn predicts, in the *Minneapolis Tribune*, "In 1999, you'll carry 'phone' on wrist . . . a miniature . . . receiver transmitter, part of the local phone system."
1958	AT&T petitions FCC for new cellular spectrum allocation; request languishes for ten years.
1962	AT&T tests a UHF cellular system in Murray Hill, NJ, for FCC.
1968	Congressional hearings on "crisis in land mobile communications." Members note that broadcast TV had been allocated 87% of the prime spectrum below 960 MHz, while mobile telephony had been allotted just 1%.
July 1968	FCC proposes to reallocate UHF channels 70–83 for mobile radio use. *Notice of Inquiry and Notice of Proposed Rulemaking* (14 FCC 2d 311).
Jan. 1969	AT&T debuts first commercial cellular service on Amtrak's Metroliner. Payphones make calls on train traveling between New York and Washington at more than 100 miles per hour.
May 1970	FCC allocates 75 MHz for wireline cellular telephone carrier. *First Report and Second Notice of Inquiry* (35 FR 8644).
Dec. 1971	AT&T, RCA, and Motorola propose 800 MHz band cellular telephone systems to FCC.

April 3, 1973	Marty Cooper, in New York City, makes first cellular call from a mobile handset.
May 1974	FCC allocates 40 MHz per market for a single wireline cellular telephone carrier. *Second Report and Order* (46 FCC 2d 752).
March 1975	FCC plans to open cellular licensing to any qualified common carrier. *Memorandum Opinion and Order on Reconsideration* (51 FCC 2d 945).
May 1978	Bahrain Telephone Company (Batelco) starts first commercial cellular phone network.
July 1978	AT&T begins testing cell phone service in limited areas around Chicago and Newark.
Nov. 1979	FCC begins setting policies for building and operating cellular telephone systems. *Notice of Inquiry and Notice of Proposed Rulemaking* (78 FCC 2d 984).
April 1981	FCC sets rules for cellular, establishing two carriers per market, one affiliated with a local phone carrier and one not ("non-wireline"). Each license allocated 20 MHz. *Report and Order* (86 FCC 2d 469).
Feb. 1982	FCC reaffirms application procedures, except AT&T is required to maintain a separate cellular subsidiary. *Memorandum Opinion and Order on Reconsideration* (89 FCC 2d 58).
June 1982	FCC accepts applications for two licenses in each of the thirty largest markets.
July 1982	*Memorandum Opinion and Order on Further Reconsideration* (90 FCC 2d 571–582).
1983	FCC grants first commercial cellular licenses.
Oct. 13, 1983	Ameritech Mobile Communications (an AT&T subsidiary) launches the nation's first commercial cellular system in Chicago.
1984–86	306 Metropolitan Statistical Area (MSA) cellular licenses issued, markets 31–306 by lottery.
1988–89	428 Rural Statistical Area (RSA) cellular licenses issued, all by lottery.

Sources: Joel West, *Institutional Standards in the Initial Deployment of Cellular Telephone Service on Three Continents,* Center for Research on Information Technology and Organizations (June 30, 1999); Tom Farley, *Mobile Telephone History,* TELEKTRONIKK 24 (March 4, 2005); Thomas W. Hazlett & Robert J. Michaels, *The Cost of Rent Seeking: Evidence from the Cellular Telephone License Lotteries,* 39 SOUTHERN ECONOMIC JOURNAL 425 (January 1993); William Horberg, *Winning the Future,* INSIDE STORIES FROM A HOLLYWOOD OUTSIDER (March 7, 2011), http://williamhorberg.typepad.com/william_horberg/2011/03/winning-the-future.html.

Its tedious cellular proceedings were a trial by fire. Ultimately the heat brought enlightenment. By 1974, regulators were searching for new paradigms:

> In the past, the Commission has treated land mobile spectrum requirements from a service perspective, allocating blocks of spectrum, usually on a nation-wide basis, to each of the twenty or so radio service categories. This method of allocation has led to parochialism among the users and inequitable situations where spectrum shortage and abundance exist side by side in the same cities. In this docket, we are proceeding to meet land mobile spectrum requirements in a somewhat different manner. Rather than allocating according to user categories or services, we have chosen to allocate by system type and to allow the market to determine ultimately how much spectrum is utilized by the various types of users.[35]

Economists had noted that the rigidities of the FCC system had led to maritime frequencies being reserved in Utah and U.S. Forest Service bandwidth being set aside in Manhattan. Radio development forcibly separated by law—spectrum apartheid, as Peter Huber calls it—was the system's most obvious and egregious structural feature. But slowly, tentatively, the agency came to see its own proceedings as hostages of "command and control." It began to "forbear," legalese for *we will delegate this decision to the market*.

For instance, instead of licensing each individual transmitter with an FCC "Radio Station Authorization" that would determine where and how the elements of the network were constructed, it decided to allow competitors to design their own cellular architectures. Mobile operators would figure out where to place base stations and towers, and how to link them. The shift was from "site licensing" to "geographic licensing," subtle but profound "permissionless innovation."[36] As the FCC's 1981 cellular plan put it:

> Flexibility to adapt to change is inherent in the cellular concept and an approach requiring any more paperwork or prior approval than is absolutely essential might destroy that flexibility. Accordingly, once a cellular service area has been established, the system operator will be able to modify its system without substantial oversight, as long as it serves the same area. Thus, the key to our regulatory structure is the geographic service area of a cellular system.[37]

To be sure, many needless limitations remained. First, the spectrum allotted was severely limited because of the rigidities of the system and the fact that, at bottom, the Commission simply did not take the emerging technology seriously. As an FCC official confidently noted, circa 1982, "It's not going to be something you and I put in the car to call home and say we're on the way home for dinner."[38]

Second, cellular's technical standard was a throwback. "The story of analog cellular radio will be written in vivid hindsight," writes technologist George Calhoun, "as one of the classic technological miscues of modern history, on a par with, say, the Zeppelin airship." The FCC's long allocation delay, Calhoun adds, was a missed opportunity. Advanced digital systems were being readied right along with the cellular rules. "In effect, cellular technology had become obsolete even as it was reaching the marketplace. It is commonplace in our era that by the time a new technology reaches the market, a better, faster, cheaper version is already well established in the laboratories."[39]

By the late 1970s, frustration over key aspects of spectrum allocation even rippled through Congress. Lionel Van Deerlin, chairman of the House Subcommittee on Communications and a former San Diego radio and television newsman, championed a "basement-to-penthouse" rewrite of the 1934 Communications Act. "We're in the middle of a technological revolution in communications," he told Congress. "If there's no sweeping revision, there's going to be pandemonium."[40] With Republican committee support, the Democrat advanced a blockbuster proposal in June 1978 that would have:

- Opened up the phone business, permitting competition with AT&T;
- Opened up broadcast television to competition from cable systems;
- Vested existing radio and TV licenses in perpetuity, following a transition period, while simultaneously providing for the "creation of new outlets to achieve greater diversity of programming sources";
- Abolished the Fairness Doctrine and the Equal Time Rule: "If Thomas Jefferson were writing the Bill of Rights today," Van Deerlin wrote, "he would make clear that the First Amendment applies to broadcast as well as print journalism";

- Eliminated "public interest" license renewals in favor of an annual license fee "based on the value of the spectrum." These monies would fund public broadcasting, replacing annual appropriations, thus increasing the medium's political independence;[41]
- Finally, the bill would have replaced the FCC with a new Communications Regulatory Commission. This agency's jurisdiction would be limited to situations where it was shown that "marketplace forces are deficient." As explained by the then Assistant Secretary of Commerce Henry Geller, "All the public interest standard says is 'We give up. Congress doesn't know [how to regulate].'" By removing the standard, the statute sought fundamental reform focused on pro-competitive criteria. *Broadcasting* magazine dubbed the bill "a deregulator's dream."

Broadcasters vehemently opposed it. Van Deerlin described the National Association of Broadcasters' position as "Keep the gold in Fort Knox." The broadcasters' champion in Congress was conservative icon Senator Barry Goldwater, who proclaimed it "impossible to support a bill which include[s] license fees based on the scarcity value of the radio frequency spectrum."[42]

In fact, Van Deerlin's proposals were savaged across the political spectrum. As broadcast industry historian Erik Barnouw noted, looking forward, "The commercial broadcasters will attack the notion of the fee to support public broadcasting and the media-access people will attack everything else."[43] Democratic and Republican FCC members objected to the "marketplace standard" as an obvious effort to reduce FCC regulatory discretion. Democrats on Van Deerlin's subcommittee, including Ed Markey (D-MA), Al Swift (D-WA), and Al Gore (D-TN), abandoned him. Left-wing activists were apoplectic over eliminating the public interest standard, condemning the bill as a "Titanic without life boats"[44] and a "multibillion dollar giveaway."[45]

By late 1979, Van Deerlin's charge had been repelled. "With the ABC Network coming down on the same side as Ralph Nader, it was hard to buck,"[46] said FCC Commissioner Nicholas Johnson. Van Deerlin was defeated for reelection in the Reagan landslide of 1980, a result perhaps not unrelated to his efforts to reform communications policy. Constituents were thought to be suspicious of a congressman who spent so much time on "Washington business."

Yet the historical trend was clear: the regime was at long last being challenged. Van Deerlin, a respected member of the House leadership, had advanced legislation that would be considered radical even today. His ambition was striking. And the ideas he floated would not be lost.

14

Deal of the Decade

The communications monoliths of yore—American Telephone &
Telegraph, Western Union and ITT—were bloated, slow-moving
megacorporations that were used to getting everything handed to
them by federal regulators. But now, young and hungry new companies like MCI were determined to wrest away a piece of their lucrative business however they could.

—James Murray, Jr.

BY 1982 THE FEDERAL GOVERNMENT was set to award cellular licenses. This
had always been the fun part, selecting lucky individuals—or megacorporations—to receive state-created largesse. When hundreds of broadcast TV
licenses were awarded during the administration of Republican President
Dwight Eisenhower, Republican newspapers had excellent luck. No paper
that had endorsed Eisenhower for president lost a license contest to a newspaper that had endorsed Adlai Stevenson, his Democratic opponent.[1]

Free Stuff Out of Thin Air

The FCC was well practiced in crafting grandiloquent documents detailing how any given assignment advanced "public interest, convenience or
necessity." These statements, required by administrative law, laid a veneer
of respectability over processes that might otherwise attract interest from

journalists or prosecutors. In one revealing episode, a surprisingly self-confident FCC staffer—tasked with writing up a justification for a license award—asked the chairman of the Commission to describe the policy grounds for the selection. The annoyed chairman responded: "You'll think of some." In another, the FCC voted to grant a company a TV license, and the staff wrote up an order of more than one hundred pages explaining it. For reasons undisclosed, the FCC reconsidered and switched licensees. The staff dutifully revised its order, using the original draft as template, producing an equally glowing public interest justification for the new winner.[2] Even a loyal FCC policy maker who sternly defended the spectrum allocation system conceded that license grants were "ritualistic, formalistic, wasteful and inefficient."[3]

When the Commission was finally ready to distribute cellular authorizations—some 1,468 licenses, two in each of 734 local franchise areas—it prepared to do so in the standard manner, via what the United States calls "comparative hearings" and other countries describe as "beauty contests." The Wise Men of the Commission, reading fat corporate applications, were to choose cellular service providers according to "public interest, convenience and necessity." While "many scholars have attempted to define" that interest, leading communications lawyers wrote at the time, they "have added little to an understanding of the real relevance of this concept to the regulatory process."[4] But every incarnation of the Commission has its favorite buzz phrases—*local ownership, educational programming, universal service, diversity of information sources*—to be melded into powerful sentences. Any investor seeking a free license is unregulated in his or her ability to hire expert counsel—including former FCC attorneys—to produce such sentences.

Economists could not understand the thrill. Starting with Ronald Coase, they saw license awards by fiat as a waste of time and money. Government was squandering value by not selling licenses at auction, like oil drilling rights or Treasury bills. Real resources were wasted as platoons of highly educated people were put to fighting over the distribution of government permits rather than making valuable goods or services. Assigning licenses by competitive bidding would get wireless services to market faster by making lengthy hearings unnecessary. Services would improve for consumers

because winning bidders would tend to be the parties who expected to produce the most economic value from licenses. And policy makers would be
cleansed because arm's-length transactions would guard against the stink
of crony capitalism. The unseemly way in which powerful interests pan-
dered for spectrum rights, Coase noted, rewarded competitors "who often
use methods of dubious propriety."[5]

In any event, wireless licenses were commonly traded after they were
assigned, and the FCC routinely approved these transfers. Whatever wisdom the Commission asserted in selecting licensees was wiped clean in
secondary markets.

Yet the traditional giveaway had significant defenders. "It does not seem
to me to be an outrageous idea," Representative Ed Markey (D-MA) intoned
in 1988, "that broadcasters—who are granted, at no cost, the exclusive use
of a scarce public resource, the electromagnetic spectrum—be required to
inform the public in a responsible manner. . . . We do not exact any monetary payment for the use of the spectrum, but we do ask broadcasters to
serve the public interest."[6] The rationale was curious. Could not policy
makers see that the public interest, as they defined it, did not require them
to dole out free wireless licenses? Far simpler—and more transparent and
enforceable—to sell the licenses with social obligations spelled out. Economist Jora Minasian, describing the objective as "socially desirable 'censorship,'" explained that "a simple solution . . . would be to incorporate a
proviso in the rights of radiation themselves. . . . Such a license could spec-
ify the required time to be devoted to certain types and quality of programs.
There is no obvious reason why this method is inferior to the present
method."[7]

One nonobvious reason is that it was clearly censorship. Loose "public
interest" requirements scampered through a constitutional loophole. By
issuing licenses gratis and renewing them without payment, influential
policy makers could indulge their preferences for particular broadcasts and
broadcasters. Profit-maximizing licensees aimed to please their patrons.
In 1965, Coase noticed that "speeches are constantly being made which
suggest that if the industry does not do something to improve its programs,
the FCC may have to take more positive action—this is what has been called
regulation by the raised eyebrow."[8]

David T. Bazelon, chief judge of the D.C. Circuit Court of Appeals from 1962 until 1978, came to see that this relationship between broadcasters and the FCC did great violence to free speech.

> Despite their tremendous influence, the [TV] networks have never developed the leverage to free the broadcast media from government influence. On the contrary, the tremendous stakes in the highly concentrated television medium make the networks particularly sensitive to the prevailing political winds at the FCC, in Congress, and in the White House. And the government has fostered sensitivity to government wishes by making clear that the failure to respond to the government's concept of appropriate program content would jeopardize the all-valuable license. I am reminded by one broadcaster who observed: "We all live or die . . . by the FCC gun."[9]

Congressional appropriations committees often proposed to end the "giveaway." In 1958 Representative Henry Reuss (D-WI) argued, "The airwaves are the public domain, and under such circumstances a decision should be made in favor of the taxpayers, just as it is when the Government takes bids for the logging franchise on public timberland."[10] He introduced a bill to assign TV licenses by competitive bidding,[11] but the leadership of the House and Senate Commerce Committees killed every effort to adopt auctions.

John D. Dingell (D-MI), the powerful chairman of the House Energy and Commerce Committee, was not coy about explaining why. In 1987, when the Fairness Doctrine was law and license auctions were yet to be tried, he tied the issues together. Removing regulation would disrupt an implicit bargain. If the Fairness Doctrine "were to be repealed," he explained, "I would be strongly moved to perhaps test [broadcasters'] dedication to competition by offering provisions to the law which might necessarily either deal more fairly with renewals or . . . requiring payments for the use of a portion of the spectrum by broadcasters, or perhaps simply by eliminating the monopoly under which they function so splendidly under the protection of a broad federal mandate which ensures . . . splendid financial returns on the use of a public resource."[12]

The U.S. courts backed Dingell's threat. That licenses were initially awarded for free opened the door to much wider regulation. Economists, who call this the "sunk cost fallacy," see no reason why the way a resource is acquired should affect the manner in which it is used. It is an excellent operational point, but federal judges have no obligation to embrace the economic logic.

Many did not. In a 1974 case heard by the D.C. Circuit Court of Appeals, a radio station owner sought to sell to a buyer who planned to abandon the station's classical music format in favor of contemporary music. The FCC considered format choice as a factor in whether to approve the transfer. The station owner asserted that the government had no right to abridge its (or its assignee's) choice of genre. But the FCC won: mandating Bach over Bacharach was legal "when it is remembered that the radio channels are priceless properties in limited supply, owned by all of the people for the use of which the licensees pay nothing. . . . Congress, having made the essential decision to license at no charge for private operation as distinct from putting the channels up for bids, can hardly be thought to have had so limited a concept of the aims of regulation."[13]

Raised eyebrows accomplished what mandates could not. During the Nixon administration, FCC Chairman Richard Wiley attempted to impose a "family viewing hour" on TV stations. For the government to thus dictate content raised constitutional issues. The legal analysis was supplied by law professor William Mayton:

> [Wiley], by various informal contacts, meetings and telephone conversations with network leaders . . . pressured the networks into adopting a "family viewing policy" that restricted prime time programming. . . .
>
> The district court, in a long and elaborate opinion, found that Commissioner Wiley had "foisted a policy on the networks" in violation of the First Amendment. On appeal, however, this judgment was vacated but not on the merits. Instead the court of appeals found that the Commission's action was not sufficiently definitive to support court intervention in an area "primarily" committed to the Commission. Ironically, the very practice at issue, the informal Commission pressures and intimidation, by that informality saved the Commission from the courts.[14]

A more circular theory of the First Amendment is scarcely imaginable: Congress chooses to give away a valuable commodity when it could have sold rights to it, and thereby breaks free of constitutional bonds.[15]

Even when constitutional subterfuge is not the goal, it is not categorically true that clarity in contracts is the best path. Precise specification and clear enforcement are often more expensive than other approaches. Victor Goldberg showed decades ago that contracts that bind two parties together may help solve coordination issues without spelling out terms in advance.[16]

This outcome is also seen elsewhere—for example, in the classic Rose Bowl ticket allocation problem explored by UCLA economists Armen Alchian and William Allen.[17] They asked why tickets to the Rose Bowl were always sold out far in advance. The universities could have raised prices and generated far greater revenues. Why did they sacrifice such returns?

By keeping prices low, top university personnel transferred benefits from the university (and its students), to themselves. Because they controlled access to the tickets, they enjoyed the power to dole out favors to friends and family. Later, Harvard economists Andrei Shleifer and Robert W. Vishny formalized and expanded the theory to explain a wide range of economic activity when agents who do not own an asset yet get to set the price of that asset. This explained, in particular, the phenomenon of "pervasive shortages under socialism."[18] State enterprises typically underprice so as to increase managers' clout. Pricing FCC licenses at zero assures a similar shortage and yields valued discretion for those able to distribute intensely sought assets.

FCC Beauty Pageants Lose Their Luster

Cellular—or, more precisely, the FCC's looming comparative hearings to award 1,468 cellular franchises—rocked this happy go-along, get-along scheme.

The Commission established a June 7, 1982, deadline for applications for the sixty licenses being issued in the thirty largest U.S. markets. The thirty wireline licenses were restricted to incumbent phone carriers. Even though these firms' service areas did not overlap, there could be more than one local exchange company in a given cellular market; hence there were forty-five of these incumbents. The single-applicant markets required no hearings as there was nothing to compare. Nor, it turned out, did those with multiple applicants: the few companies involved made deals, combined their applications, and split license rights. The FCC's administrative task (for wireline systems) was done. Phone carriers were free to launch. The first to go live was the Chicago system, in October 1983.

The nonwireline license hearings, for which the FCC received 135 applications, were different. One hopeful, Graphic Scanning (with 160,000 paging subscribers), filed applications for all thirty markets. Rivals considered this an unduly aggressive strategy that would prompt FCC pushback; no one else tried it. While few beyond the FCC's orbit knew of the opportunity, telecommunications law firms and "application mills" completed the forms for those who did. These paperwork entrepreneurs, for a fee, created voluminous documentation certifying that an applicant was capable of constructing and operating a state-of-the-art cellular network so as to serve the "public interest, convenience and necessity," even if the applicant had never previously operated a phone company—or even a business.

Eighty-five companies applied, including MCI (the upstart long distance operator), McCaw (a tiny cable system in the Northwest), Western Union (the fading telegraph operator), Lin Broadcasting (owner of a few small market radio and TV stations, with a sideline in paging), John Kluge (wealthy CEO of Metromedia, owner of broadcast TV stations), and Metro Mobile CTS ("a New York City–based startup formed specifically to file cellular applications . . . led by an ambitious, martini-drinking, cigar-smoking, socially competitive multimillionaire named George Linemann"). Upstarts, old-timers, gamblers, fixers, and dreamers.[19]

But which of them knew anything about wireless? At the 1982 press conference announcing the antitrust settlement splitting Ma Bell into parts, AT&T CEO Charlie Brown was asked who got cellular, AT&T or the Baby Bells? Brown had to turn to an aide for help. (Answer: the Baby Bells.) In negotiations with the government, Ma Bell had not thought cellular worth inquiring about, let alone fighting to retain.[20]

License applicants described the architecture of the systems they would build, what they would offer, how many would subscribe. In numeric detail they defined the business models they would deploy, how they would finance them, the vendors they would use. But the "models" were all guesses and the projections were nonbinding. Each project was a high school term paper heavy on jargon designed to appeal to bureaucrats.

One license hopeful ambitiously sought to generate information by surveying four hundred businesses in Cleveland. Sixty percent were "not interested at all" in having a cellular phone; 92 percent of these "had no need for one."[21] Years later, all of these respondents—and their kids, parents, spouses, partners, employees, neighbors, and dogs with pet trackers—would be toting cellular devices.

An arms race kicked in: a higher mountain of paper was believed to connote greater heft. Companies deposited applications that required four full-time FCC employees—not to read, but just to log and stack.[22] Each corpulent compendium included "attachments and addenda of all sorts— maps and charts and graphs and reams of appendixes, all carefully indexed into color-coated binders." Graphic Scanning sent an application that ran to 1.5 million pages. Multiply that by five, as the rules required, and then by thirty—for all markets offered—to gauge the deforestation impact. To what benefit? As an attorney who participated in the process later wrote, "The applications were essentially carefully prepared works of speculative fiction."[23]

FCC staffers stared blankly, caught in the headlights of the tractor-trailers hauling paper to the Commission (Graphic Scanning needed two). The beauty contests were getting ugly. As Commissioner Glen O. Robinson wryly observed, comparative hearings were "the FCC's equivalent of the medieval trial by ordeal."[24] The difference, as economic research has recently shown, is that the medieval ordeals were relatively efficient and environmentally friendly, save for the occasional pot of boiling oil.[25]

The massive applications filed on June 7, 1982, prompted a regulatory crisis. They were too large—literally and politically—to be deposited in a drawer. The agency would have to distinguish among "public interest" claims, technical specifications, and market projections that all looked pretty much the same, and were almost entirely irrelevant. And they were all so complicated.

Then, a strategy emerged: *coverage.*

Applications projecting larger network coverage areas were better. Coverage was quantifiable; the metric varied across applications; more was desirable; and this selection tool might be rationally related to the "public inter-est." In reality, projected coverage had little to do with what coverage would actually materialize, but that was beside the point. By relying on this simple differentiation, regulators could avoid the kind of "arbitrary and capricious" result that could be overturned by federal courts. FCC administrative law judges, tasked by the agency to determine license awards, grasped it.

As decisions began to trickle out, Murray writes, "applicants . . . were blindsided by the single-criterion trend."[26] Applications for markets thirty-one through sixty, due at the FCC November 8, 1982, had been expected to taper off, given that they featured smaller populations (and expected prof-its) than those licensed in round 1. Yet there were nearly twice as many.[27] When round 3 filings, for markets sixty-one through ninety, closed on June 8, 1983, there were even more, by far.

The top thirty licenses took a mean time of eighteen months to award.[28] But that was deceiving: the average included many markets, like Boston, that avoided comparative hearings when applicants settled the matter them-selves. The contested markets would take years, and looked as if they would engulf the entire process as new applications poured in.

Not even coverage could save the Commission.

There had to be a better way. President Jimmy Carter had asked Congress for authority to sell licenses in his 1979 state of the union message.[29] In a 1980 Commission paper, economist Douglas Webbink argued that com-parative hearings should be replaced by "auctions, lotteries or expedited paper hearings." Webbink considered auctions the best option, with lotter-ies next, preferable to comparative hearings because "they are faster and less costly," even though there was "no guarantee . . . that the winner of a lottery is the one to whom the license is most valuable." He noted that "it is important that the lottery winner have the right to sell *immediately*."[30]

The titans in Congress resisted change. They claimed the public interest standard would be undermined if regulatory discretion in license assign-ments were removed. But the 1980 elections nudged the equilibrium, and a reform budget passed in 1981 that permitted the FCC to distribute cellu-lar licenses by lottery. Lotteries were no one's first choice, but they were the reformers' second. They would let the FCC avoid lengthy hearings and get the licenses into the market. The compromise became law.

Even with the authority granted the FCC in 1981, the Commission was aware of strong preferences of the leaders of the FCC oversight committees. It trudged forward with the beauty contests. Eventually, however, "the crush of paperwork, the quagmire of comparative hearings and the increasing pressure to get the licenses out . . . proved too much."[31] In October 1983 the FCC decided to switch to lotteries.

Applications were already filed in markets one through ninety. While decisions were being made in one through thirty, hearings had yet to commence in markets thirty-one through ninety. The FCC announced that it would assign the latter licenses by lottery. With the filing windows already closed, this proved a windfall for applicants. They would save the (considerable) legal costs entailed in arguing their case at the Commission. Moreover, with random chance now the "public interest" rule, applicants had strong incentives to split the license ownership rights among themselves. This obviated the lottery and eliminated risk for the license seekers. There were settlements in fifty-nine of the sixty markets. In just one market, an annoying gadfly forced a drawing—and, to the consternation of its more cooperative rivals, won. Say what you will about the outcome, the "medieval ordeals" were over.

Still, the vast majority of markets had yet to be licensed. And now, with lotteries in place, came the deluge. Because any applicant had as much chance as any other, insider advantage evaporated. Application mills ramped up, and the word got out—helped by the marketing efficiencies of American capitalism. TV personality Mike Douglas, paid $25,000 to serve as a spokesperson, cut commercials in which he told viewers, "You and I have an equal opportunity to compete with the corporate giants for a piece of the multibillion-dollar pie!"[32] He touted cellular lotteries as "the investment of the decade . . . possibly of the century."[33]

Step right up and play a game of chance.

Total applications for markets 1–90, filed under the expectation that the winners would be chosen through FCC beauty contests, numbered 1,110, an average of just about six per license.[34] Applications for markets 91–120, due July 16, 1984, were the first ones filed with the knowledge that licenses would be awarded by random selection. These entries soared to 5,178, or 86 per license.

The lotteries, wrote FCC policy experts, brought "huge numbers of applications, each of which must be filed, logged and pre-screened prior to selection."[35] Because regulators maintained the fiction that the lottery entries were still part of a public interest determination, they described the networks to be built and the services to be offered. The submissions remained huge. "The FCC was forced to call in a structural engineer," wrote media attorney Nick Allard, "to determine whether the . . . floor could bear the weight" of all the applications.[36] While, technically speaking, only qualified cellular service providers could apply, the app mills met that barrier by getting a real communications company, like Nortel, to certify that it would build and operate a network should the applicant win a license. Members of the communications bar can create these shell companies in their sleep.

The threat to the FCC building's structural integrity posed further delays. Officials revamped rules, hoping to reduce the avalanche. Lottery application windows were shortened, fees ($200 per app per market) imposed; some entries were thrown out for errors as minor as misnumbered pages.[37] The 306 metropolitan statistical areas (MSAs) were formally completed in 1986, the 428 rural statistical areas (RSAs) in 1989.

By this time some seven years had elapsed, nearly 400,000 applications had been received,[38] and part of a federal warehouse in Pennsylvania had (in 1986) collapsed under the weight of the FCC cellular lottery ticket requests.[39] Thankfully, no one's number came up.

Romulus Engineering in San Francisco emerged as the largest of the app mills, creating more than fifty thousand applications. When the RSA windows opened in 1988, the company offered a standard package: for about $250,000 it would apply for all 428 nonwireline licenses. No specialized market knowledge or understanding of cellular was required. Romulus would prep and submit everything needed according to "FCC reports and orders, rules, regulations, technical memoranda, releases, and other guidelines required by the Commission."[40]

Investors formed blocks. A typical syndication had fifty individuals kicking in about $5,000 each, supporting applications across hundreds of markets. From "pig farmers to hairdressers,"[41] those willing to roll the dice embraced the moment. Respectable investors cringed. *Forbes* published an article called "Cellular Suckers," exposing the app mills for fleecing the

Table 14.1. Cellular License Lottery Entries (Metropolitan Markets)

Markets	FCC Closing Date	Number of Entries	Entries/Market
91–120	7-16-84	5,178	172.6
121–135	2-7-86	8,007	533.8
136–150	2-28-86	7,436	495.7
151–165	3-21-86	6,367	424.7
166–180	4-11-86	8,471	564.7
181–240	5-2-86	25,018	417.0
241–305	5-23-86	37,650	579.2

Source: Thomas W. Hazlett & Robert J. Michaels, *The Cost of Rent Seeking: Evidence from the Cellular Telephone License Lotteries,* 39 SOUTHERN ECONOMIC JOURNAL 425 (January 1993), 428. *Note:* Data are for MSAs, 91–305. An additional MSA was later established and licensed after 1986, bringing the MSA total to 306. The 428 RSA markets were assigned by lottery, 1988–89.

unwashed.[42] Wall Street preached restraint. Alas, the bluebloods lost out. By the end of the RSA lotteries in 1989, applicants were averaging at least a $5 return for every dollar invested in the application,[43] as much as $20 by another estimate.[44] The real question was: Why were there *only* 400,000 applications?

Due to an extraordinary rise in license sales prices from 1985 to 1989, when per pop license values (meaning the price of the license divided by the number of people in the franchised area) rose from $20 to $200, those who held on for a while prospered far more. By 1990, the U.S. Department of Commerce estimated the total value of U.S. cellular licenses at $56 billion to $90 billion.[45] Craig McCaw became the first cellular billionaire, buying out the pig farmers and hairdressers to form a national network, Cellular One. (In 1994, AT&T paid $12.6 billion for the company.)[46] Mark Warner, an ambitious 1980 Harvard Law School graduate, became nearly as wealthy by supplying legal expertise and relentless salesmanship to the effort. (Warner, a Democrat, subsequently became governor of Virginia and is now a U.S. senator.)

The 400,000 cellular license applications cost perhaps $1 billion to prepare and file. That princely sum was wasted. A pure random draw—say, throwing all U.S. Social Security numbers into Uncle Sam's hat and having a blindfolded secretary of commerce draw them out—would have produced faster assignments and given secondary markets a head start in forming actual networks.

These FCC lotteries took six years. As with broadcast radio and TV before them, licenses were assigned without charge, bestowing riches on the chosen. But the previous windfalls had a cover story: "public interest." The lotteries were condemned far and wide as embarrassing—"a significant step backward," a contributor to the *Federal Communications Law Journal* wrote, "in developing spectrum use, efficiency, and equity." This was not quite true, nor was it true that the application mills generally fleeced license seekers—realized mean returns were amazingly generous—but the myth was so widely believed that the North American Securities Administrators Association cited wireless lotteries as "the largest investor fraud in the nation."[47] In fact, comparative hearings had been worse, both for deterring new technologies and for corrupting the regulatory process with what congressman and broadcast entrepreneur Lyndon B. Johnson boasted of exploiting—"government between friends."[48] In contrast, naked nonsense performed a public service.

As economists and FCC auction design consultants David Porter and Vernon Smith would write, "the lotteries . . . made visible what had previously been hidden: Failure to employ competitive bidding left billions of dollars of potential revenue on the table."[49] *Business Week* grasped the point:

> It was a license to print money. In 1989, Rural Cellular Development Group in Los Gatos, Calif., won a government lottery for the right to build and operate a cellular-phone system on Cape Cod. Just 73 days after getting its construction permit, before erecting a single antenna, the group "flipped" its license—selling it to Dallas' Southwestern Bell Mobile Systems Inc. for $41 million.[50]

Say what you will, a randomly drawn ping pong ball got that Cape Cod cellular system rolling. The furious secondary market activity was expensive, needlessly so given that an auction could have taken the licenses where they were headed in one step. In 1990 alone, cellular license exchanges burned up an estimated $190 million.[51]

But lotteries, for all their absurdity, were faster than comparative hearings.[52] And bonus: they exposed a ruse. Even part-time observers could see that something needed fixing. By 1985, the *New York Times* editorial page was ready to pounce.

> The [FCC] spent years choosing operators for the top 90 cellular telephone markets from a pool of 1,200 applicants. Hardly any knowledgeable observer believes there was a rational way to decide among the candidates. All that the lengthy hearing process accomplished was to waste millions of work-hours and lawyer-dollars and to delay greatly the availability of cellular phone service.
>
> To speed the assignment of licenses in the next 30 cellular phone markets, the commission resorted to a lottery. But that produced another time-wasting paper shuffle. Some 5,000 applications had to be evaluated to insure that the would-be operators were minimally qualified. So now the commission is asking Congress to permit a more promising approach. It wants discretion to auction scarce channel space for all services save the broadcast media to the highest bidders.[53]

FCC license auctions winged their way onto the *New York Times* op-ed page in 1987, *Barron's* in 1990 (in essays by the author),[54] the *Times's* front page in 1991, and *Newsweek* and *Business Week* in 1992. These pieces savaged the FCC's lotteries for enriching front companies and (as *Business Week* had it) "the fortunate few who win licenses—from broadcaster William Paley to cellular pioneer John W. Kluge," while "the public receives nothing for the use of a valuable national resource." The Cape Cod story offered the standard object lesson: "Rural Cellular's windfall shows how the government's system for licensing the electromagnetic spectrum has gone haywire."[55]

Rural Cellular's license flip finally moved the dime. Ronald Coase's call for competitive bidding had been a dream. His University of Chicago Law School colleague Harry Kalven told him that while the problems with the existing regime were real, "the remedy appears too radical to be helpful." The superiority of market allocations was an interesting thought, but "an insight more fundamental than we can use."[56] Meanwhile, communications experts dismissed Coasian thinking for *technical reasons*. The naïve and impractical idea of an auction had been offered by those who did not understand that frequency use rights could not "be objectively specified."[57] When competitive bidding was finally suggested by a single FCC member, the estimable Glen O. Robinson, in 1976, it was treated as a joke. Rival commissioners scoffed that the proposal's odds for enactment were equal to "those on the Easter Bunny in the Preakness."[58]

Even as the Carter and Reagan administrations had repeatedly asked Congress for auction authority, that betting line seemed about right. Reagan's first FCC chairman, Mark Fowler, helped draft an auctions bill in 1985; it went nowhere. His successor, Dennis Patrick, made it his main goal to auction something off. But in pushing for legislation from 1987 through 1989, he had trouble procuring a Republican House sponsor. When he did, Representative Markey pronounced the measure "dead before arrival."[59]

The equilibrium held. When Assistant Secretary of Commerce Janice Obuchowski trekked to Capitol Hill for hearings in 1991, she dutifully made the administration's pitch for auction authority. The *Washington Post* reported that the Commerce Department had gone on offense and produced a "three-minute video poking fun at the lottery system, editing in sections of the 'Lotto America' broadcast to show the capriciousness of the current system that uses four hot-air machines that pop out numbered Ping-Pong balls to pick winners."[60] But Markey seized the high ground:

> Secretary Obuchowski, as you know, this [lottery] idea is a Reagan idea . . . developed in order to streamline the system. If you are unhappy with the lottery system, fine. Come to us. But you have to remember that the reservations I had about the lottery system went to the point that it did away with the comparative hearing. . . . My concern was that I wanted to have a comparative hearing from the get-go, and that is something that we have avoided.[61]

He derided the auctions proposal as Obuchowski's "pet rock."[62]

The congressional leadership had Markey's back. When the topic of license auctions landed in a general session of the House on September 21, 1991, it inspired outrage. "Rep. John D. Dingell was hopping mad," reported *Congressional Quarterly*. "'This is the same old, tired, hackneyed approach that my colleagues on the other side of the aisle have carried forward at the request of a bunch of unthinking dunderheads in the Office of Management and Budget,' Dingell thundered in a floor speech."[63]

Some have observed that Congress finally approved FCC auctions because the government needed the money. But it always had. During the Reagan years, intense pressure for deficit reduction had led to a bipartisan push to cut spending and raise revenues, ultimately resulting in the Gramm-Rudman-Hollings law. Still, auctions were never seriously considered; even as fervent a deficit hawk as Senator Rudman responded to the FCC's auction request by telling the agency to "go back to the drawing board." So hostile were Congress's communications committees that in 1989, John Dingell introduced a bill with a section labeled PROHIBITION OF SPECTRUM AUCTION. The measure was gratuitous: the FCC had no authority to conduct such an auction.[64]

In 1991, the perceptive economics columnist Robert Samuelson wrote that auctions were "common sense—but not yet good politics." In deficit or surplus, "Congress loves to give something for nothing."[65] This was especially true in broadcasting, where the scarcity of spectrum slots let Congress evade a constitutional barrier to regulation. Even after the FCC began auctioning cellular licenses in 1994, broadcasting licenses were still assigned by fiat. As FCC Commissioner Gloria Tristani commented in 1998, "I think that broadcasting is special. I'm all for giving broadcasters special benefits in exchange for special obligations."[66]

One rationale for the delay in auctions is that the House Democrats who killed the idea every year were uncomfortable acceding to the free-market demands of a Republican administration.[67] But during the Carter years, a Democratic Congress had no trouble throttling auction reforms requested by a Democratic White House. In 1993, however, the stars somehow realigned. Congressional angst lessened with another Democratic White House, which issued a strong plea for auctions via its commerce secretary, Ron Brown. And a new wireless licensing task loomed—for personal communications services (PCS), which would more than triple cellular bandwidth. This promised to tie up the Commission for years in useless hearings or create a new avalanche of lottery entries.

The Success of Failure

The delay in issuing PCS licenses was already a debacle. European and Asian countries had lagged the United States in first generation or "1G" analog cellular, but by 1992, Austria, Denmark, Finland, France, Germany, Ireland, Italy, Japan, the Netherlands, Norway, Portugal, Sweden, Switzerland, and the United Kingdom had all issued second-generation digital cellular licenses. These licenses were national in scope, not local, which meant that 2G would be available throughout these nations sooner rather than later. As well as the secondary transactions worked to aggregate hundreds of permits in the U.S. market, they took time, delaying the growth of national cellular networks. The Americans already trailed badly in 2G.[68]

Serious auction proposals always included explicit exemptions for TV and radio. As broadcast industry lawyers never tired of pointing out, "The exclusion of broadcasters from auction requirements is based on the long-standing reality that broadcast licensees already 'pay' for the use of the spectrum by performing a plethora of public interest programming requirements."[69] But the broadcasters, nervous of the precedent, had always killed auctions anyway. The Congressional Budget Office cited this as the "camel's nose inside the tent" argument.[70]

A key political change occurred in 1993 as broadcasters were digesting the regulatory treats won the year before. In 1992, Congress had passed the Cable Television Consumer Protection and Competition Act. Broadcasters got their money's worth: new property rights (must-carry and retransmission consent) that would prove to be worth billions annually in carriage or license fees, and cable TV rate regulation (disrupting cable program menus and briefly slowing the steady rise in basic cable viewership).[71] Having thus shot its lobbying wad, the industry remained enmeshed in Cable Act implementation at the FCC throughout 1993. "The ambitious efforts of the broadcast industry in pursuit of other priorities," wrote Nick Allard, "constrained the ability of broadcasters to effectively and openly oppose spectrum auctions."[72]

The Omnibus Budget Reconciliation Act of 1993, signed into law on August 10, 1993,[73] contained a provision allowing the FCC to assign licenses by competitive bidding. That the reform came in a budget bill was notable; the measure snuck past "the protests of the Commerce Committees . . . [which] enjoyed their traditional privilege of presiding over, through the FCC, the giveaway of hundreds of thousands of narrowly defined communications licenses."[74] Janice Obuchowski's pet rock grinned.

Once auctions were legal, Congress made them mandatory. The FCC was given one year to devise an auction plan. In July 1994, when narrowband PCS licenses were auctioned by oral outcry—*sold to the man in the big white hat!*—they brought in $160 million. The next month, interactive TV licenses (enabling TV viewers to, say, order merchandise they were watching on a show), brought winning bids of $600 million. The service was a bust; for the time being, telephones supplied all the interactivity broadcast viewers wanted.

But these were mere warm-ups. In December 1994 came the Big Kahuna: auctions for personal communications services—digital cellular, or 2G. Licenses for two new mobile operators, to challenge the existing two carriers, would invigorate markets with additional competitive rivalry. But the new licenses would not be national, instead divvied into fifty-one major trading areas (MTAs).[75] This required some deep thinking about auction design.

Both logic and experiments conducted with real bidders under laboratory conditions had revealed that various bidding formats could produce differing results. How to choose the best one? The FCC answered the question with help from interested parties and the emerging cottage industry of mechanism design. The late John McMillan described the deployment as "perhaps the biggest use of economic theorists as consultants since that other telephone-industry revolution, the break-up of AT&T ten years earlier."[76] Because scores of PCS licenses were being sold at once, and because the licenses were complementary—national networks would need to assemble dozens of pieces of the puzzle—the FCC developed a simultaneous, multiple-round auction format. Bingo.

Broadband PCS licenses, two per market (each allocated 30 MHz), went on sale on December 5, 1994. The auction lasted 112 rounds and closed on March 13, 1995. Winning bids totaled $7 billion. With charges assessed the three licensees assigned by FCC "pioneer preferences," the grand total collected came to $7.7 billion. Sprint, a long distance operator, was the big winner, paying $2.1 billion for licenses covering most of the United States. AT&T paid $1.7 billion to finish second. A new marketplace had formed that would soon render the cellular duopoly moot. Between 1996 and 1998, retail service prices would plummet, penetration would explode, and the promise of a wireless world would emerge.

Table 14.2. Economists Do FCC License Auction Design, 1993–1994

Scholar	University	Consulting Client
Paul Milgrom	Stanford	Pacific Bell
Robert Wilson	Stanford	Pacific Bell
Charles Plott	Caltech	Pacific Bell
Jeremy Bulow	Stanford	Bell Atlantic
Barry Nalebuff	Yale	Bell Atlantic
Preston McAfee	Texas	Airtouch
Robert Weber	Northwestern	Telephone and Data Systems
Mark Isaac	Arizona	Cellular Telecommunications Industry Association
Robert Harris	U.C. Berkeley	Nynex
Michael Katz	U.C. Berkeley	Nynex
Daniel Vincent	Northwestern	American PCS
Peter Cramton	Maryland	MCI
John Ledyard	Caltech	Department of Commerce
David Porter	Arizona	Department of Commerce
John McMillan	U.C. San Diego	FCC

Source: John McMillan, *Selling Spectrum Rights,* 8 JOURNAL OF ECONOMIC PERSPECTIVES 145 (Summer 1994), 146.

Economists took pride in their efforts. "I find strong evidence that the auction design was successful," wrote Peter Cramton in a 1997 assessment. "The information allowed arbitrage across similar licenses, so prices on similar licenses were close . . . [and the auction] enabled firms to piece together complementary licenses into efficient aggregations."[77] The FCC boasted in 1995 that the new competitive bidding procedures were "permitting the market, rather than lobbyists and Government regulators, to determine who gets valuable wireless licenses."[78] Chairman Reed Hundt proclaimed the PCS license sale "the greatest auction ever held" and created a large display of a $7.7 billion check to show to newspapers, many of which ran front-page photos of the financial prop, which was then mounted like a fishing trophy on Hundt's office wall. The FCC received a Reinventing Government award from Vice President Al Gore. Wrote Hundt: "I told the press that the FCC had raised more money than its total budget for its 61-year history. We were, I said, the most profitable American business in terms of return on equity."[79]

The slog had been long and hard. From its inception, wrote FCC attorney John Berresford, "every important decision about cellular was made not by businessmen and customers, but by lawyers—judges and regulators."[80] Yet under the pressure of time and technology, the rules loosened. Comparative hearings suffered from the inner contradictions of crony capitalism and the cringe-inducing reactions from those who observed the strategic use of "methods of dubious propriety." Lotteries, a useful disaster, were abandoned because they "engendered rampant speculation; undermined the integrity of the FCC's licensing process and, more importantly, frequently resulted in unqualified persons winning an FCC license," as the U.S. House report on the 1993 federal budget put it.[81] Regulatory forbearance was the last option.

Spectrum allocation reform had not yet crossed the Rubicon, but a revolutionary wireless technology had finally arrived in the market, and FCC license auctions had helped. This gave economic ideas, once dismissed as laughable, a new gravitas, and demonstrated what improvements might come with liberalization.

The smart money had been on the Bunny all along.

15

The Toaster Tsunami

(1) Deregulation: by auctioning spectrum with no rules attached and preempting all state regulation, we had totally deregulated the wireless industry.

—Reed Hundt, FCC chairman 1993–1997,
ranking his achievements

BY THE MID-1990S, MOORE'S LAW was pushing microprocessors faster and smaller, the Internet was commercialized, and disruptive entrepreneurs were spreading desktops, notebooks, and powerful new software tools everywhere. All of these trends converged in wireless, but it was deep policy reform—"auctioning spectrum with no rules attached"—that ignited the "wireless explosion."

The FCC had little idea what the marketplace had in store, and conceded the fact. "The definition of [2G cellular] PCS is a fairly broad one," explained Robert Pepper, a top FCC official, early in 1993, "and it is purposely broad because we don't know in advance which segments, or which niches, of this . . . market will be the ones that capture people's imaginations, which will capture the kinds of applications that people will want to buy and use."[1]

In this, the agency came to grips with "bounded rationality," a generic condition afflicting both man and beast. No creature—in Washington, on Wall Street, in Silicon Valley, or even at Bell Labs with its (then) three thousand Ph.D.s[2]—knew just what wireless held. Unseen were the forces that would bring texting, mobile data, broadband networks, smartphones, or tablets, not to mention the Apple App Store, Google Play, Facebook, Twitter, Yelp, Instagram, Snapchat, Spotify, Mapquest, Uber, or the rise and fall of a brisk market in ringtones, under the menacing stare of Angry Birds or the inviting buzz of Pokémon Go.

By embracing their ignorance, regulators grasped an essential truth. By leaving spectrum use undefined, they became better regulators. Delegating choices to entrepreneurs possessing superior incentives, information, and the financial wherewithal to innovate promoted economic discovery. Revolution soon followed. The flexible rules regulators chose were themselves transformative, allowing daring inventions to be tested and deployed. *Mother May I?* became *Look Ma, no hands!*

This was application of the much derided "toaster" model. And the wireless market went wild.

FCC Chairman Hundt's boast of "totally deregulating wireless" was a tad overstated. While restrictions were being stripped from mobile telephone licenses, the underlying frequency allocation process was left intact. Spectrum zoned for specific activities could not be bid into more productive employments. Liberalization was adopted, but on limited bandwidth—about 7 percent of the (most productive) bands under 3 GHz.[3] By 2006 regulators had moved to double that proportion.[4] In 2010, the FCC would call for another doubling to keep the developing "mobile data tsunami" from cresting and crashing.

The torrent of creative destruction underscored the high cost of legal rigidities. Looking back in 2010, the Federal Communications Commission stated that the FCC's historic approach "has been criticized for being ad hoc, overly prescriptive and unresponsive to changing market needs," and emphatically endorsed the critique. It called for further reforms that would put far more bandwidth into the marketplace, "sufficient, flexible spectrum that accommodates growing demand and evolving technologies."[5]

The toaster was popping, and consumer welfare was flying out.

Ameritech, the local phone operator previously and later known as AT&T, operated the first cellular system in the United States. It opened for business in Chicago in late 1983. The second system was owned by Pacific Telesis (later part of Verizon), which raced to have its Los Angeles network up and running for the 1984 Olympics.[6] The wireline carriers built out most of the top thirty markets in the next year;[7] the nonwireline entrants joined them the following year. By 1990 all 734 U.S. markets had been licensed, and most Americans could purchase service covering their homes.[8]

Service was painfully expensive. In 1986, "portable phones" carried price tags in excess of three thousand dollars.[9] Monthly bills averaged more than one hundred dollars,[10] as talk time cost well over fifty cents per minute.[11] Roaming charges were often some multiple of that. These rates generated healthy returns on invested capital—for some systems, more than 50 percent per annum.[12] The duopoly market structure, operating under an antiquated technology mandate called Advanced Mobile Phone Service, featured low capacity, weak competition, and relatively little pressure for innovation. There were no apps. Nonetheless, Americans began flocking. The Commission, in 1995, would find that "each year, cellular subscriber growth has approached or exceeded fifty percent—an amazing record of sustained growth."[13]

The agency had received petitions to allow new personal communications systems (PCS) in 1989.[14] The prospective licensees wanted to compete with established cellular operators, deploying new digital technologies. The spectrum allocation was an easy call; even regulators endorsed it. Yet it took until March 1995 to assign 30 MHz licenses in the first tranche, two licenses per market. Licenses for the other half—dished 60 MHz more—would take another decade.

Sprint, as noted, emerged as the big winner in the 1995 auction. A long distance telephone carrier launched by the Southern Pacific Railroad, it had turned nineteenth-century rights of way into conduits for twentieth-century fiber-optic lines. It then layered a mobile network atop its national web.[15] Other major winners were PCS PrimeCo, a consortium of Baby Bells, and long distance operator AT&T, by now a rival of the Baby Bells.[16]

AT&T had entered cellular by virtue of a 1994 deal with Seattle entrepreneur Craig McCaw.[17] McCaw was a visionary who saw mobile wireless as the wave of the future when the conventional wisdom saw it as at best a niche service, and at worst, science fiction. McCaw abandoned undergraduate studies at Stanford to save the tiny debt-laden cable TV system his father—

dying unexpectedly—had bequeathed to the family. No sooner were these assets solvent than he bet them like a riverboat gambler.

The lotteries had created hundreds of licenses with thousands of owners. McCaw put Humpty Dumpty back together again. He bought out "a beauty salon owner; an ambulance driver; a deep-sea diver; and a squadron of speculating suburban dentists." He "paid $700,000 to a guy nicknamed 'The Fat Man,' and $3 million to an Oregon man who lived deep in the woods in a mobile home."[18] McCaw decided to assemble "the nation's largest provider of cellular mobile telephone services, with operations in more than 100 U.S. cities."[19] Soon competition for his accumulated holdings was fierce, and the millions he had thrown at cosmetologists, dentists, and the Fat Man multiplied into billions: $12.6 billion, to be exact. AT&T wrote the check, buying back what it had lost, with nary a thought, in the antitrust settlement a decade before.

By 1996, PCS licensees were poised to enter the retail fray. Service prices went into decline. Momentum gathered with a marketing ploy by AT&T Wireless. The carrier's "digital one rate," unveiled in May 1998, featured a bucket of monthly minutes for local or long distance calling, for a fixed price.[20] Anytime, anywhere—up to the bucket's limit. No long distance or peak time charges to worry about.

"The very popularity of the service has overloaded AT&T's network at times, making it hard to place or receive calls," noted the *Wall Street Journal*'s technology columnist, Walter Mossberg.[21] Other carriers soon matched and extended the "one-rate" offer, adding free, unlimited on-net and off-peak minutes. With subscribers enjoying zero marginal cost dialing, "fixed-to-mobile substitution" kicked in. Customers began to reach first for their mobile phones, and crossed out their office numbers when handing over their business cards. "Call me on my cell."

There were huge increases in mobile call volumes. As of year-end 1994, there were 24 million wireless subscribers. Just five years later, subscribership stood at 86 million. In 2013, 336 million.[22] In 1994, callers recorded about 22 billion minutes of talk time; in 1999, 116 billion.[23] By 2004, minutes of use would grow to nearly 1.1 trillion; in 2009, to nearly 2.2 trillion.[24] Price per minute fell from fifty-one cents in 1995[25] to just three cents in

2011.[26] From 1997, when PCS licenses had been issued and new competition entered the market, prices declined 90 percent in real terms over the next twelve years.[27]

Operators dealt with the sharp decline in prices by selling far more minutes, but the terms of trade were shifting against them. Between 1993 and 1998, the average monthly revenue per subscriber for mobile carrier Air Touch fell from about eighty-three dollars to forty-five dollars, a precipitous five-year plummet.[28] The good old cellular duopoly days were over. As Sprint's CEO told an industry conference following the PCS auctions in 1996, "A wireless license used to be a license to print money and it's not that way anymore."[29]

In the process, an entire U.S. industry was encircled, starved, and left for dead: long distance telephony. In 1999, the industry had a market capitalization of $472 billion; by 2004, it was $41 billion.[30] At about that point, all three national long distance operators disappeared via fire sales to local telephone providers, MCI aided by bankruptcy reorganization of its 1998 acquirer, WorldCom. The wireless networks of AT&T and Sprint survived, but "long distance" was no longer a business.

Competition is not a rumor. Under the analog duopoly, mobile phone usage remained a novelty, and the world of third-party apps was nascent. With competitive entry, declining rates, an expanding user base, and the digital technologies to support it, innovation surged. The PCS allocation, like 2G spectrum allocations in other countries, produced astonishing outcomes. The new spectrum made available to the market visibly triggered robust new economic activity. Public policy making rarely demonstrates such clear success.

The gains to users, "consumer surplus," can be estimated to a first approximation. The calculation takes the maximum price customers would be willing to pay (revealing the value gained) and subtracts the market price customers actually pay. While market prices and the quantities—minutes of use—can be observed for cellular service, the willingness-to-pay magnitudes cannot. This can obscure the value generated.

Yet historic prices and quantities of use offer a useful guide. In cellular markets, service prices (specifically, revenue per minute of voice service) began falling continuously in the early to mid-1990s hand in hand with

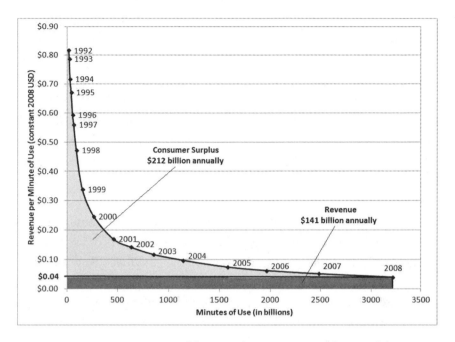

15.1. A conservative estimate of the annual consumer gains from mobile voice service in the United States, 2008. The calculations and data sources are described in Thomas W. Hazlett, Roberto E. Muñoz, and Diego Avanzini, "What Really Matters in Spectrum Allocation Design," 10 *Northwestern Journal of Technology & Intellectual Property* 93 (January 2012), 99–103.

large increases in call volumes. That makes perfect sense. The pattern graphed in Figure 15.1—where the top line comes from the prices and annual minutes of use—covers 1992 (when price per minute was about eighty cents) through 2008 (by which time the price had fallen to four cents).[31] To gauge how customers valued cellular minutes in 2008, we observe how many minutes they purchased in previous years when prices were higher. In fact, these past levels of consumption understate 2008 demand, as key trends—other than declining prices—have promoted mobile services in the intervening years:

- Handsets became cheaper, smaller, better, and featured improved battery life;
- Networks became better, increasing coverage and improving connections;

- Social mores shifted, as it became acceptable to use a phone almost anywhere, anytime—no matter how rude it might (initially) have seemed.

These influences, tending to increase mobile minutes-of-use, imply that the historic demand levels yield very conservative estimates of 2008 willingness to pay. Nonetheless, the surplus gain to such users for cellular voice minutes in that year totaled at least $212 billion over and above the $141 billion subscribers paid.

But why stop the exercise in 2008? Because that is when texting starts to gain widespread popularity, displacing voice minutes. The trend is visible to the naked eye. Figure 15.2 shows that growth in minutes of use for voice is vigorous up through about 2007; by 2008 the growth is going to texting, pictures, and videos ("multimedia messaging services"). Overall, mobile communications continue to expand, and the roughly linear trend continues unabated, if we consider one minute of voice roughly comparable to one text message.

The benefits of mobile networks are of a large magnitude. And we have thus far excluded the value of emerging mobile data networks, sharing the platform initially built for voice. This has introduced a diverse array of services and applications, including video streaming. Subscribers paid another $30 billion for data in 2008, and more than $11 billion for handsets. By 2015, smartphones had displaced the older devices and accounted for $54 billion in U.S. sales of 177 million units—more than one per family per year.[32]

In short, usage patterns in mobile markets suggest enormous value creation and—when the evolution in M2M, broadband, and the app ecosystem are factored in—innovation of vast proportion. Pointedly, these gains rest on the loosening of spectrum allocation policy, allowing bandwidth to be used as markets rather than regulators dictate. When relatively liberal licenses were authorized, relatively massive investments in networks materialized. Huge gains in consumer welfare soon followed.

Reform granting "flexible use" was key, allowing new services to be born, skyrocket to fame and fortune, and then die, making way for pretty amazing

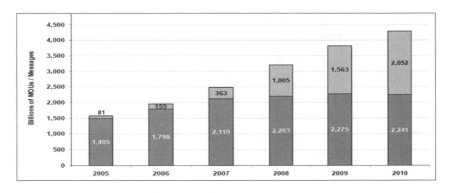

15.2. Annual mobile usage in the United States, 2005–2010: minutes of use (voice) plus text messages. Broadband Internet access not included. Data are from the Cellular Telecommunications and Internet Association.

new(er) stuff. Take the ironic path of paging. These services were launched on traditional licenses (specified for paging) and became popular with doctors and drug dealers in the 1980s. A 1991 *Business Week* article carried the headline "Jeepers, Creepers, Where'd You Get Those Beepers?"[33] Subscribership was still soaring; from 4 million in 1992 to 21 million in 1997.[34] The first FCC auction, held in July 1994, offered ten national narrowband PCS licenses of a more liberal nature; they could be used "to provide services such as voice message paging, two-way acknowledgement paging in which a subscriber can receive a message and transmit a response back to the sender, and other data services."[35] The winning bids totaled $617 million. On a per-MHz basis, proceeds were more than six times the $7 billion paid at auction for Broadband PCS licenses a few months later.

But those narrowband networks were doomed. A 1996 deregulation allowed cellular licensees to bundle paging services with digital phone service. The competition was too efficient. The stand-alone beeper business was toast.

Leaving the market free to innovate would produce cell phone improvements that makers of cameras, alarm clocks, portable radios, MP3 players, pocket calculators, flashlights, wristwatches, GPS receivers, and landline telephones would soon learn about, the hard way. As mobile phone penetration climbed to 100 percent of the adult population (by 2004, South

Koreans were stating their statistic as "percentage of population subscribing to mobile, six years of age and older"), helpful electronics crept, one by one, into the communications device.

And with that product migration, new networks formed. By 1997, cellular digital packet data (CDPD) service was being offered by such carriers as Bell South and AT&T Wireless, with coverage in 195 cities. The service, Mobitex, sent data using idle bandwidth "to take advantage of unused channel space between voice calls."[36] The system was priced for high-priority business use; an email cost seven cents and a four-page document about twelve dollars.[37] It attracted thousands of subscribers, but its chief importance was in establishing how one physical network could be shared. What else might piggyback?

Canadian entrepreneurs had an idea. In the late 1980s a start-up in Waterloo, Ontario, called Research in Motion (RIM) pondered the emergence of Mobitex, layered atop cellular services. Once the regulator had served as a prohibitively expensive middleman. But now networks in the United States and Canada (following a similar policy path) were free to deal within the spectrum spaces their licenses defined. As if by magic, great ideas appeared.

RIM's BlackBerry, delivering emails and pages along with voice calls, added a feature to the phone—a keyboard. The convenience of a mobile phone with the functionality of a personal computer became a reality by the mid-1990s. The device was a huge hit with business "road warriors," enjoying a decade during which the "Crackberry" seduced tens of millions of high-volume premium users. One famous addict, President Barack Obama, was forced to go cold turkey when, in early 2009, the U.S. Secret Service bravely informed the incoming chief executive that he would have to use the old White House phone system.

The BlackBerry business model was an architectural marvel. RIM created wireless network solutions for individuals and enterprises, all via contracts with networks. It partnered with cellular carriers around the world, meshing RIM's data distribution service with the networks of wireless operators. RIM managed this physical/virtual network, built its own handsets, and wrote its own software (operating systems and applications for BlackBerries), sending subscribers' data via transmitters and radio spectrum controlled by the operators. The company became a wireless powerhouse,

15.3. In 1998, a mobile device of the future emerges from the Canadian company Research in Motion. This is a precursor to the BlackBerry, the RIM INTER@CTIVE, supplying paging and emails to cellular subscribers. William Frezza.

with millions of users and a market valuation of more than $38 billion in 2010. Competitive forces then intervened, cruelly, with RIM losing its market to the next generation of mobile device and software innovators—Apple, Google, and Samsung.

The introduction of the BlackBerry, and myriad gadgets to follow, edged out other cellular communications, consuming precious network capacity. Operators possessed the power to welcome or block this outcome. Their interest was to configure traffic so as to maximize profit. Allowing a data network to fill in the "white spaces" where voice traffic was light could monetize underused resources, provided the inconvenience to voice customers was slight. The carrier, delegated spectrum management dominion, calculated benefits and compared them to costs.

Simplicity bred a revolution. Since the 1927 Radio Act, non-U.S. enterprises have been denied FCC licenses. While that protectionist policy still stands, liberalization of spectrum use circumvents the rule, opening domestic markets to global entrepreneurship. Upstart innovators from abroad— from as far off as Canada, or perhaps Finland or South Korea—can now shop for American bandwidth.

Even the tightly wound Japanese Ministry of Posts and Telecommunications has granted relatively wide discretion to cellular licensees. Soon after RIM's foray, another ambitious experiment came to mobile markets courtesy of NTT's DoCoMo, the leading wireless network in Japan.[38] In 1999, DoCoMo began distributing handsets with an "*i*" button. A quick tap enabled *i*-mode, a "healthy ecosystem of vendors, content, and applications."[39] It was a walled garden, cultivated under DoCoMo's rules. Preferred service providers would conserve precious bandwidth and keep e-commerce charges reasonable. Wide-ranging apps emerged, including DoCoMo's "digital concierge" to assist customers with "all the bothersome tasks of life when you are away from home."[40] Purchases were fast and simple, added to the subscriber's phone bill. The *i*-mode kept a tally of monthly purchases, mitigating "sticker shock." DoCoMo collected 9 percent of all sales.

The "wireless web" became a two-sided sensation. Thousands of vendors gained admission to the garden, while millions of customers rushed to pay three dollars a month to visit them—thirty-one million subscribers were enrolled by January 2002. DoCoMo's new marketplace became the leading Internet access platform, fixed or mobile, in Japan. Japanese mobile rivals KDD and Vodafone (later absorbed by Softbank) responded by creating versions of their own. A *Wired* reporter proclaimed *i*-mode a "pocket monster," marveling that "DoCoMo's wireless Internet service went from fad to phenom—and turned Japan into the first post-PC nation."[41]

Joseph Schumpeter saw capitalism enmeshed in constant disruption, which he regaled as the dust cloud of economic growth. Gains to society were produced in a "gale of creative destruction," bringing the "new methods of production or transportation, the new markets, the new forms of industrial organization that capitalist enterprise creates."[42] Spectrum central planning had prohibited these new forms. Now regulators were stepping back. Schumpeter, who died in 1950, might have liked what came next.[43]

Nearly every country deregulated mobiles relative to what it had authorized for broadcast radio and television. Market-driven innovation became almost instantly obvious in the United States, Europe, Japan, South Korea, and elsewhere. Others became envious. The new consumer goods and the new forms of organization were coming at a pace impossible to imagine under administrative control.

Steve Jobs, the personification of Schumpeter's vision of market combustion, seized the moment. Apple's first wireless idea was to create a Tiffany network. Company designers had been secretly at work, first on a tablet, then on a smartphone. The products were ultimately released in inverse order. Apple designers converged beauty and grace in the iPhone: a large touch screen, smooth and shiny, metal and glass. Connected to networks, the hand-held device was far more than just a pretty interface. It was the business-oriented BlackBerry on Olympic ice skates.

In 2005, Jobs quietly approached Verizon, the largest U.S. wireless network, to inquire about partnering. Apple would produce its new, gorgeous phones while Verizon delivered the cell service. But Apple's price was stiff. Verizon, the confident owner of what it considered the best U.S. mobile system, balked. "Verizon, notorious for wanting to control everything on its phones," wrote Fred Vogelstein, was "so convinced of its dominance in the wireless business that it turned down Jobs' offer of a partnership."[44]

But Verizon had rivals. Sprint offered to host an Apple mobile virtual network operator (MVNO) where Apple Wireless would retail its new mobile brand, using its devices and service supplied by Sprint's network infrastructure. But word of that offer made its way to yet another carrier, AT&T,[45] whose executives "became terrified" by the news. Elbowing Sprint aside, it agreed to pay Apple its ransom. AT&T would supply network services and yet yield its vendor unprecedented power: "total control of the design, manufacture and marketing of the iPhone."[46] The carrier then paid Apple more than retail for iPhones, including fat commissions to Apple for every subscriber using one. In return, AT&T would obtain a three-year exclusive: an American wanting to use an iPhone would have to subscribe to AT&T.

The market for radio spectrum was open for business and Apple was buying. Offers were so competitive, and demand for the iPhone so brisk, that prices were *negative:* Apple was *paid* to share AT&T's bandwidth so long as it agreed to sell its shiny new toys to AT&T's customers. Soon iPhones and myriad applications were nesting on AT&T's platform through Apple's App Store—some 1.2 million apps by 2014, with more than 75 billion downloads across all platforms. AT&T was not getting that money, either; 30 percent of sales went to Apple, 70 percent to the software developers. The app developers' annual take was soon about $3 billion.[47]

When the iPhone went on sale in June 2007, AT&T was struggling, ceding market share to Verizon. Instantly, the trend reversed. AT&T's network choked on the exponential increase in traffic. The carrier was forced to invest not only in iPhones but in the new capacity to serve them. Cell splitting and other network improvements, including a more rapid adoption of 3G technologies and aggressive bidding for new bandwidth rights (in both secondary markets and the FCC's 700 MHz auction in 2008), cost the carrier tens of billions.[48] But service quality improved, subscriber growth continued, and AT&T's financial outlook strengthened.

The iconic success of the iPhone motivated capitalists elsewhere. Google, the Rottweiler puppy of Silicon Valley, was young, cute, and already a force of nature. Keen to extend its hugely profitable search engine beyond the desktop, Google fashioned an iPhone competitor by proxy. It acquired and honed the Android operating system software to power mobile devices, licensing its package to manufacturers without charge. The default apps included its sponsor's favorite: Google Search. The firm then curated its own marketing channel for software applications, which became Google Play.

The ecosystem was too good for vendors to refuse. Samsung, Motorola, LG, HTC, and others partnered with Google in a counterstrike aimed at Apple, a company that produced its own hardware and operating system. Android, with 0 percent market share in 2007, came installed on 75 percent of the smartphones shipped globally in the third quarter of 2012.[49] In a stunningly short time, Nokia—the world's largest smartphone maker in 2007—was left gasping for air. But don't feel sorry for Nokia; BlackBerry and Microsoft's Mobile Windows were crushed too. Microsoft bought Nokia in order to relaunch its mobile market strategy in 2013. By 2015, it had written off nearly $8 billion of that investment as a financial miscalculation.[50] Consumers, enjoying the competition, were not troubled.

Meanwhile, Apple, the presumed target of Google's Android assault, flourished. Not only did iPhone sales remain brisk as new releases thrilled fanboys around the world, the iPad became the new must-have device after its launch in 2010. The firm's operating margins were the envy of the industry, the online App Store was vibrant, and the company's plunge into brick-and-mortar retail stores, designed in the image of the Gap and offering a Genius Bar fashioned after the Ritz-Carlton's concierge, sent profitability through the roof. In 2012 Apple became the most valuable company in history.[51]

Google was not far behind. In 2014, it leaped past ExxonMobil to become the second-most valuable company in the world. By 2016, it had surpassed Apple. The ascendant Android ecosystem was helping the cause. Google Play offered hundreds of thousands of third-party apps for Android devices. Like Apple, Google gave 70 percent of sales to the developer. Yet unlike Apple, Google did not keep all of the remaining 30 percent but shared it with carriers.[52] Free software for smartphone makers, extra profits for mobile networks, and billions of clicks on mobile devices for Google Search. Everyone loved Android—save Apple and Steve Jobs, who saw the smartphones as appropriating "the iPhone's look and feel." Jobs denounced Google's corporate slogan, Don't Be Evil, as "bullshit," and pledged revenge: "I will spend my last dying breath if I need to, and I will spend every penny of Apple's $40 billion in the bank, to right this wrong. I'm going to destroy Android, because it's a stolen product. I'm willing to go thermonuclear on this."[53]

When AT&T's exclusive iPhone deal expired in 2010, Verizon procured iPhones of its own. In October 2011, Sprint joined the party. By then, the third-largest U.S. carrier was in dire shape. While its market capitalization was only about $10 billion, it agreed to buy 30.5 million iPhones over four years at a price of $675 per phone.[54] The astounding $20 billion commitment, combined with Apple's 75 percent gross profit margin on iPhone sales,[55] meant Sprint was more valuable as an Apple retailer than as a wireless operator.

Between 1985 and 2014, mobile carriers invested some $430 billion in network infrastructure.[56] This response was undertaken to complement the exclusive, flexible-use frequency rights released starting in 1983. And these capital outlays, in turn, created a platform from which a spectrum market sprang.

By relaxing how it defined services, the government left entrepreneurs free to create and then reimagine networks. Innovators did not have to file petitions, register comments, commission studies, and trek to Washington to plead with regulators to open a proceeding—hoping to eventually demonstrate, over the din of public objections and against the private pull of assorted interest groups, that their new radios would serve the "public interest." Instead, they went on road shows to impress institutional investors and made elevator pitches to venture capitalists. Unlike FCC hearings, a "no" from one investor left them free to pursue a "yes" from others.

The contrast with traditional spectrum allocation was striking. In 1990, the FCC officially began a proceeding to allow FM radio stations to switch from analog to digital transmissions. Finally, in 2002, the Commission decided it was time, and mandated an industry standard.[57] But just to be safe, power levels were kept very low. Too low, the FCC later determined. Dubbing the limits "regulatory impediments," regulators found in 2010 that they had deterred "FM radio's ability to meet its full potential."[58] More-powerful signals would not result in harmful interference between stations. The restrictions were relaxed. It had taken two decades to make the world safe for terrestrial digital radio broadcasts.

The mobile marketplace, without such micromanagement, did not descend into anarchy, even as radios and networks skyrocketed in complexity. Mobile carriers, effectively private FCCs, streamlined and competitive, policed the mix of devices and applications sharing their airspace. Bandwidth-hogging uses that generated little consumer gain were constrained, killer apps that brought customers to the network were nurtured. Where the Next Big Thing required more space, the network sank additional capital to accommodate. Where overuse threatened, subscription fees were adjusted accordingly. Service buckets were sold that encouraged voice calls be made off-peak, and then that text messages be substituted, economizing on bandwidth. Cost and revenue metrics provided continuous feedback. This optimization process governed how networks were designed, built, upgraded, managed, and priced.

National mobile networks in the U.S. serve between about 50 million and 150 million subscribers each; operate between 40,000 and 100,000 base stations (cell sites) each; utilize, in total, more than fifty thousand wireless (FCC) licenses; plug into hundreds of popular devices, from smartphones to tablets to mobile hot spot modems to myriad M2M radios; host more than two million applications and countless content sources. Coordination with technology suppliers, infrastructure builders, device vendors, real estate agents (for cell sites), and spectrum brokers (for licenses), is continuous and intense.

Regulators in the United States and the European Union have long talked about "secondary markets for spectrum," generating thousands of pages of proposals and regulations on the topic. Yet with liberal licenses, the vast majority of spectrum transactions do not transfer naked bandwidth. Instead, wireless services, supplied by a carrier holding both mobile licenses and a physical network, are bought and sold. The assets are tight complements. Keeping spectrum rights and the associated network under common ownership eliminates myriad potential conflicts and coordination problems. This integration makes a wireless transaction easy as a phone call or text message. When phone users stray "out of network," they are seamlessly switched to new spectrum. A T-Mobile subscription in the United States comes preloaded with at least forty-five domestic roaming partners and far more internationally.[59]

Spectrum rights are sold at the press of "Send." And resold. Scores of virtual mobile network operators, like Tracfone, buy billions of minutes per year from carriers, wholesale, creating new retail services.[60] Boundless machine-to-machine services are developed, like OnStar, automatically calling for help when a car's airbags deploy, or Amazon's Kindle, an e-reader that downloads books or movies via the mobile network Amazon embeds by contract; the human reader is oblivious to the connection. A vast universe of M2M devices are taking hold, part of the Internet of Things. Sensors with radios are already telling central offices how to best manage trucks, railroads, or ships, routing them to avoid traffic, optimizing pickups and deliveries. Inventories are made more efficient, stocked "just in time," both in vending machines and in Walmart. Dollars are saved, energy is conserved, prices are reduced, people live better.[61]

Not so excited about radios in candy machines? Then think about your health. Those zippy computers in our pockets can—and in some cases, already do—monitor our vital signs, inform our doctor when prescribed meds have not been taken, and signal the hospital when our pulse slackens or our blood sugar soars. Mobile health applications—"mHealth"—are in their infancy but full of promise.

In general, these innovations have not been the target of specific spectrum allocations and have not had to wait for one. They are being hatched in the humdrum of competitive profit seeking, hosted by licensees with rights to create what they conjure and to eat what they kill. Vast new mobile ecosystems form; firms come, dominate, and go. They struggle in an arms race to catapult the next killer app. Mobile platforms arise in wireless without procuring specialized wireless rights, sneaking in by seducing users. Producers of content, devices, and software apps become wireless giants.

"Apple was dominant in the mobile market," wrote Walter Isaacson in 2011, and yet the firm did not have any wireless licenses or deploy a single base station. A *Wall Street Journal* article the following year explained how Apple was being challenged as the "dominant mobile platform of the future"— not by carriers but by Google, Facebook, and Microsoft.[62] The much-maligned Toaster Model overwhelms markets with waves of creative destruction. But as impressive as the wireless networks are in the United States, Canada, Europe, Japan, and South Korea, the effects of liberal spectrum policies on developing countries may be more impressive still.

Take fishermen in Kerala, a state in India with a large fishing industry. Before cellular, a boat would head out to sea, with luck snag a catch, and return to its dock. The day's bounty was sold at the "beach price," set by the local buyer. The existence of more favorable prices across the bay was not so interesting; fish spoil quickly.

But that changed, says UCLA economist Robert Jensen.[63] "Between 1997 and 2001, mobile phone service was introduced throughout Kerala." Wireless phones allowed fishermen to find, and deliver to, the port offering the highest price. Of course, when that business reality dawned, buyers had to up their offers. Jensen documented a "dramatic reduction in price dispersion, the complete elimination of waste, and near-perfect adherence to the Law of One Price." He called it the "digital provide." C. K. Prahalad, professor of business at the University of Michigan, observed, "One element of poverty is the lack of information [and the] cellphone gives poor people as much information as the middleman."[64]

"I can't imagine life without my phone," exclaimed Babu Rajan, skipper of a seventy-four-foot ship operating out of Pallipuram in Kerala. In years prior, few "could afford expensive marine radios, so if someone hit upon a massive school of sardines, there was no way to alert friends on other boats." In addition, fishermen were often stranded at sea when engine trouble struck. Now Rajan "can call his mechanic, who also carries a cellphone, to ask for emergency service."[65]

There are macroeconomic gains from improved communications. A 2001 study by economists Lars-Hendrik Röller and Leonard Waverman estimated social welfare improvements for European countries associated with the rollout of fixed-line phone systems, 1970–1990.[66] They found that fully one-third of total productivity increases resulted from expanding telephone networks. In France in 1970, for instance, there were only eight phone lines per one hundred people; by 1980, there were thirty. The splurge in infrastructure allowed the French economy to grow substantially. There were similar improvements in phone penetration throughout western Europe during the period, with associated gains in income.[67]

What about developing countries? As late as 1996, wired phone systems, utilizing technology more than a century old, were yet stunted. It was typical for much less than 10 percent of a population to have access to phone service, with multiyear waiting lists for new installations. While the United States had 55 phones per 100 persons, Tanzania and Bangladesh had just 0.3, India 0.7, China 4.5, Mexico 9.5, and Brazil 9.6.[68] As mobile networks were authorized in these and other nations, they surpassed their fixed line rivals almost instantly. Those cellular licenses went, by and large, to private competitors. The existing communications monopolies—held by the government PTTs (post, telegram, and telephone operators)—were financially far stronger. They exercised the privileges of state and were tasked with their own "public interest" mandates to advance the economy and help the poor. Yet their networks failed to serve poverty-stricken areas or rural districts or, indeed, almost anyone outside an elite power structure. To become connected, you had to *be* connected.

Wireless rocked this world. "As more and more countries abandon government-run telecom systems," the *New York Times* reported in 2008, "offering cellular network licenses to the highest-bidding private investors and without the burden of navigating pre-established bureaucratic chains, new towers are going up at a furious pace."[69] By 2002 there were more mobile subscribers, worldwide, than fixed line. By 2013, the ratio of mobile to fixed was more than six to one, and traditional phone subscribership was declining, abandoned. Journalists wondered, "Can the Cell Phone Help End Global Poverty?," noting that "a cellphone looks a lot different to a mother in Uganda who needs to carry a child with malaria three hours to visit the nearest doctor but who would like to know first whether that doctor is even in town."[70] A 2010 book on cellular service in Africa carried the title *Less*

Walk, More Talk.[71] There are several African nations in which the number of mobile users exceeds by two hundred times the number of fixed-line subscribers.[72]

In 2009, Iran's streets filled with demonstrators, taking on the Ayatollahs armed only with mobile phones. An innocent bystander, college student Neda Agha-Soltan, was killed; the martyr inspired a huge outpouring of antigovernment wrath when a video of her death went viral, largely through Twitter traffic on #Neda. The 2011 "Arab Spring," where millions took to the streets to protest dictatorial regimes in Tunisia, Libya, and Egypt, allowed protesters to mobilize via wireless. Texts and tweets organized the masses; social media and multi-media messaging via smartphones connected the leadership.[73] Applications such as Facebook and Instagram, transmitted via airwaves controlled by cellular carriers, helped dissidents organize, toppling governments in what became known as the Twitter Revolution. In 2014, the Chinese Communist Party took note. When hundreds of thousands of pro-democracy protestors took to Hong Kong streets in the Umbrella Revolution, "dependent on mobile apps to coordinate their huge, seemingly unstoppable uprising," the government apparently sought to disrupt the demonstrations with computer viruses.[74]

If the political effects are ambiguous and much contested, the global economic gains from wireless are already profound. Adam Clayton Powell III, a scholar at the University of Southern California, was stunned by what he found in Singapore in 2014. A farmer using an Android mobile app walked into cropland, held his phone out, and kept still for a minute. The phone recorded the *sound* of the field, identifying insect activity and determining which pesticides would best control them, the striking innovation offered by mPest software.[75]

Poorer countries have used wireless to leapfrog antiquated infrastructure, bounding into the twenty-first century. Economists Leonard Waverman, Meloria Meschi, and Melvyn Fuss found that the boost to growth "may be twice as large in developing countries compared to developed countries."[76] New business models adapt mobile to local conditions. Prepaid phone cards make service available in markets where few have a formal employment or credit history or, for that matter, a fixed address. Innovative services like cellular banking and microfinance are facilitated by mobile phone connections. The creation of Grameen Phone in Bangladesh by entrepreneur and 2006 Nobel Peace Prize winner Muhammad Yunus was a perfect demonstration of markets tailoring solutions to needs, "creat[ing] economic and social development from below," as the Nobel Committee put it, "to advance democracy and human rights."[77]

Waverman, Meschi, and Fuss found that the more mobiles a developing country had, the more economic growth it enjoyed. "All else equal, the Philippines (a penetration rate of 27 percent in 2003) might enjoy annual average per capita income growth of as much as 1 percent higher than Indonesia (a penetration rate of 8.7 percent in 2003) owing solely to the greater diffusion of mobile telephones." Among low-income countries, each increase of ten users per one hundred population was associated with extra annual per capita growth of 0.59 percent.[78] A country with 75 percent mobile penetration would be about one-third wealthier in a decade, relative to an identical country with just 25 percent subscribership.

A less regulated and more competitive phone medium has emerged all around the world. But perhaps the most ambitious experiment in spectrum liberalization was launched in Guatemala in 1996, one year after the agreement that ended the country's long, bloody civil war. The initial democratic elections brought serious reformers to power. Young, U.S.-educated advocates for markets and civil rights, Freddie Guzman and Giancarlo Ibárgüen gained influence in the new governing coalition and sought to privatize the state's communications monopoly, Guatel, while authorizing new competitors in fixed and mobile. The scheme proved extraordinarily liberal.[79]

A new, independent regulator, SIT, was created. (Previously, the government, in conjunction with the military, made opaque decisions.) Existing airwave users were grandfathered, and a public registration system listed the parties with rights to use given airwaves. In this way, unoccupied bands became known to all. Any individual or firm, including noncitizens, could petition the regulator to acquire rights to these frequencies. The request would be posted. If no competing claims were made during prescribed windows, the rights were assigned. If the claims were contested, rights were auctioned. Strict timelines, and arbitration rules to quickly resolve disputes, were imposed.

The spectrum use rights issued were generic, not for "television" or "cel-

No. Orden: No. Registro:

LA SUPERINTENDENCIA DE
TELECOMUNICACIONES DE GUATEMALA

Con base en el Articulo 57 del Decreto 94-96

Otorga el Presente

Titulo de Usufructo de Frecuencia

A:

Banda o Rango de Frecuencias:
Horario de Operación:
Potencia m xima efectiva de radiación:
Maxima intensidad de campo eléctrico o
potencia m xima admisible en el conforno:
Fecha de Emisión:
Fecha de Vencimiento:

15.4. TUF love: Guatemalan reforms in 1996 created spectrum property via titles to the use of frequencies (TUFs). This is a rendition of the minimalist legal document granting liberal ownership rights, notable for what it does not prescribe.

lular" or "fixed microwave." They were not licenses, associated with special privileges issued at the discretion of the regulator, but TUFs—*títulos de usufructo de frecuencia*. These conveyed property rights in the use of radio spectrum. The scheme was simple, highlighted by the minimalist TUF form. The title is defined by (a) frequency band, (b) hours of operation, (c) maximum power transmitted, (d) maximum power emitted at the borders of adjacent frequencies, (e) geographic territory, (f) duration of right. On the back of the one-page document are blank signature lines where an owner can assign the TUF to a new party, like the title to a car.[80]

The reformist zeitgeist spilled over to El Salvador, which adopted similar spectrum reforms in 1997. There, licenses were issued based on international spectrum allocations, but made permissive. Licensees were free to use bandwidth with other technologies and for other purposes; the mandated rules became default suggestions. The approach was liberal compared with that of the United States, and radical relative to approaches elsewhere in Latin America. In an economic analysis of Central American mobile markets I conducted with Giancarlo Ibárgüen and Wayne Leighton in 2007, it became apparent that the two nations adopting ultraliberal spectrum allocation regimes had—despite poor populations and generally illiberal government policies (particularly in Guatemala)—developed superior mobile networks. Other nations, many with much higher incomes, allowed less competition while imposing tighter restrictions on operators. Adjusting for other social and economic factors, mobile prices and outputs were highly favorable in Guatemala and El Salvador compared with peer countries.[81]

But the visual graphic provided by Amazon in November 2010 is perhaps most compelling proof of the advantages of such competitive policies. Amazon's Kindle, as noted, relies on M2M communications to download content. The customer is untethered; books, movies, and TV shows flow to the Kindle wirelessly. Amazon bargains with mobile data networks to supply these services. Because coverage differs across markets, Amazon displays coverage maps (supplied by data firm Mosaik Solutions) indicating where mobile downloads are available.

In late 2010, as shown in Figure 15.5, Central American e-reader users found many places lacking any high-speed data service (white spaces). Some areas had 2G availability (gray) and others had the fastest and best downloads via 3G networks (the dark shading). Guatemala was blanketed by 3G, and El Salvador was widely covered by 2G. Yet surrounding countries, including the far more prosperous Costa Rica, lagged badly. Costa Rica maintained a mobile monopoly, run by the government, even in 2010. (Two competing wireless licenses were finally assigned, by auction, in 2011.)

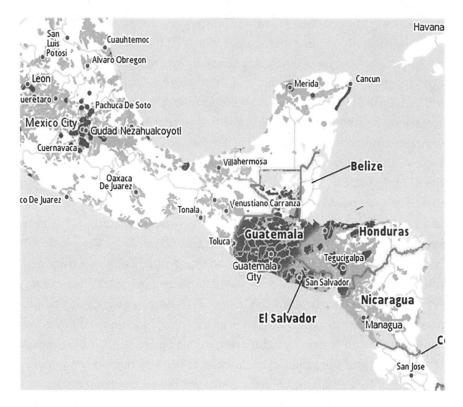

15.5. High-speed mobile data network coverage for Central America. This is a screen shot (taken November 17, 2010) showing where downloads, such as those used by Amazon Kindle owners, were available on cellular networks in the region. White spaces lack connections, gray shading indicates 2G access, dark areas offer faster 3G service. Image © 2010 Mosaik Solutions (www.Mosaik.com).

Mexico had 3G in 2010, but it was provided by just one company, TelCel, the affiliate of the Telmex landline monopoly owned by Carlos Slim, and largely restricted to Mexico City—white spaces almost everywhere else.

The Guatemalan liberalization shows how even poor countries might pull themselves into the modern world, and it supplies a critical insight. The Coasian spectrum market vision was not erratic or inept, lacking in technical understanding that would derail its practical adoption. Instead, its adoption—embraced more in some places than in others—has brought sensational results. You can read about it on your Amazon e-reader, which works pretty well in Guatemala.[82]

16

Dirigiste Backlash

After a century of wires, now everything is wireless. I see the forlorn
public phones at airports. I'm not even sure if they work anymore,
and if you tried to use one, passersby would look on you with pity.

—Robert Lucky

BY THE NEW MILLENNIUM the wireless craze was in full swing. Scholars
and policy makers widely agreed that flexible-use licenses were leading the
breakout. In a proceeding involving secondary markets for radio spectrum, launched under President Bill Clinton and concluded under President George W. Bush, the FCC found academic experts strongly endorsing
liberalization and urging more of it. A policy statement by "37 Concerned
Economists"—including Cornell's Alfred Kahn, the dean of U.S. regulatory
specialists, and Martin Neil Baily, the just-retired chairman of President
Clinton's Council of Economic Advisers—counseled that "The Commission should move decisively to broaden the rights generally granted licensees, permitting flexible use of the allocated spectrum. Licensees will then
find it profitable to pursue all productive uses of available airspace, and
market trades will make such space available."[1]

FCC license auctions had evolved from heresy to orthodoxy.[2] President
Clinton seized upon FCC license auctions as Exhibit A in Vice President
Gore's Reinventing Government campaign. "For the first time ever," boasted
FCC Chairman Reed Hundt, "the FCC truly follows a market-based ap-

proach to the allocation and use of spectrum."[3] Spectrum policy makers sought a new paradigm, seeing the "public interest" as best advanced by the release of large blocks of flexible-use frequencies. Hundt's successor, William Kennard, warned of a "spectrum drought," pushed for more generous allocations, and embraced airwave markets with a fervor worthy of Milton Friedman.

Internationally, the movement away from administrative control had farther to travel and was thus even more marked. Governments privatized their state-owned telecom monopolies, opened cellular to private competition, and let markets orchestrate network operations to an unprecedented degree. Countries around the world held mobile license auctions. Carriers were given broad authority to determine their own network architecture and business models (even as some regimes, in some circumstances, continued to mandate specific technologies). Wireless market competition eclipsed state fiat.

Some countries ventured beyond. Ambitious attempts to realize Coase's vision of property rights in spectrum were launched in New Zealand in 1989, Australia in 1992, Guatemala in 1996, and El Salvador in 1997.[4] As we have seen, these experiments provided laboratories for policy analysis. In 2002, England's Labour government issued a remarkable report in which Warrick University economist Martin Cave explained that the government's "over-arching principle is to expose all spectrum users to the opportunity cost of the spectrum which they occupy."[5] Natural economic incentives would unleash efficiencies, correcting the errors of state misallocation. "Market-based spectrum management tools, in conjunction with greater flexibility for spectrum users, are the primary means to this end."[6] Call it the Market Consensus, elegantly articulated in a study commissioned by a socialist party.

The tech boom of the late 1990s converged with the wireless craze to produce financial frenzy. England, auctioning five nationwide 3G licenses in April 2000, astounded sector analysts when it received some $34 billion in winning bids. Germany beat that total three months later, when it took in $46 billion for six licenses.[7] As financial markets plummeted in the dot com bust, the licenses—like many tech investments—suddenly looked overpriced. Yet as markets came back and their balance sheets stabilized, carriers built capacious new 3G broadband networks. Companies prized licenses because consumers were anxious to buy the services created with the bandwidth they delivered. Government cashed large checks. What's not to like?

Surprisingly much. Because at just this happy juncture, the policies hailed as vital turned fuzzy. In Europe, the backlash against the market has been generally dubbed *dirigiste,* the "centralizing tradition in French historical development, from the long reign of Louis XIV, punctuated by the rule of Napoleon."[8] The antonym to *dirigisme:* liberalism. The Market Consensus encountered new opposition.

In research on wireless telecoms policy in developed countries, published in 2010, economists Tomaso Duso and Jo Seldeslachts found that the emergence of mobile phone networks had crushed the natural monopoly presumption in the sector, supplying social momentum for pro-competitive policies. "However," they wrote, "regulatory policy . . . can exhibit a great degree of inertia."[9] When examining the pace of spectrum policy reforms across markets they found that independent regulatory commissions and powerful industry incumbents tended to slow reforms that would undercut established interests in government and the private sector.

In the United States, dirigiste backlash kept mobile allocations mired in stasis. Despite bold policy pronouncements and even a presidential executive order expressing the urgency of new bandwidth, between 1997 and 2006 the FCC made virtually no new spectrum available for mobile markets, freezing the flexible-use allocation at about 170 MHz. When the FCC auction window finally lifted with the September 2006 sale of advanced wireless services (AWS) licenses, allocated 90 MHz, a U.K. tech publication trumpeted: "Spectrum-starved US prepares to feast."[10]

The input drought slowed output growth and forced industry consolidation. Spectrum is a force multiplier, but needlessly withheld spectrum is a party killer. The decade of spectrum lag cost consumers, throttled technical progress, and delayed the emergence of the data networks that would eventually host the smartphone revolution and far more.

The point is not that more mobile spectrum is the answer to all wireless policy questions. It is that there is no good reason to keep airwaves locked up in government inventory. When flexible-use rights are issued to economically responsible parties, entrepreneurs will prowl the Halls of Science seeking strategies for disruption. Spectrum markets will be test beds for new ideas, deploying bandwidth where it is expected to provide more bang for the buck. And then, as better opportunities emerge, redeploy it all over again.

Despite its heralded successes, liberalization encountered increasing opposition in the new century. FCC economist Evan Kwerel described "those holding up more rapid progress" as "opportunists" seeking special favors, public sector agencies resisting change, "technology romantics" who see "property rights" as obsolete, and "anti-capitalists" who oppose markets, "willing to forgo trillions of dollars in economic growth that would benefit consumers."[11] From whatever source, the momentum for policy reform slowed, and regulations began slipping back to traditional administrative mechanisms. Predictable consequences have ensued.

Special Favors: The C-Block Fiasco

The fragility of the market consensus can be seen in the 1993 legislation authorizing FCC auctions, in which lawmakers indulged their nostalgia for the ancien régime: bidding preferences for "designated entities." The Commission was required to favor certain companies in counting bids. This undermined the auction's central purpose—discovering who might best use a resource—and, as implemented, led to bureaucratic mayhem. The decadelong "PCS C-Block fiasco" involved auctions, widespread bankruptcies, reauctions, and litigation all the way to the U.S. Supreme Court, keeping precious bandwidth unproductive. Commissioner Harold Furchtgott-Roth went so far as to pronounce the C-Block "cursed."[12]

And women and minorities, the "designated entities" or "DEs" Congress said it had in mind, got nothing. In 1995 the Supreme Court ruled that race and gender categories could not be favored except in circumstances unlikely to be proven under the "strict scrutiny" standard imposed by the Court. So the FCC went looking for other groups to favor as designated entities. It found "rural phone carriers" and "small businesses." The owners of these companies—typically middle-aged, upper-income white men—would receive bidding credits of 25 percent, meaning that each dollar they bid would count $1.25 in the auction.[13] Even better, designated entities' winning bids would be collected over a ten-year period, with the licensee paying only interest, at the rate of ten-year U.S. Treasury bonds (then about 6.5 percent), for the first four years. This was a substantial subsidy for such risky businesses. Most enterprises—particularly those without collateral assets, as was required to meet the FCC's definition of "small"[14]—would expect to pay at least 15 percent for capital investments.

The ill-considered aims of this program need not concern us, as they were quickly overshadowed by the incompetence of its execution. When auction participants are allowed to bid using subsidies, they tend to compete away the benefit: rather than limit their bids to capture the asset's value, they add in the government's bonus payment. The most aggressive bidders added a second strategy, overpaying and then opportunistically declaring bankruptcy, preventing the FCC from either retrieving the licenses or collecting the money promised.

In March 1995, the FCC's auction of personal communications services A and B licenses ended. No DE bidding credits were involved. Winning prices averaged 57¢ per MHz per pop (per person living in the areas covered by the licenses).[15] The auction for the next PCS tranche, the C licenses, included preferential terms for DEs. When that sale concluded in May 1996, prices averaged $1.33 per MHz per pop—more than 130 percent higher.[16] Market values had not skyrocketed. It was the bidding subsidy for designated entities that drove auction participants wild. Taxpayers bore the risks.

Under the FCC rules, winners made just a 10 percent down payment and could go back to financial institutions to raise the next year's payment. If they succeeded, they would presumably continue to pay the FCC. If not, they switched to plan B: declare bankruptcy, accuse the FCC of luring them in with exorbitant claims, and ask the bankruptcy court for protection on the grounds that the government was pursuing a "fraudulent conveyance."

Plan B was a winner. The two largest C block bidders, General Wireless, Inc. (GWI), and Nextwave, were soon in bankruptcy, refusing either to pay or to forfeit their licenses. Amazingly, the FCC's lawyers had failed to protect against this outcome. Licenses were awarded, on credit, without any security.

GWI was the first to be heard. Its trial was held in Dallas in early 1998.

(I testified as an expert for the Justice Department, whose attorneys were attempting to retrieve the FCC's licenses.) The company's lawyers argued that the FCC had tricked GWI into bidding too high by exaggerating the financial opportunities in wireless and that shortly after the auction the Commission had dumped new spectrum onto the market, reducing the value of GWI's holdings. A picture of President Clinton holding an enlarged check for more than $7 billion made it into trial as an exhibit. GWI, the defense claimed, was not a frisky bidder chasing a risky investment but a victim of wily federal government hustlers trying to balance the budget on the backs of innocent businessmen.

That the FCC had given GWI extravagant bidding credits and subsidies, only to see their new licensees cancel their payments and gain sanctuary in bankruptcy, should have been a clue as to how slick the federal bureaucrats actually were. As a factual matter, there was no surprise spectrum released after the auction but rather a decadelong drought. Yet GWI's attorneys won in a rout. The U.S. Bankruptcy Court determined that the $1.1 billion freely offered by GWI for fourteen wireless licenses had been elicited by fraudulent means, and that GWI now owned these licenses for just under $200 million.

Then came Nextwave. The largest C Block licensee, it had bid nearly $5 billion for its sixty-three licenses and then quickly gone belly-up. The mess took until 2005 to resolve. The highlight came when the FCC reauctioned Nextwave's licenses, alongside other DE licenses salvaged from defaults, in January 2001. It received $17 billion in winning bids, none of which it was able to collect. On June 22, 2001, the U.S. Appeals Court for the District of Columbia determined that the FCC did not own the licenses it had (re)auctioned.[17]

The $17 billion was refunded and the FCC entered negotiations with Nextwave. The parties eventually settled, with Nextwave keeping some licenses and returning others. In January 2005, when the FCC re-reauctioned the "cursed" PCS permits that remained under its control, the prime nationwide mobile spectrum finally became ready for useful employment.[18] In a 2009 article in the RAND Journal of Economics, Roberto E. Muñoz and I estimated that the licensing delay—depriving U.S. mobile markets of 30 MHz for about a decade—cost consumers some $100 billion. Women and minorities, the intended beneficiaries of the original policy, absorbed a substantial fraction.

Partial Deregulation

Since 1996, the Commission had intended to make additional spectrum available for broadband data services. But while EU markets welcomed such 3G bandwidth via license assignments by 2001, the United States lagged. The holdup was not due to the lack of a plan. The FCC had determined in 1996 that it would like to "re-farm" UHF TV channels 52–69 for mobile, shifting the TV broadcasts there to lower channels.[19]

Congress had endorsed the policy in the Balanced Budget Act of 1997, ordering new license auctions, with checks cashed by the Treasury by September 30, 2000.[20] These deadlines were pushed back by the FCC on at least seven occasions before being voided altogether in 2002. There was "no logistical basis for delaying the auction," Commissioner Furchtgott-Roth noted in dissent; it was a policy choice.[21]

In its first budget, the Bush administration proposed pushing auctions back still farther, from 2002 to 2006.[22] The delay, said a White House official, would be "a 'win win' for all parties involved . . . and it's good telecom policy."[23] One "win" registered the mobile networks' desire for a break from bidding because, in the 2001 recession, money was tight and balance sheets were stretched. The other "win" came from the Office of Management and Budget, which believed the delays would produce higher receipts.

But while this box score included the interests of industry incumbents and government revenuers, it omitted businesses and consumers, not to mention innovators and competitive entrants. The President's Council of Economic Advisers, having apparently failed to get the White House memo, suggested that "delays in introducing 3G products could have a severe economic impact." The CEA was right, but the auctions were scuttled anyway. "To the [wireless] industry's relief," wrote a trade magazine in 2001, "FCC Chairman Michael Powell, with the blessing of Secretary of Commerce Donald Evans, recently halted a mandate from former President Clinton that would have required all government branches to identify suitable 3G spectrum by July 30 [2001] and auction it off by September 2002."[24]

It is a cruel joke, of course, that the views of carriers should be binding. Any new license auction has three direct impacts on mobile networks—and two are bad. First, the network's competitors get access to extra spectrum, expanding their capacity and increasing competitive pressures (which is why consumers benefit). Second, when a network bids to win licenses, it must pony up billions of dollars in payments, if only to keep up with rivals. The positive impact is the third: the carrier gets access to new capacity. Sometimes the two losses outweigh this benefit. Carriers have been both pro- and antispectrum, depending on time, place, and company.

Even with its myriad problems, the old system of comparative hearings had one progressive aspect: advocates won implicit rights. If the Commission was persuaded to allocate spectrum, those who had petitioned the agency stood first in line for licenses. That gave these applicants incentives to endure through the long rulemaking process, helping the FCC find its way to the "public interest."

But the switch to auctions appropriated the queue.

Consider the plight of Carmen and Saleem Tahil, a husband and wife engineering team. They discovered that the direct broadcast satellite (DBS) band, located between 12.2 GHz and 12.7 GHz, could, via a smart angling of transmitters and receiver dishes, support new television and Internet access service. This would share the band with DirecTV, EchoStar and other satellite providers, leaving the incumbents' services undisturbed.

In 1994 the Tahils formed a company, Northpoint Technology,[25] which attracted considerable interest. Sophia Collier, one of the country's most successful female entrepreneurs (starting SoHo Natural Soda in her kitchen, she sold it for a small fortune to Seagram's in 1989, then launching multiple investment funds), was brought on as CEO. Northpoint approached the FCC to request authority to use satellite frequencies. They received an experimental permit but nothing more. Then Congress became interested and mandated that the Commission determine whether the new technology would conflict with existing services. A respected consulting firm, Mitre Corporation, rendered the verdict that the DBS band could host new video and data streams without creating harmful interference.[26]

So the upstart entrant won. In 2002, after Northpoint had spent more than $10 million in lawyers' fees, engineering studies, and lobbying costs,[27] the FCC created new multichannel video and data distribution services (MVDDS) licenses. Chairman Michael Powell and Commissioner Kathleen Abernathy were ebullient in their praise:

> Northpoint arrived at the Commission many years ago with a proposal for a new and innovative way to share the DBS spectrum. Today, thanks in large part to its fine work and diligence, that service will go forward. . . . There is little question that had it not been for Northpoint, the MVDDS service would not be ready to move forward today.[28]

But this diligence and creativity did not benefit Northpoint. The FCC, alongside its praise, rejected the company's application for licenses and instead put the firm's business plan up for sale, complete with spectrum access rights. Cable and satellite incumbents bought the rights for $119 million; Northpoint refused to participate in the auction and soon ceased operations. A decade later, MVDDS licensees had yet to develop the service.[29]

In explaining their decision, Commissioners Powell and Abernathy wrote:

> Northpoint has put significant time and resources into developing its service model as well as its Commission and congressional advocacy over a long period of time. We applaud these efforts. But . . . if Northpoint's service model is a winner, the market will reward it just as it has done for other technology companies.[30]

In fact, it was highly unlikely that Northpoint had the most efficient service model. As an upstart technology creator, it would likely be a relatively costly service provider and would logically seek a partner—a Verizon, Sears, Apple, or Walmart—that could leverage existing distribution channels. Northpoint's creative input was to discover new wireless opportunities in the government's warehouse. The FCC "applauded" and then seized Northpoint's invention, inviting the newcomer to bid against established operators to reclaim its innovation. The press reported, "FCC Gives Northpoint Half a Loaf,"[31] but Northpoint actually received a negative loaf. It invested considerable effort in promoting a new spectrum allocation, worked with regulators to achieve it, and then lost everything. The deterrent effect on future innovators is clear.

License auctions can speed rights assignments and eliminate wasteful rent seeking, producing substantial social gains. Yet as the economist Dwight Lee has shown for the economy in general, they can also reduce incentives for productive rent seeking.[32] This is a problem of partial deregulation. While DBS licensees had been given broader freedom than had terrestrial broadcasters, the rights Northpoint sought were not in the market. Were the underlying radio spectrum generically liberalized, new competitors like Northpoint would not need to lobby the FCC for a new allocation. They could buy access rights at market prices.

Auctioning license awards while leaving bandwidth locked up created a policy disconnect, outlined by FCC legal experts Henry Geller and Donna Lampert in 1986.[33] With the advent of license lotteries, entrepreneurs seeking new spectrum allocations would be forced to reveal proprietary business information to the Commission, yet they would have no protection against rivals ultimately receiving licenses and using their ideas. Geller and Lampert's solution was that "innovators should be allowed to apply for a right to use spectrum themselves at the same time they ask for spectrum to be allocated to a type of service."[34]

The Commission agreed, and in 1991 it adopted a "pioneer preference" policy. Wireless licenses would be delivered to those companies that contributed materially to the development of a new service or technology. Alas, this nice idea turned out badly.

PCS, just then in development, drew more than one hundred "pioneer preference" applications. The FCC could hardly distinguish one application from the next. Each applicant had hired professional buzzword writers of the highest order, and in the end the Commission awarded just three PCS preferences. It was unclear what the lucky winners had really contributed. (That one was partly owned by the *Washington Post* spurred several truly excellent conspiracy theories.) A federal court later held that a firm having the superior claim on technological innovation, Qualcomm, had been wrongly denied a preference, and ordered that the firm be compensated.[35]

The FCC subsequently undid the "free license" policy, charging PCS preference winners a fee for their licenses (set at 85 percent of the market price, as estimated from other sales). Congress, in the 1997 Budget Act, then abolished the program.[36] It was, all in all, an impressively rapid policy experiment, created and killed in just six years. But the failed solution had addressed a real problem. The problem remains, as the thwarted wireless entrepreneur Sophia Collier found out.

The Problem with Maximizing Revenue

In authorizing the FCC to conduct auctions, Congress instructed it to ignore the revenue implications. That was good government: the easiest way for a regulator to raise receipts is to preempt competitive opportunities. Economist John McMillan wrote: "If revenue had been paramount, the government could have offered a single monopoly license in each region— at the cost, obviously, of creating future inefficiencies."[37]

Auction bidders will offer no more than anticipated future returns, and market rivalry limits those profits. That simple economic result—a negative correlation between subsequent competition and up-front license bids— produces a screaming conflict of interest if the agency charged with making spectrum available is rewarded for bringing in high auction receipts.

Regulators clearly gain something from the political kick of high bids, but it is curious that economists often succumb to a similar temptation. Revenue became a standard metric in scholarly articles assessing the first generation of wireless license sales. Auctions producing high receipts were "successful," while countries with low prices were "fiascoes." Retail market outcomes were often overlooked.

Worse, economists recommended auction rules to restrict competition in order to goose auction receipts. The money received by the state was seen as a public finance dividend. This rested on the germ of truth that when a dollar is raised through taxes, it costs the economy more—according to empirical research, perhaps about $1.33. The extra 33¢ results from price distortions and tax collection costs. When the government sells a license, it can raise dollars without these inefficiencies. If sales are $9 billion, tax receipts can be $9 billion less, and perhaps some $3 billion in gains accrue due to this public finance bonus.

But such gains evaporate when access to spectrum is restricted by gimmicks. Yet this is just what is recommended in the form of high-reserve prices, bidding credits for weak bidders, and reducing the number of licenses for sale, in order to increase revenues. These policies are typically penny wise, pound foolish.

Reserve prices. Boats sold on eBay, for instance, typically include a "Starting Price" or "Reserve Price," below which the boat will not be sold. This makes the seller also a buyer. Unless other buyers are willing to outbid her, she keeps the asset. An influential 2002 paper by Oxford University economist Paul Klemperer makes the case for reserve prices in spectrum rights, illustrating the logic via the 3G auctions held in 2001 in Belgium and Greece. Each country offered four licenses for sale; only three were sold, as bids for the fourth fell short of the minimum. Klemperer applauds this result because the governments realized substantial revenues in each case. "It is very hard to argue plausibly that an auction deterred much entry when a license goes unsold, and there is also no obvious reason to criticise the reserve prices that these governments chose."[38]

But if a fourth network had entered the market, purchasing a 3G license at some price below the reserve, the consumer welfare gains from additional competition would have far exceeded the public finance dividend.[39] Moreover, even if no fourth competitor emerged, consumers would have been far better off if the spectrum allocated to the unsold license had been made available—free of charge—to the three winners. Instead, 35 MHz went back into the government vault and remained idle for years. This reduced competitive pressures and harmed consumers.

Bidding credits. In a 1996 article in the *Stanford Law Review*, economists Ian Ayres and Peter Cramton showed how bidding credits, by subsidizing weak bidders, could heighten competition within auctions and thus produce higher revenues for governments. The policy was endorsed on the grounds that it increased the public finance dividend.[40]

Paying companies to participate in an auction can raise license prices. Yet it also defeats the central purpose of the auction, which is to allocate rights to those who are best able to use them productively. As Paul Milgrom has written, "delays or failures are inevitable in private bargaining if the good starts out in the wrong hands."[41] In fact, bidding credits helped small businesses and rural telcos win PCS C licenses, but these parties did not build out systems, as seen. Instead they went bankrupt, squandering use of 30 MHz of prime frequencies through a decade of legal wrangling.

Restricting licenses. A 2003 study by economists Michael Rothkopf and Coleman Bazelon dispensed with reserve prices, instead recommending fewer competitive awards to begin with.[42] This idea responded to a 2002 FCC paper that had proposed granting flexibility to traditional licensees. The reform would have resuscitated dead spectrum, energizing wireless markets, but some were offended that incumbents might enjoy windfalls from receiving additional spectrum rights. (Yes, but many incumbents would actually *lose*, as new competition was unleashed.)

The authors proposed to end the "giveaway" by making the incumbents bid for the extra rights. The problem was that the new rights complemented those already held by the licensee. (Controlling two adjacent blocks of spectrum is generally much preferred to accessing the same bandwidth divvied up on separate bands.) Incumbents, able to project higher returns than newcomers, would easily outbid rivals, ensuring that few bidders would bother to participate. Prices would be low and the windfalls would remain.

Rothkopf and Bazelon's solution was for the regulator to auction new rights in batches. One hundred TV stations might bid, simultaneously, for the right to switch businesses, using spectrum flexibly. But by withholding some rights—perhaps only fifty would be offered—the FCC would force incumbents to bid against one another. Fifty stations would pay top dollar and win the option to supply more valuable services.

But the other fifty stations would be stuck with old license terms. Legacy spectrum restrictions would continue to crush innovation. The dollars gained in higher bids might well be each worth $1.33 to society, by avoiding inefficiencies of standard tax exactions, but the continuing rigidities (for auction losers) would cost society far more, wiping out any possible gains.[43] Rather than skimming additional revenues, the government would be locking in historical inefficiencies. Distorting wireless competition is one of the worst ways to raise government revenues.

The Counterrevolution in Unlicensed Spectrum

Just as FCC auctions were (finally) commencing in 1994, a grumbling was heard. George Gilder—supply-side champion, proud member of the digerati, and featured contributor to *Forbes ASAP* and the *Wall Street Journal* editorial page—turned sour. The auction, he railed, was "a major obstacle to the promise of PCS because "new radio technologies are emerging that devastate its most basic assumptions."[44] Those premises were that licensees owning flexible-use rights would harbor the spectrum resource, invest in its bounty, and deploy it wisely.

Gilder saw that approach as a "protectionist program for information smokestacks and gas guzzlers." Science had improved. It was no longer necessary that "every transmitter be quarantined in its own spectrum slot." New radios could tune across not a narrow dial but a vast one. New technologies departed from the use of high-powered signals traveling narrow, well-defined channels, doing the reverse: going "wide and weak."[45]

Gilder's attack on the liberalization program picked up steam when wi-fi routers—the now-ubiquitous radios that extend home or office broadband connections via wireless local area networks (WLANs), thus linking computers to the Internet—came to market around 2000.[46] Wi-fi uses unlicensed frequencies. In these spaces, any approved device can be deployed, with no license (or permission from a licensee) required. Wi-fi equipment accesses the 2.4 GHz band, where 83.5 MHz is set aside for unlicensed use, and the 5 GHz band, where some 555 MHz is allocated.[47]

These bands are often called "spectrum commons," a marketing term that clashes with standard legal usage. A commons is a resource owned jointly by private individuals, typically its users, who make and enforce rules to guard against "tragedy of the commons." Elinor Ostrom won a Nobel Prize in economics in 2009 by showing how such properties can thrive. In a key chapter of her classic 1990 book *Governing the Commons,* she examined group-owned resource arrangements that lasted more than one hundred years ranging from grazing lands in the Swiss Alps to water-sharing systems in Spain and the Philippines.

Each commons included mechanisms to restrict use; none was "open access." Governing councils created rules to exclude nonmembers, protecting against asset dissipation, along with enforcement mechanisms to impose those rules. These systems preserved natural resources to further the owners' interests.

Unlicensed bands are neither "open access" nor "commons." They are administrative allocations in which regulators set rules to mitigate congestion. Chief among these rules are power limits and technology prescriptions. Sometimes there are additional constraints, such as the indoor-only restriction that the U.S. applied to unlicensed 5.15 to 5.35 GHz frequencies until 2014, or bans once imposed in Britain and Mexico on commercial hotspots using the 2.4 GHz band.

An important episode in U.S. wireless history illustrates the political processes involved, with a twist: a surprisingly happy result.

In the late 1970s, the FCC received a directive from the Carter administration to find some wireless technologies suppressed by regulation, and to free them. The initiative originated with Alfred Kahn, then serving as "inflation Czar" in the Carter White House. The story is well told by Michael Marcus in his 2009 essay "Wi-Fi and Bluetooth."[48]

Marcus arrived at the FCC after earning a Ph.D. in electrical engineering at MIT and doing a stint in the U.S. Air Force. Along the way, he learned a bit about a shadowy technology called "spread spectrum." The idea dates to efforts during World War II to disguise Allied radio traffic, a campaign joined by the actress and G.I. pinup model Hedy Lamarr, who also had serious talent as an electrical engineer. Lamarr scribbled out her cutting-edge ideas for the U.S. Army, granting the Allies free use of her patent. While her specific technique never made it to market, her inventive genius, demonstrated in her description of a "secret communications system," earned her U.S. Patent 2,292,387 and a place in the National Inventors Hall of Fame.[49]

Spread spectrum abandons high-power transmissions on narrow channels dedicated to a particular link, and instead casts coded messages "wide and weak." Receivers scan across the dial, matching codes and reassembling the communications sent to them, ignoring other signals. To anyone without the codes, the transmissions appear simply a jumble. MIT's Claude Shannon, known as the "father of communications theory," likened the approach to what the human brain does at a cocktail party: it focuses on one particular conversation and disregards the din. As one writer describes it, if the older method of "frequency division" were used at the party, "each one-to-one conversation would take place in a separate room" on "your assigned frequency channel."[50] There could be no more conversations than there are rooms. With code division, many individuals can populate each room, so long as they're speaking different languages and not yelling too loudly. Potential data flow increases—by orders of magnitude.

In 1979, Marcus received the mandate to seek and destroy some "anachronistic technical regulations."[51] He noted that unlicensed bands had been authorized since 1938, with wireless remote controls being among the first uses. While power limits tended to keep applications very local, such bands could host spread spectrum technologies. But the FCC's rules prohibited such a use. Marcus had found his target for reform.

The policy initiative eventually made it to a Commission vote in 1985. Now headed by the Reagan-appointed Mark Fowler, the FCC still claimed to favor the open competition that the "liberal Democrat" Fred Kahn had championed.[52] Against strong lobbying pressure from major radio manufacturers and the U.S. Defense Department, the 900 MHz, 2.4 GHz, and 5.8 GHz unlicensed bands were made to welcome spread spectrum technologies under minimal rules, which included a one-watt power limit.

Wireless local area networks were not on the minds of 1980s deregulators, and the wi-fi standard did not appear until more than a decade later. But the permissive rules made deployment of these radios possible. Wireless local network devices were launched when broadband connections via cable modems and digital subscriber lines (DSL) began connecting residential customers to the Internet in the late 1990s. First came Apple's Air-Port in 1999. Routers with the wi-fi standard—for "wireless fidelity," a marketing term—started shipping in 2000. They were a hit with Internet users who had been using wires to link computers to their broadband connections but could now go cordless. As high-speed networks replaced slower dial-up services, wi-fi took off. And U.S. success led to international imitation. Chipmakers like Intel embedded wi-fi radios in the processors powering computing devices worldwide.[53]

Wi-fi equipment sales did much better than Marcus's career at the FCC. While he later received international recognition—the Institute of Electronics and Electrical Engineers (IEEE) gave him its Public Service in the Field of Telecommunications Award in 2013 for "pioneering spectrum policy initiatives . . . for applications that have changed our world"[54]—Marcus was demoted. He went from supervising a staff of thirty-five to sharing an office with four others in the FCC's enforcement bureau, the American version of Siberian exile. As he saw it, the motivating factor was industry backlash from the 1985 ruling he spearheaded.

Like many a banished revolutionary, Marcus left a legacy that would trigger its own rebellion. As spread spectrum wi-fi radios began to dot the land, Gilder's kickback on exclusive spectrum rights gained allies. It was curious timing. Spread spectrum had taken the world by storm before the first wi-fi-connected notebook ever downloaded a cat video—using spectrum licensed liberally.

In 1959 Irwin Jacobs, a student of Claude Shannon, began his academic career as an assistant professor of electrical engineering at MIT, moving to the University of California, San Diego, in 1966. In his spare time he founded a company, Qualcomm, with fellow wireless tech visionary Andrew J. Viterbi. Both would later have engineering schools named after them—Jacobs at UCSD and Viterbi at the University of Southern California. Qualcomm developed spread spectrum techniques in the early 1990s, using a code division multiple access (CDMA) system that was a direct rival to the 2G cellular standard, called GSM, set by European governments and favoring European suppliers such as Alcatel, Siemens, Ericsson, and Nokia. CDMA did not dominate in 2G, where GSM took 80 percent of the market by exploiting the advantages of first-to-market and government standard setting, but in 3G wireless broadband.

As wi-fi radios became popular, a tech meme developed that itself tuned out background noise—the millions and then billions of mobile subscribers who were enjoying spread spectrum technology via licensed spectrum—and heard only the conversation desired. That airwaves could be productively used when the regulator provided governance rules (protocols, power limits, and basic technical standards) seemed to reject the Coasian notion that exclusive rights were "best practice."

We can see how the argument against the liberal license paradigm jumped to the mainstream by looking at the *Economist*. The magazine gave its December 2003 Innovator Award to Ronald Coase for illuminating the efficiency of "private property rights over spectrum."[55] But by the following August it had relegated that approach to the dustbin:

> The old mindset, supported by over a century of technological experience and 70 years of regulatory habit, views spectrum—the range of frequencies,

or wavelengths, at which electromagnetic waves vibrate—as a scarce resource that must be allocated by governments or bought and sold like property. The new school, pointing to cutting-edge technologies, says that spectrum is by nature abundant and that allocating, buying or selling parts of it
will one day seem as illogical as, say, apportioning or selling sound waves to
people who would like to have a conversation.[56]

The *Economist* had something Gilder did not: a data point. Wi-fi adoption
in businesses, college campuses, and broadband-connected homes seemed
to bolster the argument that exclusive spectrum rights might block superior approaches. Stanford Law School's Lawrence Lessig (later of Harvard)
dubbed the uprising a "counter-revolution" springing not from "any opposition in principle to 'property' . . . [but] from the success of the most important spectrum commons so far—the 'unlicensed' spectrum bands that
have given us Wi-Fi networks."[57]

Yale Law School professor Yochai Benkler (later of Harvard) pressed the
case. "The most immediate debate-forcing fact is the breathtaking growth
of the equipment market in high-speed wireless communications devices,
in particular the rapidly proliferating 802.11x family of standards (best
known for the 802.11b or 'Wi-Fi' standard), all of which rely on utilizing
frequencies that no one controls."[58] The Wharton School's Kevin Werbach
challenged Coase head on: "The property approach made sense in 1960,
but is now questionable."[59]

The zeitgeist rolled over a Republican FCC. During much of this period,
the Commission was led by Michael Powell, a sharp attorney with impressive credentials both in military service and as an antitrust official in the
Justice Department. Appointed to head the FCC by President George W.
Bush, Powell quickly convened a Spectrum Policy Task Force. Its November 2002 report evinced a policy pivot. "When Powell took charge," wrote
Lessig in late 2004, "most thought the FCC would quickly launch massive
spectrum auctions. . . . But Powell's FCC quickly sabotaged this idea, in
part because technologists pushed him to see that spectrum is not like land:
that perhaps the best way to allocate spectrum is to share it." This, Lessig
noted, "has confounded liberals and free market purists."[60]

The Bush-Powell approach stalled auctions while it explored alternate
paths. The Spectrum Policy Task Force's prescription was that "the Com-

mission should pursue a balanced spectrum policy that includes both the granting of exclusive spectrum usage rights through market-based mechanisms and creating open access to spectrum 'commons,' with command-and-control regulation used in limited circumstances."

Even as it documented the shortcomings of "command and control," however, the Task Force embraced it. And not just on limited margins, but broadly. How would the Commission evaluate competing plans to achieve the properly "balanced spectrum policy"? By administrative processes, examining each allocation case by case, pretty much as Herbert Hoover, Senator Dill, and the National Association of Broadcasters had envisioned it back in 1927.

Alternative policies existed which would get more bandwidth into the market and allow competitive mechanisms to choose between the various options. Yet these approaches went unexplored—all the more curious given that the agency's two top spectrum policy experts were actively studying those alternatives. In November 2002, Evan Kwerel and John Williams released an FCC working paper detailing a means for reallocating a vast chunk of dead spectrum in prime bands (438 MHz below 3 GHz) to liberal licenses, a "big bang" that would more than double the extant mobile allocation. They noted that this flood tide of spectrum could raise all wireless boats, even those floating unlicensed:

> Future expansion of dedicated spectrum for unlicensed use could be obtained through negotiation between the manufacturers of such devices and spectrum licensees. One possible arrangement would be for a licensee or group of licensees covering a particular band throughout the U.S. to charge manufacturers a fee for the right to produce and market devices to operate in that band. . . . Competition between licensees would ensure that fees reflect the opportunity cost of the spectrum.[61]

The arguments would lie dormant for years.[62] In the interim, Powell— who boasted, "I love my DSL, but I love my Wi-Fi more"[63]—pushed the commission for more of what he thought he loved. That the products were complements, that wi-fi service augments the DSL (or cable or fiber) connection, and vice versa, escaped notice. But the deeper point is that a happy customer experience does not suggest that rearranging input markets produces additional units of happy. That we enjoy wi-fi (or baby monitors or cordless phones or remote controls) does not guarantee that more MHz should be saved for unlicensed bands, any more than loving *Seinfeld* or *Under the Dome* implies that more spectrum should be allocated to TV broadcasting. Spending on TV sets in the United States, about $18 billion in 2014,[64] continues to dwarf expenditures for WLAN equipment, about $5 billion globally.[65] Neither statistic makes a compelling case for additional spectrum, or for locking in what is already set aside. Nonetheless, it is exactly the argument that TV broadcast station owners employ to oppose liberalization—the "TV band valuation fallacy."[66]

Super technologies, such as spread spectrum, open up more intense spectrum uses, but *more* does not equal *unlimited*. Trade-offs continue and choices must be made. The Stanford electrical engineer James Spilker, Jr., explains that CDMA allows "multiple signals [to be] transmitted in exactly the same frequency channel space with limited interference between users, if the total number of user signals . . . is not too large."[67] Spectrum policy engineers Charles Jackson, Raymond Pickholtz, and Dale Hatfield agree. "Spread spectrum is a great technology . . . [but] it is not good enough to make the problem of interference go away."[68]

Of course, there are ways to deal with interference, squeezing value out of communications while mitigating its more damaging effects. One is to delegate control of given frequency spaces to competing parties to organize their use. Another is to have regulators set aside bands and permit non-exclusive (unlicensed) access, with regulators "impos[ing] constraints on how spectrum is used,"[69] as the former FCC Chief Technologist Jon Peha puts it. Either way, more bandwidth tends to ease those limits and accommodate growth.

The Bush-Powell FCC tacked hard toward unlicensed spectrum. Its flurry of activity was positive, in the sense that the Wireless Craze was blowing strong, even in Washington. But by setting spectrum-sharing rules and fixing allocations, the FCC retained the authority to structure markets. Competing claimants did not seek to outbid rivals but to outlobby officials. And in distributing nonexclusive usage rights, the Commission created disparate and uncoordinated incumbents whose future cooperation could be extremely difficult for either regulators or entrepreneurs to enlist when they wanted to deploy yet newer technologies or business models. These dynamic costs were almost completely ignored.

Among the leading unlicensed initiatives in the Bush 43 years were:

- February 2002: Ultrawideband (UWB) devices are granted "underlay" rights allowing them to access spectrum between 3.1 and 10.6 GHz, but only at extremely low power levels.[70]
- December 2002: The Commission, citing unlicensed services as a "tremendous success," proposes using TV band "white spaces"— the unused frequencies between station contours—for unlicensed devices.[71]
- November 2003: The FCC allocates an additional 255 MHz for unlicensed devices in the 5 GHz band (5.47–5.725 GHz), bringing the total to 555 MHz.[72]
- March 2005: The FCC dedicates 50 MHz at 3.65 GHz for unlicensed devices.[73] Users are required to register with the FCC, but authorizations are nonexclusive and unlimited. Radios are mandated to embed a "contention-based protocol"—for example, "listen before talk."[74]

This spasm of regulatory activity was a departure from the past. "The word 'unlicensed' does not appear in the November 1999 FCC press release announcing its comprehensive Spectrum Policy Statement," observed the Wharton School's Kevin Werbach, and the "2000 Spectrum Policy Statement extolled the virtues of market forces in spectrum policy, a code word for property rights."[75]

The policy shift was aimed at duplicating the success of Marcus's 1985 reforms. But that effort, in spaces already allocated, relaxed uneconomic restrictions. Now the FCC was requisitioning new bandwidth, planning markets, rejecting alternatives.

The FCC's most important unlicensed initiative under Powell may have been the TV band "white spaces." These frequencies, idle for decades, resulted from a TV plan that wasted wads of bandwidth and froze markets in the death grip of 1941 technology (the year the analog TV transmission standard, which ruled the airwaves by FCC mandate until 2009, was set).

These airwaves were a particularly attractive target for reallocation. Put up for auction as liberal licenses, the rights would have fetched tens of billions of dollars in license revenues. The consumer surplus generated in addition would have produced hundreds of billions more. TV stations using the channels could have been easily relocated, or moved entirely to cable and satellite platforms. Instead, regulators chose to preserve TV broadcasting and sprinkle unlicensed use rights around them, authorizing low-powered technologies.

Beginning in 2002, the FCC crafted plans for devices, predicting that they would create "significant benefits for economic development . . . providing additional competition in the broadband market."[76] The promise was for "Super Wi-Fi," or "Wi-Fi on Steroids,"[77] given that the signals traveling through TV frequencies possess favorable propagation characteristics, poking through walls while supporting broad geographic reception.

But neither the advance of technology nor the innovation in unlicensed allocations eliminated hard choices. For instance, the Wireless Internet Service Providers Association (WISPA), an organization of hundreds of mostly "small entities" supplying fixed broadband service using unlicensed bands,[78] sought to exclude consumer devices from the "commons":

> WISPA is opposed to any use of the Whitespaces for personal portable devices at this time. . . . We do NOT wish to see a spectrum issue similar to the current 2.4 GHz WiFi band . . . all but useless for large-scale, wide area deployments. . . . We believe that personal portable devices, especially in urban and suburban markets, would be best left to the higher frequency bands.[79]

Such pleas, while not fully heeded, drove regulators to place special restrictions on mobile white space devices, giving them less power and more interference-avoiding responsibilities. These constraints take their toll: bandwidth effectively shrivels, device costs rise, batteries drain. More than a decade on, the TV white spaces still sleep. While FCC rules were released to great fanfare in 2010, as of May 2016 there were but 597 fixed white space radios registered for use anywhere in the United States. (The radios must be listed on FCC-approved databases.) Meanwhile, there were zero mobile radios approved for sale (rules for such models are highly restrictive). Wi-fi on steroids has not yet threatened broadband networks or, indeed, made any impact whatever.

Yet even if it did, would the resulting services provide as much value as their spectrum opportunity cost? In its policy enthusiasm for unlicensed spectrum, and its attendant judgment that existing TV broadcasts should be protected, the government answered this question without seriously considering, let alone permitting, wireless models to compete for consumers. This replaced a market test with administrative mandate, short-circuiting key feedback loops. Even when the chosen approach visibly flops, there is no mechanism—short of a new, open-ended rulemaking—to change course. Indeed, regulators have yet to acknowledge failure; in the United States and around the world, many policy makers continue to herald the TV white space initiative as a public policy prototype.

Brave new worlds are often forecast in the political spectrum, but as Yogi Berra might have said, it is tough to make predictions, especially about the future. When they were authorized by the FCC in 2002, ultrawideband (UWB) technologies were deemed revolutionary and "received lots of favorable press from high-power technopundits."[80] These applications utilize extremely low power levels and ultrawide frequency spaces, taking the spread spectrum idea to its next level. It was presumably disruptive. An editor of the tech magazine *Red Herring* wrote: "On paper, [UWB] smokes the Bluetooth Standard." A 2002 article in *Scientific American* described UWB as a "wireless data blaster."[81] Whatever smoked, there was no fire. UWB was not a serious entrant into the wide area space, and in the personal-area-network rivalry that developed, Bluetooth crushed the upstart.[82] The best that can be said is that the UWB spectrum allocation, with its exceptionally low power limits, leaves a tiny footprint and may therefore cost little.

Not so with the FCC's 3.65 GHz allocation. The agency predicted in 2005 that issuing nonexclusive use rights for this band would "encourage multiple entrants [and] . . . stimulate the rapid expansion of broadband services— especially in America's rural heartland."[83] Fixed wireless was to provide cable operators and phone carriers a run for their money in residential and small business data services, invigorating competition to incumbent Internet service providers. In the end, the FCC rules proved a dud. The agency's

Table 16.1. High-Speed Internet Subscribers by Category
(Year-end 2014, thousands)

Technology	Dec. 2010	Dec. 2012	Dec. 2014
Total Fixed	182,065	262,564	321,305
DSL	31,637	31,106	28,612
Cable Modem	45,334	51,646	56,301
Fiber	4,993	6,733	9,180
Satellite	1,176	1,454	2,006
Fixed Wireless	587	777	988
Mobile Wireless	97,544	170,053	223,495

Source: Federal Communications Commission, *Internet Access Service, Status as of December 2014,* Industry Analysis and Technology Division, Wireline Competition Bureau (March 2016), Table 13.

data show that, at year end 2014, just 988,000 subscribers received "fixed wireless" broadband, the one category of service where some systems (perhaps not most) rely on unlicensed spectrum. This compared with 223 million "mobile wireless" customers, all served by networks dependent upon cellular licenses.

The 5 GHz unlicensed allocations doubled during the early Bush-Powell years.[84] While wi-fi and Bluetooth would continue their rapid growth in the ensuing decade, the additional allocations would play a modest role. In 2011, as a *PC World* reporter noted, "most existing Wi-Fi equipment operates in the crowded 2.4 GHz band."[85] Shipments of 5 GHz routers that deliver faster downloads by gobbling more bandwidth—including that set aside in 2003—started in 2012. In 2013, they accounted for 4 percent of global wi-fi router shipments. While they were ramping up quickly, they were not yet delivering "gateway-free access to computer networks" as a substitute for "commercial service providers,"[86] as had been touted in the FCC's 1997 allocation.

That may change. Ultrawideband technologies may make a comeback. The TV white space devices, or even the unlicensed 3.65 GHz radios, may one day support popular services. The system defect is that regulators are making the fundamental spectrum use choices. When allocations do not deliver what they predict, it causes the spectrum allocators no pain. Were opportunity costs borne by those who tout one approach, excluding others, decision-making incentives would improve.

Liberal licenses sold to high bidders have fared well in this regard, revealing the competitive forces available to help achieve the tasks at hand. In September 2006, the FCC auction of advanced wireless service (AWS) licenses generated some $13.9 billion for access to 90 MHz of spectrum. The top bidder was T-Mobile, by then a sad fourth in the national carrier rankings and slipping further due to its lack of a 3G network. That, in turn, was due to the decadelong spectrum drought. To catch up, it paid some $4.2 billion to the government, obtaining rights to about 25 MHz of nationwide AWS spectrum. It immediately announced that it would invest an additional $2.7 billion to build out 3G services. The new network was launched in New York City in 2008. Soon, the national T-Mobile network was vibrant and was a "disruptive competitor" in mobile markets, forcing better deals for customers.[87]

The 700 MHz auction in March 2008 also brought network upgrades. About $19 billion was collected for liberal licenses allotted 52 MHz in the space formerly occupied by channels 52 to 69 in the UHF TV band. AT&T and Verizon, the top winners, quickly translated their new bandwidth rights into 4G deployments, faster and more capacious than existing services. Sprint secured its 4G bandwidth by contracting with (and later buying) Clearwire, a broadband operator that had purchased thousands of 2.5 GHz spectrum rights in secondary markets, creating an estimated 133 MHz of available bandwidth. Those frequency spaces had been locked into "spectrum tragedy," as rights were highly fragmented and overly regulated, but they became far more productive via gradual liberalization.[88]

By 2014, millions of U.S. customers were enjoying next generation long term evolution (LTE) mobile services. In fact, no "4G" licenses were issued by regulators; the innovation was accommodated by the 160 MHz of spectrum allocated to new, broadly permissive, licenses sold at auction (mostly) between 2006 and 2008.[89] This allowed U.S. networks to form relatively quickly, without permission from regulators seeking to divine the "public interest."

The new growth triggered widespread innovation, and unleashed at least

$100 billion in annual consumer surplus. This can be inferred from the January 2015 prices received by the FCC in Auction 97. Winning bids for licenses, allocated 65 MHz, totaled over $41 billion. Economists have found that consumer surplus on an annual basis is probably equal to or greater than the present value of producers' surplus, a magnitude approximated by winning license bids.[90] The bandwidth released earlier is likely to be valued at least as highly per MHz as the bandwidth to come later (resource values, per unit, tend to decline as supply increases). And, here, much of the earlier access rights (for 700 MHz) constitutes prized "beachfront" spectrum with favorable propagation characteristics. Adjusting for the bandwidth differences, 160 MHz v. 65 MHz, implies that the dollop supplied in 2003–2008 was soon generating about $102 billion in annual consumer gains.

Asymmetrical Hype

No other newly allocated bandwidth comes close. That is not to say that there are not highly productive, even mission-critical wireless applications —in the military, public safety, satellite television, or supporting local area networks—supplied by other bands. But nowhere else does additional spectrum spur—according to capital inflows, infrastructure creation, and consumer purchases—such valuable new economic activity.

There is strident debate over the policy implications that follow from this assertion. My claim should be subjected to strict scrutiny. The case for more liberal licenses, for reasons given above and below, does pass that test.

But the most interesting discussion is comparative. Unlicensed spectrum has its own deregulatory success story. Harvard Business School economist Shane Greenstein, in his recent (and highly compelling) book, *How the Internet Became Commercial,* underscores the importance of reform in making wi-fi possible. "The key insight," he writes, "seems too simple to be so profound: flexible rules would enable commercial firms to put many options in front of users, and users would choose which applications gave them the most value." Due to freedom in the marketplace, many existing applications were ignored while "the vast majority of use migrated to an application [consumers] liked more—namely, Internet wireless access."[91] In essence, two competing liberalizations now vie for mindshare in the tussle over how to distribute more bandwidth access in the political spectrum.

While not all radio spectrum rights should be exclusively (or nonexclusively) defined,[92] allocations for liberal licenses and those for unlicensed bands tend to conflict. Different rules affect wireless services and the firms that supply them. That is why there is contentiousness. And the degree of flexibility afforded markets—where public policy supports the ability of consumers to migrate to applications they like more—is a key consideration in how the choice between the rival models plays out.

Yet the public debate is riddled with mischief. The standard approach features studies commissioned by rival interests. These present calculations of the social value of spectrum in alternative modes. Of course, the self-interest of corporate sponsors nicely predicts the direction of slant, but confusion over the basic economics of spectrum allocation allows the variance of the estimates to be so large.

The TV band valuation fallacy demonstrates that just because certain popular wireless applications exist, the bands allocated to such activities are not necessarily well used or wisely regulated. The overwhelming consensus, endorsed by the FCC in its 2010 National Broadband Plan, is that the 1952 TV allocation table is obsolete. Better means are available to distribute video than the old terrestrial broadcasting scheme, including popular networks subscribed to by the great majority of viewers—cable TV, satellite TV, and broadband Internet. Meanwhile, the opportunity to shift the use of TV channels to enhance mobile networks is enormous, perhaps worth a trillion dollars or more.[93]

Nonetheless, the broadcast TV industry "documents" the purported social value of off-air transmissions, seeking to protect current allocations. It cites $109 billion in TV set investments (by households) that would be "stranded" should the government allow "TV spectrum" to be used for other services.[94] It estimates that liberalization would endanger an asserted $732 billion in annual GDP gain.[95] These are fabulous numbers. In fact, no investment in TV sets—or flat-screen panels—would be stranded by a migration of TV band spectrum. Virtually no one would notice, so long as cable, satellite, and broadband Internet connections continued to work. As

for the postulated gains to the economy, they are actually negative. Terrestrial broadcast frequencies have substitutes for the delivery of broadcast TV
that are so effective that when ABC TV station owner Disney has valuable
content like ESPN, it offers it to customers not via broadcast but over cable,
satellite, and Internet. Meanwhile, TV spectrum actually has a huge opportunity cost. Were it free to migrate to support mobile networks or other services, it would generate far higher benefits for society. In offering annual
estimates of three-quarters of a trillion dollars in TV band spectrum value,
the broadcasting industry performs a useful public service in calibrating
the level of hype that routinely swirls in the political spectrum.

Indeed, competing "spectrum valuations" are regularly commissioned
by companies and trade associations to purportedly show how much better
the world would be were regulators to pump more bandwidth into their
business models of choice. Just as with the TV spectrum valuation fallacy,
estimates of the value of unlicensed bands attribute all gains from given
services (say, wi-fi or Bluetooth) to the particular spectrum set aside for its
provision. As shown in a recent paper, the key policy issue is not that the
social values—generally from about $50 billion to $150 billion annually—
attributed to unlicensed bands are dominated by the gains seen to accrue
from mobile networks largely relying on liberal licenses, generally about
$200 billion to $600 billion annually. It is that the estimates are themselves focused on the wrong margins for determining spectrum policy.[96]

In a recent patent infringement case a U.S. federal court undertook to
calculate the economic value of the gain produced by the wi-fi standard. In
that proceeding, the judge criticized the method by which a plaintiff, suing
for damages, inflated its claim. The firm had added up the sales of myriad
devices that embedded wi-fi functionality, and attributed large, arbitrary
shares of their retail prices to the wireless feature. "By this reasoning, the
court wrote, "the feature factor of the radio in a car that both transports its
occupant and plays radio would also be 50%, an absurd outcome."[97] Indeed, the approach is used in unlicensed spectrum valuation studies (particularly with respect to wi-fi offloads in mobile phones), and leads to valuation estimates orders of magnitude larger than what the court found to be
reasonable.

Even if one accepted that there were wildly large gains from—say—TV band spectrum, does that mean that more should be allocated just so? Or that markets could not do better to supply broadcast TV services using spectrum allocated differently—say, with liberal licenses, or wires, or satellite frequencies? Not only are the estimates highly imperfect and subject to gaming, they do not answer the relevant question: how can spectrum rights be defined and distributed so as to create the ultimate greatest good?

One great advantage of putting bandwidth into service via liberal licenses is that it enlists the tools of competitive economic forces to reveal the options available for competing investments. Some of the most intense current disputes in the political spectrum rage not between generic licensed and unlicensed claims, but between different flavors of unlicensed. Take the jockeying in the 5.9 GHz band. In 1999, the FCC set aside a wide swath, 75 MHz, for automobile safety. The protected applications include vehicle informatics for crash avoidance and driverless cars. But companies like Comcast (cable TV) and Google (Internet search) would prefer that the frequencies be used for wi-fi, extending adjacent bands that are already used for that purpose. They argue that the auto apps have taken too long and have yet to really take hold. Car companies like General Motors counter that they have invested heavily on the assumption of bandwidth availability, that the new technologies will soon take off, and that lives will be lost if the government changes the rules.[98]

What to do? Imposing an edict is a stab in the dark, and the lunge suffers from the traditional limitations of the centralized allocation system.

An auction would help, as some smart FCC economists have proposed.[99] If the parties put their financial resources where their "public interest" arguments were, it would clarify the price of doing one thing versus another. It could also produce some magical thinking on how to compromise and produce the best of both services, perhaps by sharing the band in an economical way. But so long as there is no effective ownership right, and there are only administrative choices, the process will be confusing and long, frustrating efficiencies that might otherwise emerge.

There are high opportunity costs in mandating allocations that cannot be easily undone by market participants. Greenstein writes, "In comparison with the old command-and-control rules [for unlicensed spectrum] . . . the movement between uses occurred more quickly." Surely. But the inside-the-bureaucracy point man who fought for the better outcome, Michael Marcus, did not get rich, he got demoted. Moreover, many unlicensed bands, with their widely dispersed, nonexclusive use rights, have thwarted an efficient migration to advanced services—as with Unlicensed PCS, or TV White Spaces, or LightSquared's attempt to build a new national 4G network using spectrum previously allocated for satellites—precisely because vital forms of economic coordination were made impossible. That part of the policy trade-off is nowhere seen in the "social value" estimates.

The Margins That Count

What matters is how a change in spectrum allocation will perform.

The *Economist* magazine observes that 1G (analog voice) came in the 1980s, followed by 2G (digital voice) in the 1990s, 3G (high-speed data) in the 2000s, and 4G (much faster broadband) in the 2010s. "Each generational change brought new frequency bands, higher speeds and greater emphasis on streaming data rather than simply transmitting voice."[100] Just so—but with opposite causality. It is not the new generations of mobile wireless that have brought new frequencies but the reverse. When new frequencies have become available in the form of liberal licenses, innovation has leapt forward. When U.S. companies joined the auction for AWS-3 licenses in 2015, it was compelling evidence that no other resources—unlicensed bands, existing licensed allocations, beams from space satellites, fiber-optic conduits, or anything else—would better supply future communications applications than the 65 MHz of de facto spectrum ownership rights for grabs. Plus $41 billion.

We face a paradox. Reforms had pushed regulators back, allowing greater scope for wireless technologies and business models to productively deploy radio spectrum. Some of the key advances for flexible use had come in unlicensed rules that pared back restrictions. The resulting enthusiasm led to demands for the government to set aside additional frequencies to unlicensed bands. But this led regulators right back into "command and control" mechanisms where administrators, not consumers or entrepreneurs, determine allocations across business models. The roadblock becomes general when each new slice of potential bandwidth is subjected to a public

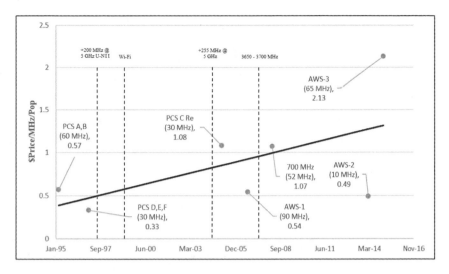

16.1. The average winning bids for U.S. mobile licenses, sold in FCC auctions, 1994–2015. Prices are displayed for the major license auctions (excluding those where bids went uncollected) and are adjusted for bandwidth and population ("per MHz per pop"). Despite large increases in the availability of spectrum and the use of wi-fi on unlicensed bands, demand for exclusive spectrum rights has remained strong. The data are from the FCC website, the U.S. Census, and Thomas W. Hazlett, David Porter, and Vernon Smith, "Radio Spectrum and the Disruptive Clarity of Ronald Coase," 54 *Journal of Law and Economics* S125 (Nov. 2011), Appendix.

interest determination as to how it should be used. The path to liberalization circled back to administrative allocation.

Some champions of unlicensed allocations have bravely attempted to predict spectrum values at the margin. Gregory Staple and Kevin Werbach, arguing in 2004 that new wireless technologies were making licensed spectrum obsolete, forecast that markets would reflect this: "The spectrum portfolios of large cellular phone companies will certainly be devalued."[101] Their prediction flowed from the premise that firms will not pay for additional inputs when superior substitutes are available for free. And it can be tested against marketplace facts: are unlicensed opportunities really eclipsing those associated with exclusive rights?

While many factors affect the prices bid at auction, including license bandwidth, frequency range, whether paired bands are offered, the auction structure, and the level of the NASDAQ Index,[102] it does not appear that unlicensed allocations have caused mobile license prices to crash. Despite the rapid growth of wi-fi, as well as substantial new unlicensed allocations, carriers still bid aggressively for exclusive rights. PCS A/B licenses brought in 57¢ per MHz per person in 1995, and prices have risen since, reaching $2.17 in 2015—an increase of about 147 percent in real terms. The trend is notable particularly because the total mobile spectrum available to the market has increased more than tenfold, from 50 MHz in 1995 to about 600 MHz in 2015.[103] Market values have held up despite the advent of new technologies, popular WLAN usage, and expanded licensed and unlicensed allocations.

Past performance is no guarantee of future results. Superior organizational modes should be free to spontaneously emerge, perplexing or displacing legacy networks. Yet such progress is stymied under top-down administrative allocations. There exist practical mechanisms to get more flexible-use spectrum into the market, and to simultaneously accommodate a multiplicity of wireless use models under conditions that reliably encourage speed and accuracy in serving technological innovation and satisfying consumer demands.

17

What Would Coase Do?

If man is not to do more harm than good in his efforts to improve the social order, he will have to learn that in this, as in all other fields where essential complexity of an organized kind prevails, he cannot acquire the full knowledge which would make mastery of the events possible.

— Friedrich A. von Hayek

It's complicated.

— Jane Adler, played by Meryl Streep, *It's Complicated*

IT IS INCONCEIVABLE THAT TODAY'S spectrum allocation system will exist one hundred years from now. Its restrictions and inefficiencies are too deep; the emerging opportunities too wide. Our policy conflicts, once quiet and unnoticed, are now public and filled with rancor. Frustrations over wasted bandwidth will grow. Wireless science will only improve; the agony of the suppressed radios, networks, and applications will only increase. At some point, the political equilibrium crafted in the 1927 Radio Act will collapse.

The splits are already evident.

Irony did not die with the Bush-Powell FCC. When President Barack Obama installed his team in early 2009, a new activist vibe was in the air. Foremost among its products was the American Recovery and Renewal Act of 2009—the stimulus. Among its spending projects was $13.28 million[1] for an FCC broadband study, which was expected to show how far the U.S. had slipped in the race to provide the fastest Internet service to its citizens. As outgoing President George W. Bush had said: "We rank 10th amongst the industrialized world in broadband technology and its availability. That's not good enough for America. Tenth is 10 spots too low as far as I'm concerned."[2]

But when the study, titled the National Broadband Plan, was released in March 2010, it did not call for a massive new government program. It instead sang the praises of wireless. Mobile networks, it claimed, were disrupting the Internet landscape yet were choking on their own success. The runaway popularity of smartphones and the "mobile data tsunami" had created a desperate need for more bandwidth for mobile. To supply it, the Commission proposed allocating 500 MHz to new liberal licenses—a doubling of available mobile spectrum—by 2020.

The NBP report documented that the FCC was taking six to thirteen years to approve spectrum allocations. In the current environment, waiting a decade for spectrum would frustrate innovators, rob consumers, blunt broadband networks, and compromise the future. It was past time to open up the spigots.

An "Incentive Auction"

The primary innovation for achieving that was a two-sided auction to repurpose the TV band. Incumbent stations would indicate prices at which they would sell their licenses back to the Commission, while mobile operators bid for rights to use the vacated frequencies. In a flash, spectrum markets would swell. As this book is being prepared for publication in late 2016, that competitive bidding process is under way.

This strategy is rich with implication.

First, whereas Michael Powell had earned praise from Lawrence Lessig for enabling a "triumph of technology over ideology" in veering toward unlicensed allocations, the FCC under its new head, Julius Genachowski, tacked back toward Coase. This triumph belonged to economics. The FCC based its case for pro-consumer reforms on the blessings of market-based spectrum deployments.

Second, the strategy to buy back TV licenses exposed the system under which Wise Men—undaunted by Hayek's (or Meryl Streep's) warning about the limits of centralized knowledge—dictated spectrum use. Nominally, broadcasting licenses are created in the "public interest." They can be canceled (or not renewed), reconfigured, or reassigned whenever the government deems it appropriate. But with the National Broadband Plan, regulators conceded this system was a chimera. Having determined that TV band airwaves would be better employed some other way, they were yet hamstrung in making the necessary changes. Policy makers, the FCC avowed, should explore novel ways of "expanding incentives and mechanisms to reallocate or repurpose spectrum."[3]

Hence the FCC would bid for its own licenses, in what it dubbed an "incentive auction." It had floated this idea before. In 2001, it "suggested that broadcast licensees . . . might accept compensation from wireless companies in return for vacating their assignments." But Congress was not amused. Senator Ernest Hollings, head of the Commerce Committee, declared the plan "outrageous. . . . The spectrum is a public resource that is to be used according to the laws prescribed by Congress, and not based on the dictates of industry."[4]

The problem the FCC sought to solve in 2010 had been in existence far longer. As Ronald Coase and his colleagues wrote in 1962, "The most conspicuous feature of the present [FCC] allocations procedure . . . is its rigidity in the face of changing conditions. One aspect of this is the long delay which occurs before the FCC is able to make a decision. . . . As [FCC chairman, 1957 to 1960] Mr. [John] Doerfer said, 'The practical problem is taking it (the right to use a frequency) away from them, and there isn't anybody that wants to take something away from somebody and . . . pull the rug out from underneath him.' "[5]

In an important pivot, the National Broadband Plan capitulated to reality. "Public interest" regulation would be achieved through market mechanisms —in this instance, reverse auctions to help free bandwidth. While there had been intense political and ideological opposition to the "windfalls" associated with payments to incumbent licensees,[6] the incentive auction seemed to magically silence this criticism. Nixon goes to China, Obama embraces spectrum markets.

Yet plenty of other criticisms developed. Many questioned whether the "spectrum crunch" was real and whether the mobile market needed new bandwidth. One study, distributed by the National Association of Broadcasters, attacked forecasts used by the FCC for falsely (in the authors' view) "predicting a skyrocketing demand for mobile spectrum. . . . When spectrum demand forecasts are inaccurate, governments may make inappropriate policy decision[s]."[7] Marty Cooper, the "father of the cell phone," also weighed in: "We have never had a spectrum shortage and never will."[8]

When the policy debate prompts a thinker of Marty Cooper's caliber to view the issue this way, it's high time to reorient the discussion. In fact, there is a persistent spectrum shortage in input markets. Operators ready to bid billions for bandwidth are ignored, as legacy allocations quarantine them in uses worth little or nothing. In output markets, conversely, spectrum shortages (in input markets) translate into higher prices, slower data speeds, and poorer quality service. There are always workarounds to ration bandwidth: cell splitting, technology upgrades, wireline substitution, data caps, or price increases. But these are costly and become inefficient options when spectrum is artificially truncated. Consumers get less than they are willing to pay for.

These arguments are so straightforward that even the FCC voiced them. Chairman Genachowski became the most prominent "Crunchie" fighting the "Anti-Crunchies," leading the Commission to publish "spectrum exhaust" forecasts. In October 2010 it predicted that mobile carriers would need 822 MHz by 2014, about 275 MHz more than then available.

One can appreciate the exercise. The goal is laudable. But the government's numerical estimates are artificial concoctions. At best, they are FCC talking points. At worst, they are used to support traditional illiberal allocations that lock in one vision of the future on the basis of government-calculated "need."

In the NBP, the FCC promised to release 60 percent of its goal, or 300 MHz, by 2015. Little-used satellite telephone frequencies were to be allowed to supply terrestrial mobile services (90 MHz); other spectrum would be shifted from government blocks (65 MHz); and the "D license" for public safety networks, unsold in a 2008 FCC auction (bids did not hit the FCC reserve price, and the spectrum has been wasted for a decade), would be put up for sale again (10 MHz). But the biggest reallocation would come from the television band, scheduled to account for 120 MHz in the market by 2015.

The FCC's target date came and went; the goal was unmet. And the dollops that were parceled out—adding up to about 100 MHz—came from processes that far predated the National Broadband Plan. WCS licenses, auctioned in 1997, were useless for mobile services due to restrictions; an FCC proceeding to relax the rules was opened in 2007.[9] The first licenses assigned under the NBP's "hurry up offense," in the February 2014 AWS-2 auction, were created in 2004.[10] The AWS-3 licenses put up for bid later in 2014 began their reallocation in 2000.[11] Moreover, as we discuss next chapter, 40 MHz was taken off the NBP's rollout when regulators changed their mind in February 2012 about satellite licensee LightSquared.

The FCC is blessed to recognize the dysfunction in legacy spectrum allocation but cursed to repeat it. Alfred Kahn decried policy makers' efforts to micromanage the process by which, under the Telecommunications Act of 1996, competition came to local telephone service. Extensive rules were crafted to direct how telco entrants could lease existing networks as "stepping stones" to building their own competing systems. (Those steps, it turned out, led nowhere. Phone rivalry came via wireless and cable TV voice-over-Internet offerings, neither of which was regulated or a "stepping stone.") Kahn found regulators conflicted by their career interests, which encourage them to impose complex schemes as a job security program. His 1998 book was titled *Letting Go—Deregulating the Process of Deregulation.*[12]

The U.S. Transition to Digital Broadcast Television, 1985–2009

In 1997's *Defining Vision: The Battle for the Future of Television*, the late *New York Times* reporter Joel Brinkley traced the first half of the U.S. transition from analog to digital television. In engrossing detail, Brinkley described the process that led to a world-class digital standard for U.S. high-definition television (HDTV), leapfrogging the analog alternative used by Sony's pioneering system in Japan.

But the central premise of *Defining Vision* was wrong. Yes, "cunning, conceit, and creative genius collided in the race to invent digital, high-definition TV," as the book's cover teased. But the United States did not jump ahead by producing a better broadcasting standard. Today, many believe that the European, Japanese, and Chinese versions of HDTV are superior. The very discussion is beside the point, because digital video programming in the United States rides virtually everywhere over wires, satellite connections, and mobile networks, bypassing terrestrial transmissions. The shiny nugget in Brinkley's book is that the entire digital TV transition was a political ruse, a two-decade delay tactic by TV broadcasters to block the reallocation of "their" spectrum.

The threat to broadcasters first materialized around 1985, when cellular began its brisk ascent and various interests—including public safety organizations and cell phone makers—asked the FCC to allocate additional licenses. They targeted unused UHF-TV frequencies hosting channels 60–69. In order to block this, TV station owners needed a "public interest" argument to mask their protectionist agenda. "They had to show Congress and the Federal Communications Commission that they needed the channels for something, *anything*," wrote Brinkley. So they came "up with a scheme: Let's tell everyone we need those channels for high-definition television."[13]

National Association of Broadcasters President Eddie Fritts, in a December 1986 speech, declared high-definition TV the wave of the future—and steadied his lobbying aim. The FCC "appears predisposed to give Land Mobile [cellular] users the available UHF-TV frequencies," warned Fritts, eliminating "HDTV as a free, over-the-air service to the nation" before it even arrives.[14]

In 1987, the FCC put the spectrum reallocation on hold, opened a formal inquiry, and appointed an Advisory Committee on Advanced Television (ATV) Service. Broadcasters, having thwarted the "spectrum grab" by mobile, spent the next decade sabotaging ATV to avoid having to actually deploy it, as it required an expensive upgrade with no known source of extra revenue. The FCC's 1987 standards committee was just the ticket to the delay it sought. "Advisory committees are typically zoos," gleefully opined NAB Vice President John Abel. "There are so many ways to slow things down."[15]

The resulting twenty-year regulatory dance was about spectrum allocation, not technology. Even Brinkley, a Pulitzer Prize–winning investigative journalist, missed the lede. Policy makers were not about to explain it to him. They had their own history to rewrite.[16] In June 2009, when the long and winding ATV road had reached its end, Congressman Edward Markey (D-MA) declared:

> After my hearing in October of 1987, the FCC created the Advisory Committee on Advanced Television Service to investigate how the U.S. would create High Definition TV, but the panel initially pursued an "analog" standard for the nation. When researchers . . . suggested the real possibility of achieving a digital standard for HDTV, I aggressively advocated for such a switch and successfully convinced the FCC in 1990 to begin pursuing a digital standard. This was a game-changing moment.[17]

In fact, at those hearings in 1987, Markey said not a word about a "digital standard," nor did he reference "digital television." Instead, the push was for industrial policy, as the congressman (now senator) made clear:

> It's going to be jobs, plants closing, people out of work, and a large debate, and either we put the Government and labor and the electronics industry and the university together today, and we put together a policy and we understand what our strategy is, or we are going to go down the same road that we have gone with automobiles and VCR's and every other consumer product.[18]

In 1987 Japan was the archenemy, the thief of high-tech manufacturing jobs. Markey's remedy—proposed by the NAB—was to freeze spectrum in place, reserving it for a new generation of TV sets that would pull production back to America. The argument was comical even then. When the NAB put together a high-profile demonstration in January 1987, members of Congress were treated to "high definition" TV via an NHK system linked to an HD Sony receiver—all from Japan.

By the time congressmen rushed to the microphones in 2009 to take credit for the digital transition, there was no U.S. TV industry. Indeed, there was no TV set industry anywhere. Flat panel display (FPD) production—including TVs, cellular phones, computers, camcorders, and other devices—had superseded it. Moreover, Japan proved a paper tiger. In 1993, Japan hosted 95 percent of FPD output; by 2005, just 10 percent. South Korea and Taiwan were each up to 40 percent, with China at about 7 percent. U.S. workers made their last flat panel display in 1999.

Thanks to free trade, Americans grabbed their HDTVs from the most efficient suppliers. And while just 42.5 percent of U.S. households subscribed to cable TV service in 1985, at least 89 percent had cable program packages in 2009.[19] The vast majority of households took their HD content from such nonbroadcast services, "over-the-top" Internet channels, or Blu-ray disks. One policy payoff was the June 2009 "analog switch off" in which all stations assigned channels 52–69 had to switch their transmissions to lower channels. That there were enough unused channels to accommodate them testified to how much of the TV band had been wasted.

But the new digital standard forced licensees to use all 19.4 mbps allotted them for broadcast, even though most TV programs (excepting HD formats) used but a small fraction of the flow. This prevented a station from economizing by, say, transmitting just one standard signal and using the remaining five-sixths of its channel for, say, low-power in-home networking. FCC rules mandate that every signal blast out high-powered emissions across all 6 MHz. Who had inspired that bright pro-pollution idea? Why, the broadcasters, who demanded inflexibility, seeing the requirement to consume all 6 MHz as a winning tactic in the game of spectrum keep-away.

Conventional political wisdom had it that cutting off Aunt Minnie's favorite broadcast TV station would wreak electoral havoc. The mythical Minnie—old, set in her ways, and certain to vote—would not subscribe to a video service or buy a digital TV converter. Instead she would go apoplectic when her Lawrence Welk reruns disappeared and shuffle over to her local congressman's office in a state of panic.

As late as September 2008, former FCC Chairman William Kennard described the analog switch-off, then scheduled for February 17, 2009,[20] as a "train wreck waiting to happen" and warned George W. Bush's successor of another "Katrina moment."[21] Congress, terrified, allocated $2.2 billion to subsidize digital set-top boxes that would allow pre-2004 TV sets to receive the new signals. The money went out by voucher, each U.S. household eligible for two $40 coupons. These were redeemed at retailers, who offered DTV receivers at prices starting, conveniently, at about forty dollars.

17.1. DTV vouchers: subsidizing the killer app of 1951. This card, issued by the
Department of Commerce, provided a forty-dollar credit for the purchase of a
set-top digital television receiver. Two were distributed to each U.S. household that
applied for them, an effort to smooth transition when analog TV broadcasts were
discontinued after June 17, 2009. In the event, few noticed. Almost everyone was
watching cable, satellite, or broadband video.

Fear of Aunt Minnie thus drove Congress to dedicate federal funds to
keeping viewers watching the same channels where *I Love Lucy* debuted in
1951—channels that had been rendered obsolete in their old use but were
now extremely valuable in different ones. Broadcast TV was such an unap-
pealing option for most Americans that even "free" off-air receivers were
overpriced. Despite Congress dangling some $9 billion (two times $40 times
110 million households) in cash and prizes, and then issuing 64 million
vouchers to the 34 million households that applied for them, only 54 percent of the DTV coupons were redeemed. Policy makers ultimately doled
out just $1.4 billion.[22]

The transition to digital television was no Katrina. When the final deadline arrived, on June 12, 2009, hardly anyone noticed. Aunt Minnie did not
storm the capital; she yawned, flipped off her cable- or satellite- or telco-
connected TV, and took a nap.

For the billions Congress was willing to advance to stave off a revolt, it
could have bought basic tier cable, telco, or satellite subscriptions for the
last ten million U.S. households lacking such service, clearing the entire
TV band—all 67 channels (402 MHz). I proposed this in a 2001 paper,
"The Digital TV Transition: Time to Toss the Negroponte Switch." The title

refers to Nicholas Negroponte's observation that while we were born into a world where phone calls traveled on wires and TV came via wireless, we would die in a world just reversed. It's not entirely on point—satellite TV arose to broadcast video programs, while voice-over-Internet via wires is alive and well—but you get the drift.

The DTV transition was a golden opportunity for spectrum allocations to change with the times, leapfrogging legacy lock-ins. Indeed, many hundreds of stations applied to the FCC for permission to switch off their analog broadcasts early to save electricity costs (reflecting the fact that while stations did not pay for spectrum, they did pay for transmitter power). By June 2009, more than one-third of analog stations had already gone dark.[23]

Stopping the Madness: Overlays for Market-Based Reform

In the political spectrum, the digital TV transition is considered an astounding success. Leaders jockey to take credit for it. Yet it was advanced as a decoy, and worked, thwarting productive improvements in wireless for years. By the time the TV band reallocation was completed, another was past due. The number of off-air TV channels had gone from eighty-one to forty-nine, but emerging networks were still hungry. "Cellular companies are straining within their bandwidth restrictions," wrote former FCC Chief Economist Gerald Faulhaber in 2006, while "large amounts of bandwidth are currently occupied by VHF and UHF television broadcasters, even as the audience for broadcast-delivered TV shrinks."[24]

Some quietly suggested that the best plan might be no plan: just grant TV licensees full flexibility. Station owners could sell to mobile carriers, letting market forces push spectrum where customers wanted it. Yet when this idea was broached in 1987 by Lex Felker, a high-ranking FCC official, the NAB immediately shot it down. The NAB's top lobbyist told Felker: "Don't you dare award flexibility; we'll fight that to the hilt!"[25]

That harsh reaction was borne out elsewhere. On May 9, 1996, the chairman of the Senate Commerce Committee, Larry Pressler, unveiled a "discussion draft" of the "Electromagnetic Spectrum Management Policy Reform and Privatization Act."[26] He warned that, thanks to regulatory rigidities, "a vast array of new spectrum-based products, services, and technologies will go unrealized for the American people." To jettison an "antiquated model

Table 17.1. A Spectrum Warehouse: The TV Band Slowly Slims

Year	TV Channels	MHz	Spectrum Allocation Event
1953	81	486	1952 TV Table plus 1953 UHF add-on
1982	67	402	84 MHz reallocated, 50 MHz for cellular
2009	49	294	DTV transition, 108 MHz reallocated, 70 MHz for cellular via liberal licenses
2019 target	29	174	FCC incentive auction announced 2010

[wherein] the Government—not consumers—decides who uses frequencies, what they are used for, and how they are used," the draft laid out key reforms:

1. *Exhaustive allocation and licensing.* The FCC would allocate all requested bands for use, assigning licenses via auction. Existing users would be undisturbed, protected from interference.
2. *Full flexibility.* Licensees could supply any services or technologies. "Simply put, frequencies should be treated more like private property."
3. *Privatizing public allocations.* The federal government, which commanded "nearly one-third" of airwaves, would "relinquish one-quarter of its spectrum stockpile."
4. *TV band overlay licenses.* Stations would receive their digital TV licenses as established, but all vacant TV bandwidth would be allocated to overlay licenses sold at auction. These would convey primary rights to use unassigned channels, and secondary rights for TV channels in use (by broadcasters). Bargains would become possible, with overlay owners paying TV stations to share spectrum (perhaps by terminating their broadcasts). Gains would be split in a "market-based alternative to a Government mandated and dictated transition policy."

The Pressler proposal ignited such a torrent of opposition from broad-casters as to become instantly radioactive. It was never introduced as actual legislation. In November 1996, Larry Pressler (R-SD) became the only in-cumbent senator to lose reelection.

But the overlay idea was a good one. And "proof of concept" had already been established. The personal communications service licenses auctioned in 1995 were overlays that granted spectrum use rights in the 1.9 GHz band to new mobile operators. This addressed a classic holdup problem. While European countries had assigned 2G digital cellular licenses, 1989–1992, U.S. authorities were flummoxed by four thousand microwave users in the band targeted for PCS licenses (which would support 2G). The in-cumbents, while private firms, claimed to provide public safety services and threatened that, if they should have to switch frequency assignments, people would die. The standoff consumed years. Finally, the Commission sold the PCS licenses at auction, grandfathering the microwave users, but allowing the PCS entrants to pay them to move. They took the money. Im-passe resolved. No one died. In fact, with the retail cellular price declines that followed, crime rates fell—causally attributed, in part, to the increased diffusion of mobile phones.[27]

Talking people into cooperating with other claimants for radio spectrum, when it is against their institutional interest, is extremely difficult for civil servants who wield only limited tools and whose rewards for success will be nil—if not negative. Overlays allow regulators to let market forces do the heavy lifting. By the time a staffer in the FCC's Wireless Bureau can fill out a travel form requesting authorization to visit a Texas oil company that is refusing to give up its license for radio service to an off-shore drilling rig in the Gulf, an overlay owner has made the trip, given the company free satellite phones, and put the old frequency space to new and more produc-tive use.

Overlays are straightforward. The government agency issues rights to unused spectrum and secondary rights to adjacent occupied frequencies. Incumbent interests are untouched. But the new rights enable bargains tapping fresh collaborative energies. Entrants and incumbents are able to explore options for rooting out existing rigidities, unleashing higher valued services. If they can spy such opportunities, and then craft an agreeable split of the gains from trade, they walk away happy. Contentiousness over shifting spectrum use becomes a game to create lucrative innovations. Co-operation becomes profitable, holdouts costly—reversing normal administrative polarities. Consumers, getting new and improved services, are delighted.

The policy is adaptable to diverse situations. Consider advanced wireless service licenses—also overlays—auctioned in September 2006. The rules left in place scores of radio assignments for federal government agencies using AWS frequencies, from the U.S. Forest Service, the Federal Aviation Administration, and the Department of Justice to the Bonneville Power Administration (Department of Energy) and the army, navy, and air force. These institutions, which had long resisted spectrum reallocation, appeared immovable. Yet the largest AWS winner, T-Mobile, was deploying its AWS bandwidth to offer 3G service in New York City just twenty-one months after the auction—lightning-fast turnaround by FCC standards.

Policy makers supported this relocation process through the Commercial Spectrum Enhancement Act (CSEA) in 2004. The law allowed private parties to underwrite the costs incurred by public agencies in conserving radio spectrum. Before that, a company that discovered a way to make a $5 million upgrade for a government user, yielding bandwidth worth $25 million, had no way to execute the transaction. The CSEA created a mechanism.

As AWS licenses were prepared for auction, the Department of Commerce's NTIA sent a questionnaire to the federal users in the band.[28] It asked them to estimate (a) how many months it would take to switch their radio operations out of the band; and (b) how much it would cost. The initial agency responses were generally (a) *long,* many indicating six or seven years, and (b) *high.* Yet they offered a starting point. On review by the Office of Management and Budget, the total band-clearing cost was set at about $1.5 billion. The FCC then held the auction, setting reserve prices at just above the estimated relocation cost.

License winners saw relocations as urgent. Because owners were now in place—"residual claimants" who would gain from speedier movement—the six- or seven-year time frames suddenly seemed way too long. T-Mobile, with the most to gain, was most aggressive. It filed a raft of deployment plans with the NTIA, receiving many "red lights"—the bandwidth would not be available for several more years. Negotiations broke out. T-Mobile pushed to meet with agency officials, exchanging information, offering compromises that would allow spectrum sharing during a transition period. Barriers began to crumble.

The end came far faster than planned. Rather than a seven-year clearing window, 3G services were using the spectrum within two years, and virtually all of it within five. The relocation costs came to about $1 billion—$500 million below projections.[29] "At the same time, federal incumbents received modernized systems in other frequency bands," noted the FCC. "The experience . . . proves that relocation can be a win-win-win: for incumbents, for the U.S. Treasury, and, most importantly, for the American public."[30]

Under traditional reallocations, spectrum-using agencies regard the status quo as comfortable, change as risky. Commerce Department officials or FCC policy makers, tasked with pushing for more commercial spectrum use, are only academically engaged. They are not rewarded for speed or punished for letting "deadlines" slip. Nor can they go to financial markets to raise capital for new technology deployments, or pay incumbents to move on.

In AWS frequencies, for example, the Department of Justice was using discontinued analog radios. All the replacements were now digital. This should have made the transition easy: the DOJ now gets new, improved equipment. But Justice was uninterested in the opportunity; the upgrade stalled because, technically, the agency was entitled to obtain the *same* technology. The overlay owner, however, grasped the big picture and seized the moment. T-Mobile pushed the OMB to interpret the CSEA more broadly. It worked: the Justice Department was allowed to replace antiquated analog phones with better digital ones. Truth, Justice, and the American cell phone subscriber all benefited.

Mother Knows Best

Nobel-winning economist Eric Maskin explains the science of mechanism design—creating markets that perform efficiently—with a parable.[31] A mother has two children and one cake. She seeks to divide the cake into equal pieces. She knows that no matter how carefully she measures and carves, argument will ensue. Both kids will noisily object when their sibling gets "the big half."

This is a classic border dispute. The solution is well known. Long before its formalization in game theory, mom designed a mechanism to align her kids' incentives with her own. She selects one child to cut the cake; the other then chooses his slice. The cutter is in control of her choices; the chooser is in control of his. Both diligently pursue the "big half," and Mom is off the hook. It is a Solomonic solution to a thorny informational problem.

The cake cutting delegates actions to decentralized parties and, done correctly, elicits cooperation. This method also works in reallocating radio spectrum. Were Mama FCC to make the cut and award bandwidth slices, both parties would complain. Silence conveys acquiescence; loud protest is the dominant strategy. With overlays, the parties—grandfathered incumbents and new licensees with secondary rights—control their fate. Entrants bid for new rights and then are allowed to bid for complementary rights held by incumbents, who choose to stay or go—and who now have strong incentives to seriously contemplate the full consequences of either.

The 2010 National Broadband Plan, and Why It Rejected Overlays

When, in 2009, the National Broadband Plan Task Force invited suggestions, I submitted a proposal to reallocate the entire TV band using liberal licenses.[32] Overlays would be allocated all forty-nine DTV channels; the suggested policy was seven national licenses, each for 42 MHz, sold at auction. No single firm could win more than two of the licenses. Existing full-power TV stations, all 1,800 across 210 TV markets, would be untouched and could continue broadcasting indefinitely. Overlay winners, which might include TV stations, could bargain with incumbents, buying their rights and using the spectrum thus acquired for any purpose.[33]

If policy makers feared that "free over-the-air" TV would disappear and that this would harm the "public interest," there was a cheap solution: the ten million households that had not yet migrated to the postbroadcasting world (by subscribing to cable or satellite) could be extended a "limited basic" subscription, free of charge, for some years. I opposed this policy. There is as much reason to tax broadcast TV—as with the United Kingdom's annual $250 color TV license[34]—as to subsidize it. But if such a scheme were needed to free up the TV band, it would be the bargain of the century. By holding a reverse auction, where firms bid for "broadcaster of last resort" status, the FCC could select the most efficient operator(s) to supply video to ten million Aunt Minnies. Rivalry would probably drive the price to $3 billion or less.[35]

The National Broadband Plan rejected the overlay approach and chose to protect over-the-air broadcasting while targeting 120 MHz (of the existing 294 MHz) to switch over to flexible-use licenses. The means chosen for the transition, described briefly above, was an "incentive auction." The approach does elicit some important measure of voluntary support, but two key elements of coercion remained. First, TV stations that do not sell their licenses to the government are forced to accept new channel assignments. This is meant to mitigate costs by countering broadcaster intransigence:

> Incentive auctions present a more efficient alternative to the FCC's overlay auction authority, in which the FCC auctions encumbered overlay licenses and lets the new overlay licensees negotiate with incumbents to clear spectrum. These piecemeal voluntary negotiations between new licensees and incumbents introduce delays as well as high transaction costs as new licensees contend with holdouts and other bargaining problems.[36]

Second, the FCC's offer to TV stations was made one time only. Failure to exit under the FCC's terms indefinitely locks TV broadcasters into . . . TV broadcasting. Given the fading economic returns there—stations make money from producing shows seen via cable, satellite, and web distribution, but not much from off-air transmissions themselves—the option to escape was seen to facilitate cooperation. Hence the term *incentive auction*. While all auctions rely on the incentives of buyers and sellers to reveal values, this program offers both carrot and stick. In 2014, the FCC commissioned a report to increase broadcast stations' participation in the auction, emphasizing an offer they shouldn't refuse:

Unique Opportunity
The FCC has the unique ability to unlock spectrum value through its authority to repurpose broadcast spectrum for wireless use and reorganize the 600 MHz Band on a nationwide basis

- Broadcasters cannot repurpose their spectrum for wireless broadband on their own
- Only the FCC can reorganize the UHF spectrum into nationwide contiguous spectrum for wireless broadband
- A private sale of spectrum is not an option
- The FCC has no other Incentive Auctions planned or expected[37]

The rigidity of this structure is perverse, and made more so by the fact that the large majority of stations will not, under the FCC plan, be purchased. Leaving the remaining stations in a more crowded space, stranded in their grandmother's business model, eliminates most broadcasters' interest in a speedy transition.

As for overlays, it is true that negotiations between entrants and incumbents could prove tricky. The Commission has, historically, scattered TV stations across a huge block of bandwidth and endowed them with just enough spectrum rights to be dangerous. These licensees cannot deploy innovative services or emerging technologies, but they can block others from doing so.

Suppose, for example, the Commission creates an overlay right to use 15 MHz of TV band frequencies in a given market. Suppose, further, that the license is encumbered by two grandfathered TV stations, accounting for a total of 12 MHz, and the value of the 15 MHz, free and clear of those obstacles, for use supplying mobile broadband services, is $100 million. Because the value of spectrum is nonlinear—it is generally worth more, per MHz, in larger contiguous blocks—the 3 MHz of vacant spectrum (not encumbered) is worth just $10 million. Let's further assume that each TV station is worth $4 million if limited to broadcasting. With the two stations in place, the total value of the 15 MHz is $18 million ($10m + $4m + $4m).

The efficient upgrade removes the broadcasting, raising band value from $18 million to $100 million. The $82m gain from reconfiguring the band means that the overlay winner will seek to deploy it in mobile services (or, with liberal rights, anything more valuable). To do that, it must induce cooperation by sharing its gains with the TV incumbents. Haggling ensues. If one company owned both stations, the bargaining would be relatively straightforward and would likely strike a price somewhere between $8 mil-

lion and $90 million. Transactions would be neither free nor instanta-
neous, and in some instances hard bargaining might deter deals. But the
costs of delays are now borne by the participants; sacrificing profitable trading hurts both.

When different companies own the two stations, the problem intensifies:
each rival seller might hold out for the "big half." With more TV stations
the situation is worse. Efficient deals may be sabotaged by transaction costs
that result from the legacy allocation system, which has carved rights into
uneconomic slices and scattered them among unorganized parties. Now
the challenge is to put Humpty Dumpty back together again. The FCC's
modern stance is best summarized as: The TV licenses we've been distributing since 1939 have messed these airwaves up so badly that only we can
salvage the situation.

The government could resort to coercion without an incentive auction,
simply adjusting its "public interest" determinations. Indeed, the National
Broadband Plan states the official view that too much spectrum is set aside
for broadcast television. But the FCC concedes that it cannot, in any practical time frame, alter licenses without financially compensating interested
parties. Hence, the "incentive" auction.

Overlays don't have to outrun the bear. In this race between imperfect
policies, they must simply outperform the FCC's reallocation management.
I argued in 2009 that overlays would repurpose more spectrum more efficiently and faster, and that markets would thereby gain the flexibility needed
for a bountiful flow of productive adjustments. They allow decentralized,
self-interested parties to craft profitable bargains, and usher in efficiencies
observed in negotiations already observed.[38]

The NBP ultimately decided against overlays because "bidders typically
pay significantly less for encumbered spectrum."[39] This is an unproven assertion; whatever the price reductions due to encumbrances, more spectrum
rights can be put up for sale via overlays. Yet even if the claim were true, it
focuses on the wrong goal. Moving spectrum rapidly into more productive
use far outweighs the benefits of higher auction receipts. And the delays
attendant to the bargaining between overlay owners and TV incumbents
have been overstated. In fact, delays are probably longer with FCC—not
market—reallocations. This has been observed in cellular license secondary

markets where holdup opportunities are plentiful but where, nonetheless, more than fifty thousand wireless licenses have been aggregated into four national networks.

The incentive auction incurs transaction costs, too. FCC Chairman Tom Wheeler writes, "I have often defined the complexity of this multi-part simultaneous process as being like a Rubik's cube."[40] Law professor Christopher Yoo observes that "the enormous sucking sound you hear at the [FCC] is the incentive auction pulling every single member of the staff into its ambit, essentially consuming almost every bureau and every part of the organization."[41] Policy makers must determine how much spectrum is targeted for repurposing, how TV incumbents are to be relocated, how much time to give TV stations to vacate "cleared" bands, and more. Crafting such rules costs the FCC more than $100 million each year, more than one-fourth its total cash resources.[42] And these are in the tiniest category of the costs imposed. The far larger issues concern how much bandwidth is inefficiently left out of the reallocation process.

What ensues is already a fascinating experiment in public policy. To be clear, the FCC is to be lauded for tackling a major irrationality in the political spectrum, and for embracing ambitious means for the task at hand. It has tried to shift the arc of regulatory history and has assembled an able team of communications experts and auction theorists to accomplish its goal.[43] Formally kicking off on March 29, 2016, the FCC's incentive auction rages on as this book goes to press. The Commission hopes to complete the process in early 2017. But it is not too early to make a few notes.

The National Broadband Plan Task Force was established in early 2009 and published its report in March 2010. It aimed to repurpose 120 MHz of the TV Band for mobile markets via an auction scheduled for 2012 or 2013, with the transition achieved (broadcasters out, mobile networks in) by 2015. Actual results have varied.

As the incentive auction began, the agency's spectrum target was reduced to 100 MHz. But the two-sided auction requires both sides to match. The initial stage of the auction, completed in July 2016, found that $88 billion was needed to pay for broadcast TV licenses and related expenses, while just $23 billion was bid for new mobile licenses. That was not good news for those rooting for an expeditious transfer or a capacious reallocation.

The gap must close to zero for the auction to close and for new frequency use to be unleashed, and the way to achieve that is to reduce the targeted reallocations. In the second stage, for instance, the FCC set a 90 MHz goal. New bids, taking more months to collect, are unlikely to eliminate the $65 billion deficit. More stages will surely be needed.

Assuming that reverse and forward auctions do finally clear and some spectrum is allowed to shift from TV to liberal licenses, the process is far from over. TV stations will have thirty-nine months to switch off the channels being cleared. Bernstein Research analyst Paul de Sa, formerly (and then later, again) an FCC official, wrote in 2014: "It is hard to see how [TV band] spectrum gets deployed before the early 2020s."[44] The strategy to speed bandwidth into the market, improving on the standard six-to-thirteen-year delay, has been found to produce . . . about a six-to-thirteen-year delay. And not that much additional bandwidth, particularly troubling given policy makers' stated position that the incentive auction is one-time only, and that remaining TV stations are to be—according to announced policy— rigidly confined (in broadcasting) for years to come.

It is not established that overlays, the counterfactual, would have achieved more. But I believe the case is strong. They surely could have been implemented far faster, being an "off the shelf" regulatory template not requiring a new 2012 statute,[45] nor nearly so complex an auction design. Moreover, the overlays would create flexibility for future efficiencies. Conversely, the FCC's path cedes authority to administrative diktat by freezing two-thirds of the TV band in its prehistoric allocation.

As for the workability of the overlay approach, the FCC has implicitly conceded the issue. In December 2014, the Commission announced that many of the flexible-use licenses it offered for sale in the incentive auction would themselves be encumbered by grandfathered TV stations.[46] Some licenses would incur substantial "impairment," meaning that interference from broadcasts would reduce the utility of the new bandwidth licensees would gain. Regulators announced that they will bestow price discounts to compensate.[47] The rationale for the switch: if regulators had to remove all incumbents everywhere, intransigent parties would have gained great bargaining power. Accepting "impairments" relieves the blockage.

Overlays came to the rescue of their policy rival.

To its credit, the FCC conceded that other strategies beyond incentive auctions might be needed. It would be appropriate, it wrote in the NBP, to "explore alternatives—including changes in broadcast technical architecture, an overlay license auction, or more extensive channel sharing—in the event the preceding recommendations do not yield a significant amount of spectrum."[48]

It is time to dust off Plan B.

18

Hoarders Anonymous

The main reason for the misuse of spectrum is that it is usually given away for free. Nothing is more subversive of social equality than for the state to distribute a limited resource as a privilege.
—Ithiel de Sola Pool

SPECTRUM RIGHTS CONTROLLED by government agencies are endemically underused. Michael Heller notes that municipal governments, which control prime airwaves for public safety services, "mismanage what they have, using over ten times as much spectrum per call than do private cell phone carriers, while providing much less reliable service—as attested by the tragic failure of the New York City's Fire Department network in the aftermath of the World Trade Center attacks."[1]

Treating fire and police departments as buyers of communications, not suppliers, could remedy this situation. It would allow public agencies to take the best competitive offers instead of making them use homemade networks—buggy, expensive, and limited in functionality. Were they to hop onto commercial systems, public safety agencies could seize the advantages of state-of-the-art technologies and economies of scale, enjoying far superior communications for less cost. When a crisis hits and first responders must commandeer bandwidth, civilian users could be given lower-quality voice calls (modern cellular systems dynamically alter signal clarity to manage traffic flows already), be reduced to texting, or receive slower Internet access, saving capacity for emergency personnel. And by contracting with a mix of networks—wired, wireless, and satellite—first responders could ensure that their radios still work if a particular system goes down.

The greatest pushback to this approach involves the mission-critical wireless systems used by the U.S. Department of Defense. Military planners often resist transparency, both for strategic national defense reasons and to protect bureaucratic turf. While the former justification is the more compelling, both must be dealt with. We do want to place a premium on readiness; it may be worth leaving some spectrum underutilized to gain a margin of safety. At the Pentagon, planners have good reason to prepare "back-up plans for their back-up plans."[2]

But there are trade-offs, and military decision makers systematically overshoot. We saw this in the very first days of radio, when the U.S. Navy believed the military should be vested with control over the entire spectrum. Overconsumption is a predictable consequence when costs fall on others, externalities that cannot be corrected through voluntary transactions.

A 2012 report by the President's Council of Advisors on Science and Technology (PCAST) underscored the problem: "Federal users currently have no incentives to improve the efficiency with which they use their own spectrum allocation."[3] Bureaucrats reflexively hoard resources, if only to amass chits to barter later. This reprised conclusions from previous government studies done under the John F. Kennedy, Lyndon B. Johnson, and George H. W. Bush administrations.

The White House estimates that nearly 60 percent of prime spectrum is set aside for federal government use—either exclusively or "shared" with private users who are effectively barred from pursuing productive activities. Duke University economist Leslie Marx objects that the government's accounting substantially understates the amount of spectrum it consumes.[4] The public is thus denied even a rudimentary level of transparency: how much bandwidth is being utilized, and thus how much might be available for alternative employments. Stanford economist Gregory L. Rosston summarized the state of play in 2014 when he spoke of "the two main areas of spectrum users that most observers claim are inefficient: the federal government and commercial licensees without full flexibility."[5]

In 1998, the British government moved against spectrum inefficiencies by imposing administered incentive pricing (AIP). A spectrum manager sets charges, based on estimates of market value, for government agencies assigned wireless rights, in the hope that having to pay these fees will force agencies to consider the cost of their occupancy.

Perhaps well motivated, this policy has been largely a bust. In 2005, a "spectrum audit" conducted for the British government by economist Martin Cave "found that government departments and entities held approximately 50 percent of the spectrum below 15 GHz with significant evidence of underuse."[6] A 2009 audit by regulator Ofcom gave the AIP idea favorable marks but could find only three modest actions that seemed to result from the decade-old policy.[7] The Ministry of Defense released no spectrum. In contrast, in the United States (without AIP) between 1998 and 2008, hundreds of public radio links were moved out of the 1.7 GHz band to clear 90 MHz of spectrum for advanced wireless service licensees, rights issued to mobile carriers as overlays.

The British practice has been to set prices low so as not to force changes, which defeats the purpose of the incentives. Proposals exist for compensating agencies, or agency officials, with bonuses when they do manage to—through the departments they manage—economize on bandwidth. These financial rewards might motivate decision makers, but the prices are still arbitrarily set and might—if they have any impact at all—create perverse new incentives. Administrators who undermined an agency's mission so as to collect a bonus for saving on spectrum would merely reverse the misallocation problem. Of course, a system that allowed select bureaucrats to get rich by selling public assets would predictably create political backlash, again undermining reallocations and buttressing the status quo.

It is a demanding task to shift public moneys around to compensate either individuals or programs, bypassing the legislative committees that control overall budgets. The paradigmatic experiment in this realm involves lotteries for public education. Almost all U.S. states run games of chance, with the profits typically dedicated for public school budgets. But academic research shows that the lottery money does not boost educational spending. Regular appropriations are simply reduced to reflect the new funding source.[8]

Today, U.S. policy makers have latched onto the concept of administrative "spectrum sharing," an idea that is both trivial and dirigiste. It is trivial because virtually all spectrum is shared. The question is: What parties make the guest arrangements, under what rules? How that is answered determines incentives and accountability. In economic terms, the most intensely shared spectrum spaces are those hosting mobile traffic over wireless networks. Not only do millions of end users access airwaves, sharing politely, but the carriers obtaining licenses and organizing the various services contribute to the project by investing heavily in complementary infrastructure ($33 billion in the United States "to build out and upgrade their networks" in 2013 alone).[9] That the licenses held by such network operators are billed "exclusive use" is perfectly misleading. Exclusive frequency rights holders pay billions of dollars for the inputs that they hope to share, supplying wireless services to countless others.

The authors of the 2012 PCAST Report, conversely, envisioned spectrum sharing as the exclusive purview of government agents. Regulators are tasked with identifying instances of excess capacity and—while protecting existing services—working with public agencies engaged in wireless communications to promote new radio applications. Nice thought. But the recommendation simply ignores the underlying reality that "government lacks proper incentives to manage spectrum efficiently."[10]

PCAST starts with an admission: relocating public wireless systems, clearing bandwidth of obstructions, takes too long. Therefore, it reasons, the strategy of moving inefficient incumbents should be abandoned. Leave the debris in place and work around it, using more robust technologies to fend off interference. Yet this leaves the authorities who originally misdirected traffic, wasted bandwidth, and created the existing roadblocks to supervise the new additions. Neither the information they possess nor the incentives they face are fundamentally altered. It is a redo under legacy rules.

Moreover, the assertion zips right past the historical facts that bands have been cleared by more efficient means, including the use of overlays in PCS and AWS: that secondary markets have reassembled thousands of tiny FCC license shards into efficient national mobile networks; and that spectrum markets facilitate bandwidth sharing through countless roaming agreements (where subscribers to one company seamlessly access the spectrum of others), wholesale mobile operator agreements, and an expanding eco-system of M2M (machine-to-machine) wireless services. All of these out-comes are created by simple economic transactions, some parties stepping aside to let others use more bandwidth.

The PCAST Report goes backward, reviving the old strategy in which government bands are frozen and regulators must approve new wireless technologies case by case. That these potential new entrants are now to be shoehorned into existing government bands resurrects the administrative allocation model—even while blaming that model's recognized failures on old thinking: "This report argues that spectrum should be managed not by fragmenting it into ever more finely divided exclusive frequency assign-ments, but by specifying large frequency bands that can accommodate a wide variety of compatible uses and new technologies that are more effi-cient with larger blocks of spectrum."[11]

The FCC's traditional allocations have wasted wads of spectrum, and slicing rights too "finely" is one of the key sources of mischief. But the en-terprise on which the report embarks is committed to even finer fragmen-tation. That is what is entailed in "specifying . . . a wide variety of compatible uses" and then granting nonexclusive access rights to users. First, the FCC is determining what types of radio technologies will be permitted, at what power levels and with what interference rules. It then preempts market-based activity to restructure those decisions. Even when positive sum gains exist from buying out existing users, including government incumbents, those opportunities will be squandered because spectrum users hold only nonexclusive rights. With control effectively exercised by public sector agents, the resources are unavailable to entrepreneurs wishing to pay for them.

The PCAST approach builds confidently on the TV white spaces pro-ceeding—the 2002 FCC initiative permitting unlicensed devices to use the buffer space between stations in TV band. But that policy, in the middle of its second decade, has yet to produce a hint of consumer benefit. Rather than leverage this approach, it might pay to try something different. Mark McHenry, whose company Shared Spectrum has produced some very cool measurements of radio spectrum usage, notes that "TV bands are among the worst, if not the worst, for spectrum sharing," by which he means the sort of shared use authorized in the white spaces proceeding. "The TV bands

... are subject to a number of overly restrictive sharing requirements."[12] But the deeper truth is that the TV band hosts lots of low-valued radio emissions that clutter it up. It might be perfect for sharing, just not the way they're doing it. The oppressive "sharing requirements" are man-made, as is the mandate that most television stations continue broadcasting.

The U.S. Department of Commerce issued a plea for help in 2013, proclaiming that "Uncle Sam Wants You to Help Us Design a Spectrum Monitoring Pilot Project":

> While clearing spectrum bands to make way for new wireless services has been a viable approach for many years, options for relocating incumbent operations are dwindling, getting more expensive, and taking longer to implement. Given this, NTIA has been working with the Federal Communications Commission, other federal agencies, and industry stakeholders to explore ways to share the spectrum without displacing existing systems in the same bands.[13]

But the constraint—*without displacing existing systems*—is brutal. Frequently, the best outcome is for a fabulous new network to be built, with its new owners (and then eventual users) compensating existing systems to clear out and make way. In any event, using bureaucrat-on-bureaucrat negotiations to craft sharing rules that protect incumbents has been thoroughly vetted. Its current popularity is the triumph of hope over experience.

Suppose private companies stepped into the Commerce Department's shoes. That could be done by the sale of an overlay. In 2013, FCC member Jessica Rosenworcel recommended that such rights be used to repurpose AWS-3 spectrum, part of which was encumbered by federal users: "I propose we auction . . . [an] exclusive right to negotiate with federal incumbents. By creating a source of agency, it will create opportunity for specific parties to negotiate with federal users and come up with creative ideas for near-term testing, sharing and even long-term relocation. Given the real statutory constraints, I think this is an elegant solution."[14]

It is elegant, and the Commission—after at first rejecting the suggestion —adopted it. As with "impairments" in the incentive auction, the move bent policy toward reality. Top-down directives bottle up allocations with no sense of urgency. Overlays are "hunting licenses" where a hungry hunter gets to eat what she kills. Theory and evidence suggest the latter out-perform.

The PCAST Report focuses on what are sometimes called spectrum "silos" and "stove piping" as the source of wasted bandwidth. Silos do exist, and are hugely problematic. But they are a result of traditional "public interest" allocations and legacy telecommunications policy, wired and wireless, with its enforcement of "competitive apartheid, characterized by segregation and quarantine," as explained in a 1999 legal treatise.[15] Market transactions—when permitted—piece together vast bundles of bandwidth. Yet some economists have recently challenged this notion, arguing that exclusive rights "can stifle third-party innovation" because "third-party innovators face a threat of *hold-up*."

> A company that comes up with a new mobile device or business model needs to convince the owner of the spectrum to let it develop its idea. . . . If the new development threatens the owner's existing business, it is particularly unlikely to be allowed. And, if the innovation requires the assent and coordination of multiple spectrum owners, it is even more difficult to get the owners all to agree. The potential for this type of coordination failure is sometimes referred to as the *tragedy of the anticommons*.[16]

No doubt such tragedies occur; the political spectrum is littered with examples. Yet spectrum markets have, on the contrary, proven an antidote, creating a warm, fuzzy place for third-party innovators. The "owner of the spectrum" has been driven by competitive forces to become a change agent, coordinating complex complementary activities, including large-scale network investments. The smartphone revolution was hosted on mobile networks possessing exclusive frequency rights, but the radio devices were invented and sold by independent vendors while the apps they host are developed by thousands more.

To categorically argue that exclusive rights uniquely impose transaction costs invokes the Nirvana Fallacy, relying on the supposition that there exists an alternative regime that eliminates such expense. But no regime is either perfect or free to operate. Is there a realistic counterfactual—say, imagining that the FCC had planned out mobile services using traditional licenses and/or unlicensed bandwidth—that compels us to believe that greater, more valued third-party innovation would have arisen?

The history of traditional allocations, where administrators did take charge, suggests: not likely. Indeed, connecting rights fragmentation with the tragedy of the anticommons leads to the opposite of the policy position asserted. The quoted passage includes a citation to authority referencing the seminal 1998 *Harvard Law Review* paper by Columbia law professor Michael Heller. This work described the situation in 1990s Moscow real estate following the collapse of communism. Great buildings were left vacant while brisk retail activity occurred on the sidewalks out front. Heller, working for the World Bank and stationed in the city, first thought that property rights had not been assigned to the buildings, but that turned out to be wrong. Ownership rights had been established but were given to too many parties, and were in conflict. To rationalize resource use, they needed to be combined. Fragmentation was costly, holdouts ensued, and superb opportunities were squandered.

In his 2008 book *The Gridlock Economy* Heller applied his analysis to radio spectrum. Without effective owners, he argued, the market cannot coordinate new and better arrangements. Broad ownership rights in frequencies facilitate this process; tiny slivers of usage rights, as with nonexclusive unlicensed bands, often do not. He noted, for instance, that when you buy an automatic garage door opener, the device is bundled with the right to emit a very weak radio signal from the remote device. This comes courtesy of a specific allocation of unlicensed bandwidth, and can work well for the immediate purpose contemplated. But opportunity costs are unaccounted for. Moreover, changing that wireless use, as demands and science evolve, is not so easy. "You and all your neighbors might be happy to sell your rights to garage-door opener spectrum and get in exchange access to some next-generation technology," he writes. "But in a spectrum commons, there is no way to make that deal." The lost opportunities may be unnoticed, as "the gridlock side of unlicensed spectrum remains invisible."[17] But the social costs can be huge.

The LightSquared debacle is highly instructive.[18] Since 2003, the FCC had authorized the satellite L Band (at 1.6 GHz) to offer terrestrial land mobile—wireless voice and data—service.[19] The satellite telephone "silo" had proven to be a graveyard for multibillion-dollar investments, while the emerging 3G wireless market was booming. The liberalization made perfect sense. By mid-2012, the company owning the L Band licenses, LightSquared, had sunk some $4 billion in building a projected $14 billion national LTE (4G) network. But the FCC dramatically shifted its policy, withdrawing its authorization for cellular services when interference complaints were lodged by parties using the adjacent GPS (global positioning service) band, including airlines, tractor makers, GPS receiver vendors, and the U.S. military. Within weeks, LightSquared was in bankruptcy reorganization and its network build-out was dead.

Five aspects of the conflict were salient. First, LightSquared's new LTE phones would not spill emissions beyond its license borders; the interference complaints were due to the fact that the existing GPS receivers were using the quiet frequencies next door to the GPS band (in the L band) to improve locational accuracy. The anticipated increase in LTE traffic would add noise. FCC official Mindel De La Torre noted that GPS users had "been driving in the left lane [the L Band] with impunity, but now it looks like the left lane might actually have traffic in it, [so] the GPS community is yelling bloody murder."[20]

Second, there was a border dispute, but there was no property. The FCC approved GPS devices and LTE devices; then unilaterally undid the rights extended to the latter. Formally, this represented a change in the "public interest." As a practical matter, the outcome was due to lobbying from parties such as Delta Airlines, Caterpillar, the Federal Aviation Administration, and the Department of Defense. LightSquared, an upstart, was swamped.

Third, even drawing the borderline to favor GPS service should not have stopped the emergent LTE network. That advanced new nationwide carrier would deliver on the order of $100 billion in (present value) consumer gains going forward. The GPS devices that would suffer modest performance declines could adopt filters and other upgrades to mitigate the damage; these would have cost a tiny fraction of the benefits. Certain mission-critical GPS radios, such as embedded in crash-avoidance systems on commercial airliners, could have been replaced with newer, better equipment. LightSquared offered to put up $50 million for this task; larger investments might have been necessary. All plausible estimates pale in relation to potential social benefits from an additional nationwide mobile network.

Fourth, the socially useful bargain could not get done. Without spectrum owners in the GPS band, actual or de facto, LightSquared had no responsible party to deal with. Rights were simply too fragmented. Forced to do business with the FCC, LightSquared was ultimately jilted by a fickle partner. Vast social efficiencies could not save the company.

Fifth, LightSquared did manage to fix one L Band mess. Originally, regulators had "interleaved" transmission channels assigned to rival satellite companies, leaving very "narrow roads" mixed up in alternating assignments. This forced endemic border disputes. To tidy up this jumble was one of LightSquared's first orders of business. It negotiated with other licensees, paying its counterparts to trade spaces, thus creating a large, contiguous band—40 MHz—to use for LTE. It paid one band cotenant, Inmarsat, some $490 million.

In 2011, the FCC was delighted. "Next generation broadband systems require large, contiguous blocks of spectrum," it opined, and "much of the L-band spectrum will not be suitable for broadband without such coordination."[21] Historic allocations had created worthless slivers, and market bargains had salvaged them—so said the regulators:

> LightSquared is making significant efforts to rationalize narrow, interleaved bands of L-band spectrum, held by several international operators, into contiguous blocks that will support next-generation broadband technologies for both mobile satellite and terrestrial use. . . . The Commission has recognized that these types of operator to operator arrangements, especially in the L-band, should be encouraged and are preferable to "regulations based largely on hypothetical cases."[22]

The L Band came together. But it was then again rendered a wasteland, devastated by extreme rights diffusion in the neighboring GPS band. Ironically, while mass-market GPS users would handsomely benefit—as subscribers to mobile broadband services—from an additional network, they cannot organize to seize their better options. The tragedy of the spectrum anticommons rages on.

PART FIVE

BEYOND

The liberation of the human mind has never been furthered by . . .
dunderheads; it has been furthered by gay fellows who heaved dead
cats into sanctuaries and then went roistering down the highways of
the world, proving to all men that doubt, after all, was safe.
                                                    —H. L. Mencken

THE FUTURE OF WIRELESS IS bright and exciting. There are ways to let it
happen.

19

The Abolitionists

If something cannot go on forever, it will stop.
—Herbert Stein

THE RADIO ACT OF 1927 was a legislative bouquet for commercial broadcasters, policy makers seeking to politicize communications, and Hooverite visionaries wishing to advance the administrative techniques forged in the 1887 Interstate Commerce Act. There have been many attempts to improve it. The Landis Report, delivered to the incoming Kennedy administration in late 1960, concluded that the FCC had "drifted, vacillated and stalled in almost every major area." Finding that the agency "seems incapable of policy planning, of disposing within a reasonable period of time the business before it, of fashioning procedures that are effective to deal with its problems," it recommended administrative reforms.[1]

On leaving the FCC in 1963, Newton Minow still found the agency dysfunctional. Invoking his famous Las Vegas speech, he wrote:

The Commission is a vast and sometimes dark forest where we 7 FCC hunters are often required to spend weeks of our time shooting down mosquitos with elephant guns. In the interest of our governmental processes . . . better marked roads have to be cut through the jungles of red tape. Though we have made many substantial improvements in recent years, the administrative process is a never-never land which we call quasi-legislative and quasi-judicial. The results are often quasi-solutions.[2]

In December 1968, the Rostow Report, by the outgoing Johnson administration, cautiously suggested greater reliance on competition in communications markets. This theme was pushed further with the emergence (at long last) of cellular in the 1980s. The idea was revisited in the Commerce Department's 1988 study *Telecom 2000*[3] and elaborated upon in 1991 in its comprehensive *U.S. Spectrum Management Policy: An Agenda for the Future*.[4] In 1999, FCC Chairman William Kennard issued a report arguing for substantially revising the structure of the Commission so as to "create more efficient spectrum markets . . . and increase the amount of spectrum available for use, particularly for new services."[5]

The reformist agenda has gone farther. Perhaps the most comprehensive case for abolishing the agency was put forward by lawyer and former MIT engineering professor Peter Huber, whose compelling 1997 book was entitled *Law and Disorder in Cyberspace: Abolish the FCC and Let Common Law Rule the Telecosm*. My 2001 article in the *Harvard Journal of Law and Technology* outlined how key functions of the FCC could be spun off to a Spectrum Court to enforce borders in frequency space. More recently, Dorothy Robyn (summarizing an Aspen Institute discussion) has recommended that "Congress should establish a Court of Spectrum Claims to be housed within the Court of Claims."[6] Huber had recommended relying on federal courts for these functions, but having a specialized body (like courts that decide patent or bankruptcy questions) overcomes the objection that the subject is too technical for off-the-shelf judges.[7] In any event, the move toward "fact-based adjudication"[8] of disputes could usefully depoliticize spectrum rules.

My 2008 article in the *Journal of Economic Perspectives*, "Optimal Abolition of FCC Spectrum Allocation," proposed using overlay auctions to supplant regulatory discretion, on the theory that the "public interest" is better served by letting competitive forces reveal market prices for spectrum use. I also proposed that government agencies generally not be awarded frequency rights but be funded to procure wireless services from competing suppliers.

Eloquent champions for FCC abolition have appeared in popular forums. In a 2004 essay for the tech website ZDNet, cybernews reporter Declan McCullagh called the FCC "an agency that does more harm than good."

His bill of particulars included censorship of the electronic press—the Commission was "hard at work, trying to figure out how to muzzle Howard Stern and make a national example of Janet Jackson's right breast"—and suppressing new technologies. "The Internet has transformed from a research curiosity into a mainstay of the world's economy," he wrote, "in less time than it took the FCC to approve the first cell phone licenses." He noted—citing Huber—that the Commission's vital administrative functions could be spun off to other agencies and that monopolization concerns could be remedied under antitrust law.[9]

Readers responded to McCullagh with an avalanche of abuse. While most of the complaints showed little understanding of FCC regulation, one important objection concerned how disruptive this transition would be. That is why reforms should grandfather incumbents, McCullagh suggested, while expanding flexibility in order to embrace new technology: "Amateur radio operators would be granted homesteading rights to the spectrum they currently use. Title could be awarded to the American Radio Relay League, the national association of ham radio buffs, which likely would police use of it more carefully than the FCC currently does."[10]

In 2007, media columnist Jack Shafer launched a broadside in *Slate*, "The Case for Killing the FCC and Selling Off Spectrum."[11] "Technology alone can't bring the spectrum feast to entrepreneurs and consumers," he wrote. "More capitalism—not less—charts the path to abundance." A year later, Lawrence Lessig took up the cudgels in *Newsweek:* "It's Time to Demolish the FCC."[12] His primary concern: "We need to kill a philosophy of regulation born with the 20th century, if we're to make possible a world of innovation in the 21st." He delivered a concise critique of the agency: "Its commissioners are meant to be 'expert' and 'independent,' but they've never really been expert, and are now openly embracing the political role they play. . . . Lobbyists spend years getting close to members of this junior varsity Congress." Lessig saw the agency instinctively protecting the status quo. "And you can't fix DNA." On the FCC's ashes would be created the "Innovation Environment Protection Agency (iEPA), charged with a simple founding mission: 'minimal intervention to maximize innovation.' "

In its 2010 National Broadband Report, the Commission itself documented how agency procedures had stifled markets by sandbagging spectrum allocations. By its own standards, however, the FCC has been unable to improve upon the six-to-thirteen-year "processing delay" that it labels as destructive. Simultaneously critiquing and defending its historic performance, the agency offered that "in general, a voluntary approach that minimizes delays is preferable to an antagonistic process that stretches on for years. However, the government's ability to reclaim, clear and re-auction spectrum (with flexible use rights) is the ultimate backstop against market failure and is an appropriate tool when a voluntary process stalls entirely."[13] Alas, the goal it set for 2015 has already been missed, and may not be met by 2020. Prompting the question: What happens when government action "stalls entirely"? Following Herb Stein, *if something is unacceptable, eventually it will not be accepted.*

It is not clear what adhesive would affix Lessig's iEPA to its noble mission, but the idea of moving beyond the formula of 1927—"public interest, convenience and necessity"—is intriguing. I have previously suggested a similar move: replacing "public interest" with a "consumer welfare" mandate, analogous to the guidance imposed on antitrust enforcement agencies by the courts that govern them.[14] These and other proposals remain to be fleshed out. They will be, once the momentum for reform crests. That may take a while. Entrepreneurs denied, like Northpoint's prime mover Sophia Collier—who, already flush from her forays into business and finance, abandoned wireless and became an artist in northern California—have rich opportunities in other fields and do not loiter in the FCC's start-up cemetery.[15] Meanwhile, the interests directly benefiting from capacious new liberal licenses, mobile carriers, find their enthusiasm taxed both by the increasing competitive pressures experienced when their rivals secure these same rights and by the billion-dollar payments required to win their own rights at auction. Conversely, firms seeking government set-asides, either under traditional licensing rules or via additional unlicensed allotments, evade the tax. The tilt in lobbying incentives alters policy outcomes, which then serve to buttress a status quo regulatory scheme in which political jockeying delivers the goods.

Another impediment to reform is that FCC regulation is often popular with vocal, well-organized constituencies. FCC enforcement actions undertaken to punish broadcast licensees for Howard Stern's quips or Janet Jackson's fleshy parts were pushed by protests from thousands of Americans disturbed by what they saw or heard. Disturbed Americans, it turns out, constitute a sizable voting bloc. Even the mythical Aunt Minnie commands fear. The idea that there is some "adult supervision" (as one critic of McCullagh's essay put it) brings comfort to the nervous, even if the actual regulatory outcomes are perverse.

"Public interest" regulation means never having to say you're sorry. Even when the government protects and nurtures oligopoly, ensuring low quality programming, it gains kudos for objecting to the "vast wasteland." Even when benign regulatory neglect on the "toaster model" produces communications services that deliver progress as never before, the abdication of adult supervision leads pundits and populace to cringe in horror. In sum, there is no abolitionist groundswell.

Yet it would be wrong not to notice that change has come incrementally. And the "wireless explosion" follows foundational reforms in both licensed and unlicensed allocations. The United States has often led liberalization efforts, but not always. Some countries have dared to go much farther, testing the waters for auctions (in New Zealand in 1989, and India in 1991) and crafting explicit private property rights in spectrum (Guatemala in 1996). The United Kingdom, following the landmark Cave Report in 2002, tried sweeping spectrum liberalization, and while it missed a 2010 deadline to strip away all use restrictions in the great majority of "beachfront" airwaves, it offers a template that could prove attractive to future reform endeavors. Sweden has succeeded in allotting large dollops of bandwidth via flexible-use licenses under "a dedicated route towards increased liberalization." The template includes minimalist, technology-neutral licenses, employing auctions to assign rights, encouraging the buying and selling of licenses in the marketplace, and allocating unlicensed bandwidth when there "is little risk of harmful interference and there are no other impediments."[16] Norway has employed a similar method.[17] The approach is not radical, but— and—it offers a possible template for others.

It is important to delineate the specific object of our discontent: the case-by-case spectrum allocation system. While decrying the constraints and delays inherent in a "command and control" regime, regulators continue to rely on the administrative allocation system to decide how markets should look. They erroneously conflate particular wireless applications, devices, or businesses with an efficient spectrum rights regime. It would be better if their attention were put to devising regimes that allowed competitive forces to discover what products and use models work best.

Just as the historical popularity of TV broadcasting did not erase the inefficiencies of TV band spectrum allocation, the productive use of licensed spectrum does not constitute a categorical endorsement of any particular bandwidth allotment or business model. Mobile networks from the early days of cellular have morphed into distant 4G cousins, and organizational migration should continue. Future mobile networks may use licensed spectrum but look more like deployments in unlicensed bands. Device makers and multiple network carriers may pool spectrum rights (as is done with roaming or virtual mobile networks) and share infrastructure (as cellular tower companies host multiple network base stations). The transition of video from terrestrial broadcast spectrum has already entered its third generation, having graduated from cable and satellite systems to broadband (fixed and mobile) networks, which now stream vast quantities of popular programming. Neither the application nor the service model uniquely determines the optimal spectrum allocation.

The policy debate focuses almost without fail on which existing wireless networks use existing allocations, pointing (depending on the advocate or interest) to more allocations for whatever is said to already work. More sophisticated discussions add projected scientific breakthroughs, pounding them into existing regulatory formats. But what is essential, and virtually universally missed, is how some policy options allow the arguments to prove themselves without requiring the regulators to guess right.

Take mobile TV. There was considerable interest in the service in the United States by the early 2000s, but there were no mobile TV authorizations. There were, however, liberal licenses that allowed flexible use of airwaves. Three American firms sought to use them to innovate. Crown Castle purchased rights for 5 MHz in the 1.6 GHz band; Aloha Networks won licenses allotted UHF TV channels 54 and 59, 12 MHz total; Qualcomm laid claim to permits for UHF TV channel 55, yielding 6 MHz. The firms were pioneering, as *Business Week* put it, "one of the hottest areas of tech: TV over mobile phones."[18]

Crown Castle developed Modeo, beta-tested in New York City starting in late 2006.[19] Aloha Networks developed Hi-Wire, offering a twenty-four-channel service in Las Vegas, Nevada, starting in 2007.[20] Qualcomm's MediaFLO, also launched in 2007, was the showcase. The company spent $1 billion to develop chips capable of squeezing nearly twenty TV networks (including CNN, CNBC, MSNBC, Fox News, CBS, NBC and others) into the channel space previously hosting just one. Broadcasts were received on handsets marketed by mobile carriers. Service rolled out with a nationwide advertising blitz.

Alas, few subscribed. By late 2010, all three experiments had folded, and two (allocated frequencies adjacent to those used by mobile operators) had sold their licenses. The spectrum "re-farming" helped AT&T, which bought both the Aloha and Qualcomm holdings, to handle the huge data demands of iPhone users. This unremarkable news story was the product of utterly remarkable spectrum reform:

> *Qualcomm Sells MediaFLO Spectrum for $1.93B.* Qualcomm said today that it has agreed to sell its 700MHz US spectrum licenses to AT&T for US$1.93 billion, with the operator stating that the transaction "will bolster AT&T's ability to provide an advanced 4G mobile broadband experience for its customers in the years ahead." The spectrum was used for Qualcomm's aborted FLO TV mobile broadcast business.[21]

Mobile TV was not such a bad idea. YouTube videos are close to a "killer app" in wireless. Broadcast and cable TV shows are widely viewed on smartphones, tablets, and notebook computers, driving a "mobile data tsunami." But a stand-alone service, using dedicated spectrum, proved inefficient at supplying it. Instead, mobile networks have increased speeds for data downloads, enabling TV to piggyback. The Qualcomm-AT&T transaction epitomized the shift. Markets were open to new ideas and new forms of organization. Investors financed them. Consumers chose among them. When other spectrum uses became more promising, resources were diverted to make room. The failure of one path to mobile video helped enable superior alternatives.

As in other high-tech markets, services will evolve in surprising ways, and the outcomes may stun participants in today's spectrum wars. Perhaps frequency owners would find that combining rights into vast frequency bands to supply plug 'n' play radio access, sending all messages "wide and quiet," outperforms the competing networks (and bands) of yesteryear. This is sometimes called the "private commons" solution. How would we know that that approach yields social improvement? The key is the manner in which the usage is determined and then tested against other approaches, and the process by which it can be adapted or abandoned as competing options unfold. Market survival is a powerful force for value revelation.

Flexible spectrum use rights assigned to responsible parties—neutral with respect to technology, services, and business model—are the key ingredient. The frequency allocation regime must be administratively efficient in making spectral inputs widely available for productive use. It must be transparent, with rights assignments made through arm's-length transactions such as competitive auctions. Governments should shop for spectrum rather than squirreling it away, secretly nibbling on their stash. And it must be sufficiently adaptable that wireless deployments made in one period can be reconfigured in the next.

It is easy to get confused in contemplating how liberal reforms might work in the airwaves, but it should be much easier today than it was when the discussion began in the 1950s. The idea for a robust spectrum market has been credited, with considerable justification, to Ronald Coase. Yet this attribution has an interesting footnote that turns my previous query—*What would Coase do?*—into something of a trap. This was elucidated in an impressive essay by law professors Thomas Merrill and Henry Smith in the 2001 *Yale Law Journal*.[22] They noted that Coase, early in his 1959 FCC article, spoke about treating spectrum like land, delimiting frequency access by awarding ownership of various spaces. Later in his essay, however, Coase abandons that notion, saying that we should not waste time defining boundaries. "Whether we have the right to shoot over another man's land," he wrote, "has been thought of as depending on who owns the airspace over land." But, he continued, "it would be simpler to discuss what we should be allowed to do with a gun."[23]

Merrill and Smith explain Coase's conflict as a classic divide in property law. Depending on circumstances, social coordination can be furthered in two ways: either by awarding ownership rights or by delineating use rights.

In some instances, and particularly with real estate (Coase's analogy for spectrum), it makes sense to define a space and then delegate decisions over its use to the owner, an agent with an active interest in maximizing the resource's value. Even so, the owner does not possess all possible property rights. Some limitations are needed to help support social coordination. Airplanes can fly over land, for instance, and not pay a toll to the owners of the real estate down below.[24]

Under an alternative method, rights awarded for personal activity are crafted individually. That is the "bundle of rights" approach Coase shifted to in saying it was "simpler to discuss what we should be allowed to do with a gun." It cannot possibly be what he really meant for radio spectrum. The process by which the government defined each and every wireless use right—for every service, technology, or business model—was the very object of Coase's emphatic critique. Awarding rights on this basis drives entrepreneurs back to *Mother May I?* whenever a new idea strikes. That creates an endemic holdup problem, slays incentives for innovation, and constitutes the core of what Coase was rejecting.

As for abolishing the spectrum allocation agency, it is worth remembering that in 1977, when Fred Kahn went to head the Civil Aeronautics Board, he had no intention of abolishing that CAB. (He would have preferred an appointment to chair the FCC, but that slot was taken. If Fred had gotten his wish, perhaps we would be paying far more for air travel on a government-protected cartel of carriers but enjoying better and cheaper wireless service.) That was the year that the late Milton Friedman saw the CAB and FCC as two peas in a pod, hidebound organizations that harmed consumers, each deserving the death penalty:

> If we could abolish the Civil Aeronautics Board tomorrow, there is no doubt whatsoever that our aircraft industry would be in an extremely healthy state, with . . . much less concentration of power. The same is true in the broadcasting and television industry. Why do we have three major networks? Because the Federal Communications Commission . . . has held back . . . every new invention.[25]

Friedman could scarcely have guessed that the CAB would be gone by 1985. Or that ten years later, the Interstate Commerce Commission would join it in the dustbin of history. At some point the FCC will follow. The CAB lived to be 47 years old, the ICC a ripe old 108. In 2027, the FCC's spectrum allocation system is scheduled to celebrate its 100th birthday. It is surely not too soon to start thinking about a retirement party.

20

Spectrum Policy as if the Future Mattered

> Liberating spectrum from the control of government is an important first step to innovation in spectrum use. On this point, there is broad agreement, from those who push for a spectrum commons to those . . . who push for a fully propertized market. . . . Innovation moves too slowly when it must constantly ask permission from politically-controlled agencies. The solution is . . . removing these controllers.
> —Lawrence Lessig

A CONSENSUS ILLUMINATES the path ahead.

To be clear, compromises and footnotes are essential. There can be no "fully propertized market," as limits are part of any realistic regime—in land, oil deposits, intellectual property, or radio spectrum. Users of traditional wireless licenses cannot simply be brushed aside; their rights, as a matter of practicality and efficiency, must be vested. Existing unlicensed bands, where in popular use, cannot be uprooted. It is appropriate to focus on new margins, pouring liberal rights into the marketplace. If this is done well, waves of innovation will roll, and industry watchers will scratch their heads trying to figure out whether the emerging service models are best described as "exclusive use" or "spectrum commons." The "controllers" step aside and competitive rivalry will take things from there.

Not all bandwidth rights need be defined and auctioned. Very localized uses may best be left alone. Real property owners—the only parties affected—

exercise dominion. A suggestive example is that a homeowner does not need permission (or a mandate) to switch her 2.4 GHz cordless phone to a 1.9 GHz model to protect her 2.4 GHz wi-fi access.[1] Very low-powered devices often peacefully coexist with exclusively assigned wide-area spectrum rights.

Gerald Faulhaber and David Farber explained in a 2002 paper that various compromises or intrusions could occur with owned spectrum, but the mystery is why anyone would believe otherwise.[2] Private land is subject to a variety of covenants and easements. This tweaking occurs in property law, where rules constantly evolve through the resolution of disputes. The challenge is to channel the conflicts so that the parties with the best information and incentives can tackle them, capturing efficiencies and establishing helpful precedents. The argument for property rights in airwaves often takes land as its model, but the closest analogy is to radio spectrum itself. The actual performance of "public interest" governance is what led Leo Herzel and Ronald Coase to propose alternatives. Eventually, when it came time to award licenses for cellular service, regulators tiptoed towards flexible, market-driven assignments of spectrum rights. The results offer proof of concept for the Herzel-Coase Theorem.

Some policy experts believe that liberal property rights are still difficult to define in radio spectrum, and that it would be ill-advised—even dangerous—to embrace an ownership regime until the details are thoroughly worked out. The fear is that incomplete airwave contours will trigger endless disputes. The evidence suggests otherwise. Not only are the basic paradigms supporting market allocation of spectrum already tried and tested, with bountiful economic results despite fuzzy borders, it has been demonstrated that the greater social costs accrue from slow-walking liberalization, extending the shortcomings of administrative allocation. Compared with this exercise, where disputes drag from one century into the next, defining exclusive rights is a snap. Spectrum markets are not exotic but active.

Ownership rights are never exhaustively defined.[3] Once rights are clear enough to outperform legacy regulation, striving for perfect rules is a fool's errand. Moreover, while quarrels over interference are cited as the basis for preferring extensive rulemakings to define spectrum rights upfront, the key to harmony lies not in exactitude but in incentives to cooperate. Regulators awarding rights have weak motivation to resolve problems efficiently. In their efforts to prevent border disputes, they often separate users into spacious, barren silos. As in the old TV band, where more than eighty channels were set aside to accommodate just three or four broadcasts, regulators happily create white spaces and then compound the error by protecting them for generations.

This avoids one source of "cacophony," but it overcorrects by imposing a more potent and costlier silence. Confused by combatants pronouncing doomsday—"a drawn-out battle of claims and counter-claims," as a government research study dubbed it[4]—spectrum administrators retreat into excessive conservatism, stifling wireless services that create generous value at the cost of barely perceptible conflicts. Recall Coase: "It is sometimes implied that the aim of regulation in the radio industry should be to minimize interference. But that would be wrong. The aim should be to maximize output."[5] Regulatory professionals still, more than half a century later, are flummoxed by this simple insight.

Markets discover boundaries. This approach was vindicated in U.S. mobile services, where the FCC distributed more than fifty thousand licenses, willy nilly, with different franchise sizes defined by multiple maps. This idiosyncratic public policy (almost all other countries issue regional or nationwide licenses) created an extraordinary number of borders, each of which presented potential conflict. Rights were only approximately defined, and made permissive, while secondary market trading was permitted. Private actors then assembled the dispersed fragments into national networks— just four, at the regulators' latest reckoning. Wireless technologies jumped the narrow slots originally assigned, resolving potential squabbles. Interference issues continue to arise on adjacent frequencies, but mobile carriers— motivated to solve these problems—routinely settle them privately, avoiding the overhead of the regulatory system.[6]

Should we lament the exit of "mom and pop" licensees—thousands of whom won cellular rights in lotteries and could not wait to sell out to their corporate big brothers? Not if we care about consumers, technology adoption, device makers, or mobile app innovators—or mom and pop, who enjoyed the windfall. The potent network economies that emerged from consolidation yielded far more abundant opportunities at much lower cost than afforded by atomistic and uncoordinated rights holders.

Often, *good enough* beats *ideal*. The FCC defined the AWS-3 licenses for sale in 2015 rather curtly:

> 65 megahertz total
> Block A1: 1695–1700 MHz (5 MHz)
> Block B1: 1700–1710 MHz (10 MHz)
> Block G: 1755–1760/2155–2160 MHz (10 MHz)
> Block H: 1760–1765/2160–2165 MHz (10 MHz)
> Block I: 1765–1770/2165–2170 MHz (10 MHz)
> Block J: 1770–1780/2170–2180 MHz (20MHz)[7]

There is more detail in an eBay product description.[8] That this cryptic overview encourages private corporations to offer billions of dollars suggests that conflict and uncertainty do not cripple liberal licensees. This is despite the fact that regulators concede that they have never thoroughly defined "harmful interference,"[9] leaving frequency rights less than crystal clear.

Spectrum regimes around the world embrace the idea that decisions formerly made by governments should be ceded to the competitors in the marketplace. As the International Telecommunications Union, a Geneva-based offshoot of the United Nations, wrote in 2009:

> Diverse countries have been modifying their telecommunications regulations . . . with the objective of promoting the provision of new and innovative services [and] the reduction of prices. . . . Traditional regulatory systems focused primarily on the specific means of telecommunications or on the specific service provided by the operator. . . . Such distinctions are no longer practical. Therefore, reforms . . . have consistently focused on two key elements: the introduction of the principles of technology and service neutrality, and the establishment of greater flexibility in key aspects of their existing regulatory frameworks.[10]

Technological and service neutrality. These policies have been adopted with incomplete success, and actual practices are in some conflict with boasts made in international forums. But the trend is unmistakable. This inflection in the historical course of the political spectrum should be pushed to its limits.

Markets over Ideology

Society gains from a diversity of wireless networks, applications, technologies, radio devices and service models. But how to determine the optimal combination of spectrum uses, or jiggle it when circumstances change?

Superior mixing and matching depends on reliable information about cost and performance, not only for what we deploy but for what we forgo. Prices for spectrum, no less than for cell towers or smartphones or software engineers, help operators to treat valuable resources with appropriate care. When mobile carriers seize wi-fi options using unlicensed spectrum, off-loading network traffic where subscribers are able to use fixed network connections, market prices can become more vital, not less. Substituting one approach for another is a search for conservation, saving what is precious in favor of consuming what is less so. Without competitive bids for spectrum, the data revealing these relative values are murky. The proper balance is elusive.

Property rights are not business-model destiny. Private ownership of copyrights, for instance, does not foreclose Creative Commons licenses that grant wide access to content, free of charge. Exclusive owners of software code thrive while open-source models flourish, often competing in the same or overlapping markets. Corporations developing proprietary systems may—for competitive, profit-seeking reasons—relinquish their exclusive rights and share their wares, free of charge. Intel, Xerox, and AT&T's Bell Labs all ceded key technological innovations to rivals for zero or nominal fees. Nokia, the world's leading cell phone maker at the time, gave away its Symbian mobile operating system software, following IBM's lead in releasing Linux code (which it had spent over $1 billion to develop) to all users, without payment. These "strategic forfeitures" were undertaken to promote platform development, as law professor Jonathan Barnett has explained in fascinating detail.[11]

Similarly, flexible spectrum rights, exclusively assigned, can and do support a vast array of use models. The development of spectrum markets, though rudimentary (one hopes) by tomorrow's standards, is sufficiently advanced to have given birth to "wireless" giants like Apple, Amazon, Facebook, Google, Uber, and Twitter, creators in the digital economy that contract for frequency access in the marketplace. And this scarcely scratches the surface of the deep ecosystem below. An orthogonal illustration is supplied by this politically charged advertisement for a U.S. mobile operator, a creation of secondary spectrum markets: "DOES YOUR PHONE COMPANY SUPPORT THE TEA PARTY—OR PROGRESSIVE CHANGE? Stand up for social justice—with your phone. Switch to CREDO Mobile, America's only progressive phone company."

Working Assets Wireless, owner of CREDO, bought its spectrum and network access from Sprint. The company urges customers to "JOIN THE ONLY MOBILE COMPANY THAT GLENN BECK CALLS 'EVIL.' " CREDO then invests its profits in causes such as the "Mother Jones Investigative Fund, Planned Parenthood, Rainforest Action Network and the ACLU."[12] No donations to the University of Chicago Law School, where Ronald Coase taught, have been reported, but the spectrum liberalization ideas developed there might well be considered substantial in-kind contributions to CREDO.

Property Rights to Streamline Innovation

Today, ambitious wireless entrepreneurs imagine the impossible. One outlandish idea is to bring cheap, ubiquitous, high-speed Internet access to the entire planet, from the tiniest island in Indonesia to the remotest village in Namibia to the highest outpost in the Himalayas. In fact, there's competition for this fantasy. A venture headed by serial visionary Elon Musk, SpaceX, is attempting to deploy hundreds of low-earth-orbit satellites. Google X's Project Loon launches stratospheric air balloons. Meanwhile, Facebook plans a worldwide network supported by drones. These wild aspirations are economically straightforward: there may be money to be made in extending the applications that ride on mobile devices to the "bottom billion." Some ventures will surely fall short, while others—perhaps including those not yet envisioned—may soar.

But these wireless dreams will surely be toast without sufficient radio

spectrum. In some regions, flexible-use spectrum rights allow these ambitious Internet entrants to bargain for access to bandwidth controlled by
existing licensees. Loon fortifies emerging LTE networks, sharing mobile
operators' flexible spectrum rights.[13] Yet elsewhere, bandwidth requests
must join the bureaucratic queue. That's where the dreams go to die.

There will be no pure solution to the problem of spectrum allocation.
Models are clear, but the world is a mess. Nor will there be complete "deregulation." As Coase noted in 1959, "How far this delimitation of rights
should come about as a result of a strict regulation and how far as a result
of transactions on the market is a question that can be answered only on
the basis of practical experience. But there is good reason to believe the
present system, which relies exclusively on regulation and in which private
property and the pricing system play no part, is not the best solution."[14]

Progress has been made. Our practical experience now extends to vast
wastelands *and* wireless toasters. There has been much puffery about how
regulatory jawboning may shame bad licensees, public and private, into
overcoming their wicked ways to provide heroic outcomes. But history has
exposed the wizard behind the curtain. The failure of administrative planning in radio spectrum, generally followed by the comic ritual of "holding
their feet to the fire," has been shown to yield "public interest" payoffs that
invariably benefit interests far more than the public.

Scientific expeditions to the frontiers of wireless are giving us wonderful
new communications options, but their implications for regulation are
often misunderstood. "The Aim of Science," wrote Bertolt Brecht, "is not to
open the door to infinite wisdom, but to set a limit to infinite error."[15] Here,
economics might help. Technology has not put an end to spectrum scarcity.
Just as breakthroughs are relentlessly made to supply fatter conduits, entrepreneurs ruthlessly arise to stuff them full of ever more communications.
Supply creates its own demand, and then some. Access to prime radio spectrum has generally become more contentious, not less.

We have long heard the counterclaims. The idea that scientific advance is
moving us beyond the constraints of nature continues to be heard, often in
the service of campaigns hoping to push additional spectrum allocations
away from one group of competitors to favor another—a telltale sign of the
conflict that the argument's own premise should moot. Since at least March

1941 this "end of scarcity" refrain has been heard. That was when *Radio News* broke the exciting story that as "FM becomes universal, there will be no physical limit on the number of stations in one town. The interference problem is solved."[16] Well, not quite then. And not quite yet.

Innovation should challenge incumbents; may the most efficient ecosystem win. But that victory is compromised when regulators seek to divert usage with new, unpriced allocations that effectively lock in wherever regulators leave them. The requests for additional unlicensed bands are submitted to the authorities, not to bidders or resource owners playing with their own stakes. This *is* the top-down allocation system.

It is the driver of consensus dissatisfaction.[17] U.K. economist Martin Cave calls the approach "arbitrary and unsatisfactory."[18] To improve matters, the FCC's top spectrum policy experts, Evan Kwerel and John Williams, proposed that contests over rules and bandwidth set-asides should be decided via competitive bidding. In 2008, the FCC dispatched economists Mark Bykowski, Mark Olson, and William Sharkey to devise an auction format that included offers from both licensed and unlicensed partisans. They wrote: "The allocation between licensed and unlicensed use . . . is based on the FCC's judgment, which in turn relies on information provided by interested parties who seek to use the spectrum. One method of reducing the incentive that parties have to exaggerate the value they place on a given regime involves creating a market for such rules."[19]

The authors designed a process for companies seeking liberal licenses, say, wireless carriers, to bid against companies, such as software firms, that desire more unlicensed bandwidth. Under this scheme, a license-exempt band would be created out of the spectrum allocated to a license if the sum of the unlicensed spectrum bids (for that license) exceeded the top bid made by a rival seeking exclusive rights.[20]

This approach pursued a promising path, even as it was truncated in scope. The FCC need not design special licenses or auctions to accommodate coalition bids. Flexible-use permits allow high bidders to determine business models. And under existing FCC auction formats, industry consortia have already bid for, and won, wireless licenses. Moreover, the assignment of frequency use rights to third parties, including radio manufacturers (the use model for unlicensed spectrum), is routinely a part of how liberally licensed frequencies are used. By analogy, patent pools and joint ventures routinely share property rights in this manner.[21]

This type of commons, however, would benefit from an important reform. Under current rules in the United States and most countries, licenses typically come with build-out requirements. These rules, meant to counter spectrum hoarding, mandate that license buyers must create networks within a given time frame, say five years. If the buyer does not make appropriate progress in the construction project, the rights revert to the government for reassignment. Many competing business models, including unlicensed "parks" that supply bandwidth for low-power wireless devices—Bring-Your-Own-Network—might be effectively excluded by rule. Build-out rules, which are ineffective in practice (defining build-out is more complicated than it sounds, and compliance is often lax), should be relaxed.

As noted previously, some of the most contentious allocation contests juxtapose interests seeking different flavors of unlicensed. One controversy features a satellite phone operator who aims to create "a private Wi-Fi channel and charge for access to it," opposed by corporations wanting to freeze the initiative to protect adjacent 5.8 GHz unlicensed frequencies. The latter group fears that enhanced traffic will diminish performance of existing wi-fi service.[22] Meanwhile, General Electric battles aircraft maker Boeing. GE proposes that its medical devices be permitted to transmit data (including patients' vital signs) over 2.4 GHz unlicensed frequencies, while Boeing claims that this activity will interfere with its use of the bandwidth "to test the safety of planes."[23] Life or death in a hospital bed, versus life or death in the sky—which does the "public interest" favor? In another face-off, oil companies want to use airwaves dedicated for educational institutions to reach deepwater oil rigs because "the only schools in the Gulf of Mexico are schools of fish." The educators counter with a science lecture of their own: sharing their offshore channels will create harmful interference because "wireless signals tend to travel farther and faster over the warm Gulf water, causing greater interference on shore." Moving over to channel 37 in the TV band, we find radio astronomers—who listen to deep space—howling over having to share bandwidth with automobile cruise control systems. The car controls do not emit much power, but it's fairly easy to knock scientists off a receiver tuned to Galaxy 459X7Y.[24]

And then there is the brisk controversy over LTE-U (long-term evolution —unlicensed), part of 4G carriers' efforts to improve data offload via wi-fi links. Strong protests from the Wi-Fi Alliance, representing cable operators and tech companies, claim that the innovation tends to hog unlicensed frequencies for mobile subscribers, reducing wi-fi functionality for others.[25] Some see the conflict as simply technical, but scientific research seems to follow economic self-interest. "Recent tests to see whether LTE-U technology interferes with Wi-Fi signals prove conclusively that LTE-U poses no problems whatsoever for Wi-Fi networks," reports a trade journal, "and also that LTE-U will drown out Wi-Fi, depending on which party is to be believed."[26] It would be remarkable were it otherwise.

Under current law, the FCC does not know the optimal solution to these conflicts. Nor do I. But the agency will nonetheless, after lengthy deliberation, impose its guess. And it will not permit market transactions to undo its decision, as the fragmented, nonexclusive access rights it distributes in unlicensed bands cannot be easily reconfigured. An auction between the opposing parties, however, would help to pluck the relevant costs and benefits from the dark.

Competition in a Pastoral Setting

One objection raised against this competitive solution is that the values created by "spectrum commons" cannot be captured in auctions. It is "akin to asking users of public parks to bid against developers to decide how land is to be allocated."[27] But the chosen analogy cuts exactly the other way. First, when a government agency sets aside unlicensed bandwidth, *it is* bidding against other "developers" who would seek to use the airwaves differently. It simply does so in a nontransparent and monopolistic manner that suppresses competing bids. Second, when open auctions are used for wireless licenses, the parties making offers—say, mobile networks—shoulder the task of aggregating the disparate demands of millions of future subscribers (many of whom are not even born yet). The process encourages extensive research and careful calculations, given the risks involved for bidders, and puts prices on public display. This information improves decision making for all parties, including governments, by exposing opportunity costs.

Finally, the idea that public parks are an analogy to unlicensed band allocations perpetuates a common misunderstanding in the political spectrum. As the FCC has written, "A mechanism based on markets . . . will be most efficient in most cases. However, government may also wish to promote the important efficiency and innovation benefits of a spectrum commons . . . much as it allocates land to public parks."[28]

The "however" crisply defines regulatory confusion. Public parks sit on land allocated through a system of private property rights. A market where basic resource rights are preempted in favor of "public interest" determinations is something quite distinct. That is what led Ronald Coase to characterize the spectrum allocation system as equivalent to a Federal Land Commission (FLC).[29] The analogy was extended by New York University economist Lawrence J. White, in a thought experiment about how a Federal Land Use Commission (FLUC) might hinder the productive use of real estate.[30] (Acronyms are a leading output of the spectrum allocation system, as well as the debate beyond.)

Land, a key input into parks, is not held in abeyance, rights dribbled out case by case, as government administrators impose command and control to plan where all the public facilities (now and possibly in the distant future) should be located. Instead, ownership rights are defined, and the assets made widely available. They are largely distributed, transferred, combined, and subdivided by marketplace transactions. Bids are registered not just by private developers but also by nonprofit organizations pursuing social objectives. These include governments providing green spaces, parkland, and other amenities. The latter institutions have multiple ways to bid —eminent domain, for instance, requires compensation while mediating holdout problems—and zoning regulations may be enforced. But the market for real estate does not funnel each choice through a narrow administrative spigot; rather, it cedes property rights to decentralized actors.

This yields a wide variety of land-use modes, including the quintessential "commons" produced by New York City's amazing Central Park.[31] Land markets enabled the City of New York to acquire the necessary rights and to understand how much it costs to keep them and what it would take to acquire more. By creating Central Park and determining its parameters, the city "bids" against rival (potential) landowners.[32] While White argued for property rights in frequencies, he declined to call this "privatization." He sought to avoid the mistaken impression that "public ownership of spectrum is not part of the concept."[33] In fact, "public entities own land, buildings, vehicles, etc., alongside private ownership of the same types of property." And the provision of both public and private goods is the better for it.

Reforms

There is a droll, probably apocryphal story about a graduate student who fell asleep during Milton Friedman's macroeconomics course at the University of Chicago. Friedman hurled an eraser at the snoozer and bonked his target. The young scholar awoke with a start. "I apologize, Professor, for falling asleep during your lecture," he said. "But the answer is—*reduce the money supply.*"

The spectrum policy answer here—if you've dozed off—is *auction overlay rights.* The method strategically introduces new spectrum use rights, empowering competitive forces by dispensing with centralized micromanagement. It enables entrepreneurs who then achieve the cooperation needed for progress. It protects legacy systems but not needlessly; existing users face market incentives to accommodate the future.

The reform paves the way for advanced "spectrum sharing" by allowing gains from trade to flow to those parties whose active assistance—say, by upgrading technology, revamping networks, adopting bandwidth saving applications, switching to less contested bands, or exiting the market— contributes to a consumer-pleasing outcome. Government studies may assume that these forms of sharing are too complicated to arrange, but the white flag is hoisted due to a dubious choice of tactics. Hammering out the rules for cohabitation in forced spectrum sharing campaigns sparks instant resistance and can consume decades, as the 2012 PCAST Report concedes.[34]

The nub of an economic agenda for reform would include:

1. *Auctioning overlay rights to bands allocated to traditional licenses.* These new liberal authorizations would vest existing wireless users in their current activities but would create new, fully flexible rights

to employ all idle frequencies. Moreover, the licensed newcomers (who might also be oldcomers, or any party choosing to bid) would enjoy secondary rights to use the spaces occupied by incumbents, enabling bargains. Spectrum sharing would unleash gains from trade, as entrants compensated legacy interests to make way for pretty amazing new stuff.

2. *Auctioning overlay rights—"hunting permits"—for bands assigned to public agencies.* While similar in form to overlay licenses for bands with private licensees, overlays for these bands require special measures. In particular, government departments (and their decision makers) must be able to make deals. The companies seeking to increase the value of their new rights will search for ways to make transitions happen, and U.S. regulators should provide supporting rules and institutions, for which templates (used in AWS-1, for instance) exist.

3. *Holding incentive auctions.* The FCC's National Broadband Plan called for considering overlay auctions should incentive auctions prove disappointing. Given the delays and limitations in the latter, it is best to shuffle the ranking. But the demoted Plan A might still prove useful. We should learn from the FCC's policy experiment and grasp lessons that may allow capacious spectrum reallocations to be achieved more expeditiously. These methods, combined with various overlay techniques, could develop more robust regulatory approaches.

4. *Instituting blanket liberalization of new and existing wireless licenses.* It is not necessary to reinvent the wheel, to devise more complex rules for transmitters or receivers (with extra dimensions defined by regulators), or for U.S. policy makers to ask Congress to rewrite Title III of the 1934 Communications Act so as to allow private ownership of frequencies. Liberalization can occur by tweaking existing rights. The flexible-use template that is well established in many countries' mobile licenses is an off-the-shelf model upon which policy makers can build. Where licenses still prohibit networks from providing innovative services, technologies, or business models, deregulation can supply breathing space. One option

is to invite licensees to request relaxation of use rules, with presumptive approval for any noninterfering activity. Complaints lodged by protesting parties would be limited to border disputes (which would exclude "public interest" arguments seeking to limit competition, like radio stations wishing to prohibit local programming by satellite radio vendors) and adjudicated under efficient arbitration rules with strict time limits. ("Baseball arbitration," where opposing parties submit their preferred solutions, and the arbitrator picks one, is a potential fast-track option.) This might lead regulators to deny requests to comply with the mandate, but applicants generally prefer a rejection—which can be legally challenged at the next level—to a multiyear proceeding-in-progress.

5. *Requiring government entities to purchase commercial wireless services.* Public agencies, including first responders, should buy wireless services from competitive suppliers. Contracts would include quality-of-service agreements that established priority for public safety communications, added capacity during emergencies, tapped redundant networks for backup, and delivered custom functionalities. Competitive provision, not least because it would lead to spectrum sharing with mass-market civilian customers, would demonstrably outperform the results achieved when public agencies try to create their own networks.[35] A bonus is that radio spectrum will be treated as a resource, not wantonly stockpiled. "So as not to distort spectrum usage decisions," writes former Pentagon official (and economist) Dorothy Robyn, "the government should subsidize the desired social good (i.e., public safety) directly and then let the relevant group acquire with spectrum or spectrum-based services in the market."[36]

6. *Using these spectrum reforms to salvage greater civilian use of military bands.* Military frequencies represent special targets of opportunity, as they are a vast, underutilized holding. They also present unique challenges, including the need for secrecy in national security missions. The same market reforms applied elsewhere can bring economic rationality to the armed forces' spectrum alloca-

tion choices, but it will take more strategic thinking to implement such changes.

7. *Shifting the methodology for creating unlicensed allocations to markets.* Interests favoring set-asides for nonexclusive spectrum access can and should bid for liberal licenses, and such licenses should not preclude "non-network" applications. Where a compelling determination is made that bids for such business models will suffer from public good (free rider) problems, and be inefficiently underprovided, the handicap should be remedied by explicit subsidies. Prices are then transparent, not hidden, and competing interests have the opportunity to demonstrate rival valuations in arm's-length transactions. In such a manner, new bandwidth can pour into unlicensed employments, with rules set by responsible economic agents responding to opportunities for new efficiencies under evolving conditions.

8. *Implementing complementary policies.* Many reform proposals have outlined policies to support spectrum liberalization.[37] I will mention just one: national policy, in the United States, could do more to overcome the NIMBY (not-in-my-backyard) problem encountered in building cellular towers and base stations. Metropolitan jurisdictions routinely block new radio facilities, sometimes out of concern that the radiation from such transmitters is harmful to human health. Whatever radiation threat exists is associated with mobile phones being held close to the user's brain. People who are nervous about cell phone radiation are advised to use the speaker function, which "drastically reduces RF [radio frequency] exposure."[38] The power emitted by phones *increases* when base stations are fewer; radios amp up to reach the more remote tower. Hence blocking nearby towers ironically exacerbates exposure to radio emissions. With coming 5G technologies, mobile carriers will attempt to radically densify their networks, adding vast numbers of "small cells" to ramp up capacity and reduce latency (the lags in between interactive communications). But they must first scale the roadblocks erected by local governments. Holdups here are

endemic, and federal efforts in the United States to impose a "shot clock" on municipal approvals (enacted in 2011) recall some official descriptions, circa the 1970s, of the results of the Vietnam War: an "incomplete success."

Toaster Liberation

A strong meme in the spectrum policy debate posits that spectrum rights determine how services develop and markets evolve, and the regulator's task is to choose the best outcomes for society. But the unique correspondence between rules and results holds only if we're doing public policy wrong. Truly flexible rights, distributed in ways that avoid tragedy, release competitive forces to discover new efficiencies without the need for n-year regulatory rulemakings. This simple reality brings helpful context to radio spectrum allocation.

Change is controversial. When Ronald Coase broached the idea of auctions for spectrum rights, it was received as a "big joke" whose odds of adoption—even decades later—were equal to those of "the Easter Bunny in the Preakness." These dismissals were accompanied by a phalanx of regulatory defenses. The FCC's chief economist in the 1950s, Dallas Smythe, declared it impossible to define wireless rights and sell them. That wisdom was little challenged even though licenses had traded in secondary markets since before the Radio Act of 1927.

Coase might have been naïve about the politics, but his economics were spot on. Not only could rights be defined and sold, competitive bids could be used to speed, and improve, services for consumers. Auctions of wireless assignments eliminated many of the obstacles holding back spectrum-based services and enabled elegant solutions for other challenges. From electricity to water to pollution allowances to fishing rights,[39] newly constructed markets have fashioned superior alternatives to command-and-control regulation. Today, economists and systems engineers are at work designing ever more ambitious bidding mechanisms, revealing hidden values, improving resource use, and saving the planet. Many of these innovations— according to Caltech's Charles Plott, a central figure in this revolution— emerge from "spectrum auctions."[40]

The foundational idea was dismissed in 1960s Washington not only because incumbent licensees and bureaucrats had their own agendas but because—as policy disruptor Tom Whitehead noted—the president of the United States had his.[41] This was perhaps the only time the matter of spectrum allocation got such high-level attention. As Bobby Baker, Lyndon Johnson's top Senate staffer (later imprisoned on corruption charges), wrote of his boss: "It was no accident that Austin, Texas, was for years the only city of its size with only one television station. Johnson had friends in high places. . . . LBJ demanded, and received, the opportunity to pick and choose programs for his monopoly station from among those offered by all three of the major networks."[42]

The regulatory system that accommodated this sordid record is trumpeted, even decades later, as a noble enterprise protecting the public. In 2013, FCC Chairman Tom Wheeler bemoaned broadcast video content and suggested that "maybe the industry was in need of another Newton Minow 'Vast Wasteland' moment to use the bully pulpit of the post to call them to the angels of their better programming natures when it came to violence or indecency."[43]

The "better angels" were unleashed, but not when broadcast triopolists were ostentatiously flailed by an agency that simultaneously protected them from cable TV competition. Progress came when the model flipped. With deregulation, unlicensed cable program networks created exponentially greater variety in video choice. Network information sources beaming bland "News from Nowhere" in three dull shades were replaced by a raucous rivalry of 24/7 services ranging from CNN to Fox News to Comedy Central to Vice. Diversity further widened as the unregulated Internet came to sit atop emerging broadband networks. Better angels did make it to television—*Touched by an Angel* and *Buffy the Vampire Slayer* were both hit series—but it was a very earthly set of reforms that set aside "public interest" rulemakings and made possible *Sex and the City*, *Dexter*, and C-SPAN.

Depoliticizing the Political Spectrum

Spectrum allocation has a long history of paradox. The best tool for understanding it is not the physics of radio waves but in the economics of public choice, which explains how special interests craft political coalitions and ally with regulators to distribute favors that bless the anointed while shorting entrepreneurial risk taking.

The "deregulation wave" of the 1970s changed history and many of its positive externalities blessed the political spectrum.[44] The Open Skies reform broke the government-backed monopoly of Comsat in satellite communications in about 1975. Rules blocking cable TV to protect broadcast were relaxed in the late 1970s. By the late 1980s, cellular licenses had introduced a more open regime in which competitive firms would decide what services to offer and what business models and technologies to employ. By the 1990s, mobile networks had eclipsed broadcasting as the preeminent wireless sector. The cozy spectrum allocation club became overrun with new members. They were chasing the future. In 1997, a confident Peter Huber wrote:

> It appears that old-style broadcasters will carry the regulatory baggage of the 1934 Act for another decade or so. Early in the next century, however, this dismal regulatory era will finally come to an end. Broadcast spectrum will be dezoned. *Roseanne* will have to compete for airtime with the more civil, uplifting, and profitable expressions of ordinary people talking on wireless phones.[45]

Yes and no. By the 2000s, cellular networks had gained international traction, bringing modern information services to billions, helping many up from poverty. More liberal spectrum usage rules allowed carriers to offer text, data, and video services atop voice networks; generate daring new ecosystems populated by smartphones, tablets, netbooks, and dongles; and provide connectivity for millions of applications. These emerging platforms are giving birth to machine-to-machine services that can disrupt markets, alter social intercourse, topple governments, and even extend human life. Crime rates fell with the introduction of cell phones, and mHealth innovations are pushing patients to take their meds while monitoring their vitals.

Social media gave a booster shot to global democratic movements. Spectrum policies gave markets room to roam, limiting the "controllers." Broadcast TV, tightly licensed and subject to the Fairness Doctrine and Equal Time and the Public Interest Standard, led no revolutions. Under traditional au-

Guglielmo Marconi, *Inventor*   Herbert Hoover, *Regulator*   Steve Jobs, *Entrepreneur*

20.1. *Who Doesn't Belong, and Why?* Steve Jobs picture from Annette Shaff / Shutterstock.com.

thorizations, services were preordained and innovation was lost. With liberal licenses, up popped green shoots.

The Internet of Things is just revealing its shape and scope. Tech writer Vivek Wadhwa predicts that, using "sensors and the apps that tech companies will build, our smartphone will become a medical device akin to the *Star Trek* tricorder."[46] These coming advances in science, culture, and economics stretch far beyond radio spectrum policy. But they are related.

As the dreams of visionaries grow, the drag imposed by anticompetitive spectrum regulation becomes all the more damaging. Wireless is a key component of the drive for a better world, so it becomes increasingly curious that society would seek to slow its progress. Wireless technologies of freedom have opened up new vistas; we can see the future from here. The political spectrum ought to stop blocking the magnificent view.

NOTES

Introduction

Epigraph: Quoted in Deirdre N. McCloskey, *Two Cheers for Corruption*, WALL STREET JOURNAL (Feb. 27, 2015).

1. Steve Mirsky, *Einstein's Hot Time*, SCIENTIFIC AMERICAN (Sept. 2002). A search of the Albert Einstein Archives in Jerusalem, however, turns up no record of the comment. It is possible that the trope is the product of a hijacking by cat lovers. "Guglielmo Marconi was asked how his new gadget, the radio, differed from a telegraph. 'A telegraph is like a dog,' Marconi replied. 'You pinch the tail at one end, and it barks at the other. My radio is just the same, except there is no dog.'" Peter Huber, LAW AND DISORDER IN CYBERSPACE (New York: Oxford University Press, 1997), 13.
2. Martin Cooper, *Antennas Get Smart*, SCIENTIFIC AMERICAN (July 2003).
3. *Cisco Visual Networking Index: Global Mobile Data Traffic Forecast Update, 2011–2016*, Cisco White Paper (Feb. 14, 2012), 1.
4. Quoted in J. Gregory Sidak, *Old Regulations Never Die: Featherbedding and Maritime Safety After the Titanic*, 1 CRITERION JOURNAL ON INNOVATION 201 (2016), 203. Sidak presents an interesting political explanation of the Radio Act and subsequent legislation mandating radio operators on U.S. ships.
5. The Radio Act of 1912, 47 USCA § 55. The Department was split into the Department of Labor and the Department of Commerce in 1913. The authority granted by the 1912 Radio Act went with the latter.
6. James L. Swanson, MANHUNT: THE 12-DAY CHASE FOR LINCOLN'S KILLER (New York: HarperCollins, 2007). John Wilkes Booth was a well-known actor, familiar to the Ford's Theatre audience left horrified at the scene of the crime, and to the soldiers with whom he was soon chatting during his escape.

7. Jamie Turner, *Are There REALLY More Mobile Phones than Toothbrushes?* 60 Second Marketer (Oct. 18, 2011); Yue Wang, *More People Have Cell Phones than Toilets, U.N. Study Shows,* Time (March 25, 2013).

8. Michael A. Heller, *The Tragedy of the Anticommons: Property in the Transition from Marx to Markets,* 111 Harvard Law Review 621 (Jan. 1998); Heller, The Gridlock Economy (New York: Basic, 2008).

9. Thomas W. Hazlett & Brent Skorup, *Tragedy of the Regulatory Commons: LightSquared and the Missing Spectrum Rights,* 13 Duke Law & Technology Review 1 (Dec. 2014).

10. This is a common theme in Public Choice analysis. "Olson's Law," as dubbed in a popular 2014 book, is the proposition that "the larger the group, the less it will further its common interests." Mancur Olson, The Logic of Collective Action (Cambridge, MA: Harvard Economic Studies, 1965), 36. See also, e.g., James Buchanan & Gordon Tullock, The Calculus of Consent (Ann Arbor: University of Michigan Press, 1962). The term "Olson's Law" is used in John Micklethwait & Adrian Wooldridge, The Fourth Revolution: The Global Race to Reinvent the State (New York: Penguin, 2014), 111.

11. Richard A. Posner, *Taxation by Regulation,* 2 Bell Journal of Economics and Management Science 22 (1971).

12. The bar claims that its slogan was coined by piano performer Mr. Whitekeys.

13. Ithiel de Sola Pool, Technologies of Freedom (Cambridge, MA: Belknap Press of Harvard University Press, 1983), 10.

Part I. Welcome to the Jungle

Epigraph: Ithiel de Sola Pool, Technologies Without Boundaries (Cambridge, MA: Harvard University Press, 1990), 46.

Chapter 1. Dances with Regulators

Epigraph: Joseph A. Schumpeter, History of Economic Analysis (New York: Oxford University Press, 1954), 90.

1. Newton Minow, Equal Time: The Private Broadcaster and the Public Interest (New York: Atheneum, 1964), 51–52.

2. Ibid., 67.

3. *Minow Wins Peabody Award,* Milwaukee Sentinel (April 19, 1962), 4.

4. Minow (1964), 49.

5. Ibid., 57.

6. Ibid., 59.

7. Henry Geller confirmed this outcome in an email to the author dated Jan. 22, 2013. Geller "worked at the Federal Communications Commission (FCC) at several intervals from 1949 until 1973, serving as General Counsel for six years (1964–70)." Museum of Broadcast Communications (visited Jan. 29, 2013).

8. "Like drug dealers on the corner, [TV broadcasters] control the life of the neighborhood, the home, and, increasingly, the lives of the children in their custody. Unlike drug dealers, they cannot be chased away or deterred: they claim a constitutional right to stay." Newton N. Minow & Craig L. LaMay, ABANDONED IN THE WASTELAND: CHILDREN, TELEVISION, AND THE FIRST AMENDMENT (New York: Hill & Wang, 1995), 18.

9. *Inquiry into the Impact of Community Antenna Systems, TV Translators, TV "Satellite" Stations, and TV "Repeaters" on the Orderly Development of Television Broadcasting, 26 FCC 403 (1959), 427–429, 437.*

10. Sloan Commission on Cable Communications, ON CABLE: THE TELEVISION OF ABUNDANCE (New York: McGraw-Hill, 1972), 25.

11. Stanley Besen & Robert W. Crandall, *The Deregulation of Cable Television,* 44 LAW & CONTEMPORARY PROBLEMS 81 (Winter 1981), 85.

12. *Carter Mountain Transmission Corp.,* 32 FCC 459 (1962), aff'd, 321 F.2d 359 (D.C. Cir. 1963), cert. denied, 375 U.S. 951 (1963).

13. *Community Antenna Systems* (1959), 432.

14. Stuart N. Brotman, *From the "Vast Wasteland" to "Net Neutrality,"* Brookings Institution Tech Tank (Jan. 29, 2015).

15. Minow (1964), 55.

16. Franklin Fisher, Victor E. Ferrall, Jr., David Belsley, & Bridger M. Mitchell, *Community Antenna Television Systems and Local Television Station Audience,* 80 THE QUARTERLY JOURNAL OF ECONOMICS 227 (May 1966).

17. Second Order and Report, 2 FCC 2d 725 (1966).

18. See Rolla Edward Park, POTENTIAL IMPACT OF CABLE GROWTH ON TELEVISION BROADCASTING (Santa Monica, CA: RAND Corp., R-587-FF, 1970); Gary Fournier & Ellen Campbell, *Shifts in Broadcast Policy and the Value of Television Licenses,* 5 INFORMATION ECONOMICS AND POLICY 87 (Jan. 1993).

19. Minow & Lamay (1995), 5. While the passage appeals to the public interest standard set forth in the 1934 Communications Act, that legislation simply repeated the language of the 1927 Radio Act.

20. Newton Minow, *How Vast the Wasteland Now?* Speech, Gannett Foundation at Columbia University (May 9, 1991), and cited in Thomas Krattenmaker & Lucas A. Powe, Jr., REGULATING BROADCAST PROGRAMMING (Cambridge, MA: MIT Press, 1994), 299.

21. Krattenmaker & Powe (1994), 299.

22. Newton N. Minow, *A Vaster Wasteland,* ATLANTIC (April 2011). James Warren, *Never Mind the "Vast Wasteland." Minow Has More to Say,* N.Y. TIMES (May 7, 2011); James Fallows, *Worth Watching: Newton Minow, 50 Years Later,* ATLANTIC (May 2011).

23. *In re: Establishment of Domestic Communications-Satellite Facilities by Non-Governmental Entities, Docket No. 16495,* 38 F.C.C.2d 665 (Dec. 22, 1972).

24. Brian Lamb, Tullock Lecture, Information Economy Project, George Mason University (Oct. 4, 2006). The transcript was clarified via email communication with Brian Lamb (Jan. 23, 2013).

25. R. H. Coase, *Evaluation of Public Policy Relating to Radio and Television Broadcasting: Social and Economic Issues,* 41 LAND ECONOMICS 161 (May 1965), 167.

26. Federal Communications Commission, *In the Matter Regarding the Applications of Cowles Florida Broadcasting, Inc. (WESH TV), Daytona Beach, Florida for Renewal of a License, Dissenting Opinion of Commissioner Glen O. Robinson,* Docket No. 19168 (June 20, 1976), 1976 WL 31608 (F.C.C.), 48.

27. See Clarence C. Dill, RADIO LAW, PRACTICE, AND PROCEDURE (Washington, DC: National Law Book Co., 1938), 89; and Thomas W. Hazlett, *The Rationality of U.S. Regulation of the Broadcast Spectrum,* 33 JOURNAL OF LAW & ECONOMICS 133 (April 1990). This process is explained in greater detail below.

28. Laurence F. Schmeckebier, THE FEDERAL RADIO COMMISSION (Washington, DC: Brookings Institution, 1932), 55.

29. *Mark Fowler: Interview,* REASON (Nov. 1981).

30. Ibid.

31. Steve Daley, *News Judgment from a Toaster,* CHICAGO TRIBUNE (Sept. 24, 1987).

32. Georgie Anne Geyer, *How Broadcasters Misuse the First Amendment,* CHICAGO TRIBUNE (March 22, 1996).

33. Toby Miller, CULTURAL CITIZENSHIP: COSMOPOLITANISM, CONSUMERISM, AND TELEVISION IN A NEOLIBERAL AGE (Philadelphia: Temple University Press, 2007), 13.

34. *Jeopardy!* Game #6424 (July 19, 2012).

35. Susan Crawford, *More than an Appliance: Verizon, the FCC, and Our Digital Future,* HUFFINGTON POST (Aug. 21, 2012).

36. Peter J. Boyer, *Under Fowler, F.C.C. Treated TV As Commerce,* N.Y. TIMES (Jan. 19, 1987).

37. Lili Levi, *In Search of Regulatory Equilibrium,* 35 HOFSTRA LAW REVIEW 1321 (2007), 1324.

38. *NBC v. United States,* 319 U.S. 190 (1943).

39. "*National Broadcasting* was a declaration of independence for regulatory agencies." Charles Murray, BY THE PEOPLE: REBUILDING LIBERTY WITHOUT PERMISSION (New York: Crown Forum, 2015), 74.

40. *NBC v. United States* (1943), 213.

41. Ibid. (footnote omitted).

42. *NBC v. United States* (1943), 216–217.

43. A "tragedy of the commons" in biologist Garrett Hardin's classic formulation. Hardin, *The Tragedy of the Commons,* 162 SCIENCE 1243 (Dec. 13, 1968).

44. Jim Chen, *Conduit-based Regulation of Speech,* 54 DUKE LAW JOURNAL 1359 (2005), 1403.

45. Matthew L. Spitzer, SEVEN DIRTY WORDS AND SIX OTHER STORIES: CONTROLLING THE CONTENT OF PRINT AND BROADCAST (New Haven: Yale University Press, 1987).

46. "Widely considered the most knowledgeable communications law and policy expert, Geller is legendary for his subtle grasp of the issues and ability to suggest innovative solutions." Robert M. Entman, *Communications Policy: Putting Regulation in Its Place,* in Marcus Raskin & Chester Hartman, eds., WINNING AMERICA: IDEAS AND LEADERSHIP FOR THE 1990s (Boston: South End, 1988), 261.

47. Henry Geller, *Broadcasting and the Public Trustee Notion: A Failed Promise*, 10 HARVARD JOURNAL OF LAW & PUBLIC POLICY 87 (1987), 90.

48. Ibid.

49. See presentations by Gigi Sohn (of Public Knowledge) and Norman Ornstein (American Enterprise Institute), *The Gore Commission, Ten Years Later: The Public Interest Obligations of TV Broadcasters in Perfect Hindsight*, George Mason University Information Economy Project (Oct. 3, 2008).

50. Minow & LaMay (1995), 26.

51. Alfred E. Kahn, THE ECONOMICS OF REGULATION: PRINCIPLES AND INSTITUTIONS (Cambridge, MA: MIT Press, 1988), xxiii.

52. While U.S. rules always allowed cellular licensees more flexibility than broadcast television licensees, the formal deregulation of mobile technology choices dates to 1988. This lagged the initial licenses, distributed mostly by lottery during 1984–1989.

53. Christopher Rhoads, *AT&T Inventions Fueled Tech Boom, and Its Own Fall*, WALL STREET JOURNAL (Feb. 2, 2005).

54. Drew DeSilver, *CDC: Two of Every Five U.S. Households Have Only Wireless Phones*, Pew Research Center (July 8, 2014).

55. Federal Communications Commission, *National Broadband Plan* (NBP) (March 2010), 78.

56. Ibid.

57. Roy Mark, *Spectrum Auction Delay Hits Fast Track*, INTERNET NEWS (May 3, 2002).

58. Federal Communications Commission, *Report of the Spectrum Efficiency Working Group*, Docket No. 02-135 (Nov. 15, 2002), 29.

59. As the FCC's chief economist and chief technologist in 2000–2001 labeled the regulatory system in Gerald R. Faulhaber & David J. Farber, *Spectrum Management: Property Rights, Markets, and the Commons*, in Lorrie Faith Cranor & Steven S. Wildman, eds., RETHINKING RIGHTS AND REGULATIONS: INSTITUTIONAL RESPONSES TO NEW COMMUNICATIONS TECHNOLOGIES (Cambridge, MA: MIT Press, 2003).

60. On the impact of the McLuhan metaphor, see Nicholas Carr, THE SHALLOWS: HOW THE INTERNET IS CHANGING THE WAY WE THINK, READ, AND REMEMBER (London: Atlantic, 2011).

61. *Amendment of Parts 2 and 22 of the Commission's Rules to Permit Liberalization of Technology and Auxiliary Service Offerings in the Domestic Public Cellular Radio Telecommunications Service*, Report and Order, 3 FCC Rcd 7033 (1988).

62. *Implementation of Sections 3(n) and 332 of the Communications Act*, Second Report and Order, 9 FCC Rcd 1411, ¶¶ 90–92 (1994).

63. Omnibus Budget Reconciliation Act, Pub. L. No. 103-66, 107 Stat. 312 (1993).

64. Thomas W. Hazlett, Roberto Muñoz & Diego Avanzini, *What Really Matters in Spectrum Allocation Design*, 10 NORTHWESTERN JOURNAL OF TECHNOLOGY & INTELLECTUAL PROPERTY 93 (Jan. 2012), 100.

65. Todd Spangler, *CES: Genachowski: More TV Spectrum Could Be Reclaimed Within Two Years*, BROADCASTING & CABLE (Jan. 7, 2011).

66. Margaret Kriz, *Supervising Scarcity*, NATIONAL JOURNAL (July 7, 1990), 1660.

67. Blair Levin, *When an ROI 500 Times Better than Goldman Isn't Enough: Reallocating Our Focus on Reallocating Spectrum,* BROADBAND & SOCIAL JUSTICE (March 14, 2012).

68. 14 F.C.C.R. 19,868–19,870 (1999); NBP (2010), 85.

69. Bruce M. Owen, Jack H. Beebe & Willard G. Manning, Jr., TELEVISION ECONOMICS (Lexington, MA: Lexington, 1974), 12.

70. Most regulatory tracts identify the prime frequencies as extending from about 100 MHz to 3 GHz. But higher frequencies are becoming increasingly valuable with the march of science.

71. George Gilder, *Auctioning the Airwaves,* FORBES ASAP (April 11, 1994).

72. Lawrence Ausubel and Peter Cramton write: "The best sites for offshore wind farms on the US Outer Continental Shelf are scarce. To make the best use of this scarce resource, it is necessary to implement a fair and efficient mechanism to assign wind rights to companies that are most likely to develop offshore wind energy projects." Lawrence M. Ausubel & Peter Cramton, *Auction Design for Wind Rights,* White Paper (Aug. 9, 2011).

73. James B. Murray, Jr., WIRELESS NATION: THE FRENZIED LAUNCH OF THE CELLULAR REVOLUTION IN AMERICA (Cambridge, MA: Perseus, 2001), 50.

74. Ibid.

75. Ithiel de Sola Pool, TECHNOLOGIES WITHOUT BOUNDARIES (Cambridge, MA: Harvard University Press, 1990), 42.

76. Nancy L. Rose, *After Airline Deregulation and Alfred Kahn,* 102 AMERICAN ECONOMIC REVIEW 376 (May 2012), 378.

77. Minow (1964), 36.

Chapter 2. Etheric Bedlam

Epigraph: Glen O. Robinson, *Regulating Communications: Stories from the First 100 Years,* 13 GREEN BAG 2D 303 (Spring 2010), 317.

1. *Red Lion Broadcasting Co. v. Federal Communications Commission,* 395 U.S. 367 (1969), 375–376 (footnotes omitted).

2. The phrase was used in an early U.S. Navy report, quoted in Ronald Coase, *The Federal Communications Commission,* 2 JOURNAL OF LAW & ECONOMICS 1 (1959), 2. The reference to "ether" (sometimes "aether") reflected the nineteenth-century belief, since discredited, that radio waves rippled through a gaseous substance in the atmosphere.

3. Testimony of Charles Ferris, *Hearing Before the Subcommittee on Telecommunications and Finance of the Committee on Energy and Commerce on H.R. 1934,* U.S. House of Representatives, One Hundredth Congress (April, 7, 1987), 62.

4. Testimony of Newton Minow, *Hearing Before the Subcommittee on Telecommunications and Finance of the Committee on Energy and Commerce on H.R. 1934,* U.S. House of Representatives, One Hundredth Congress (April, 7, 1987), 64.

5. Ellen P. Goodman, *Superhighway Patrol: Why the FCC Must Police the Airwaves,* WASHINGTON POST (Aug. 6, 1995), op-ed page.

6. Quoted in H. L. Mencken, THE PHILOSOPHY OF FRIEDRICH NIETZSCHE (1907), Amazon Kindle Edition (location 1,975 of 4,224).

7. U.S. Comp. St., 1916, §§ 10100–10110.

8. Dean Lueck, *The Rule of First Possession and the Design of the Law*, 38 JOURNAL OF LAW & ECONOMICS 393 (Oct. 1995).

9. Radio Control, Hearings Before the Committee on Interstate Commerce, United States Senate, Sixty-Ninth Congress, First Session 118-119 (1926).

10. Thomas W. Hazlett, *The Rationality of U.S. Regulation of the Broadcast Spectrum*, 33 JOURNAL OF LAW & ECONOMICS 133 (April 1990), 146.

11. Erik Barnouw, A TOWER IN BABEL (London: Oxford University Press, 1966), 185–186.

12. William Ray, FCC: THE UPS AND DOWNS OF RADIO-TV REGULATION (Ames: Iowa State University Press, 1990), 126–127.

13. *Aimee Semple McPherson*, Angeles Temple (visited July 28, 2015).

14. *History—Sister Aimee and Her Castle*, HIDDEN LOS ANGELES (visited July 27, 2015).

15. Aimee Semple McPherson, *Aimee: The Life Story of Aimee Semple McPherson* (International Church of the Foursquare Gospel, Los Angeles, 1979).

16. Ray (1990), 127.

17. Herbert C. Hoover, *The Urgent Need for Radio Legislation*, 2 RADIO BROADCAST 211 (Jan. 1923).

18. Herbert C. Hoover, THE MEMOIRS OF HERBERT HOOVER: THE CABINET AND THE PRESIDENCY, 1920–1933 (London: Macmillan, 1952), 143.

19. Herbert C. Hoover, Opening Address, Fourth National Radio Conference Proceedings (1925).

20. *United States v. Zenith Radio Corporation*, 12 F.2d 614 (N.D. Ill. 1926).

21. *Hoover v. Intercity Radio Co.*, 286 Fed. 1003 (App. D.C. 1923).

22. Douglas B. Craig, FIRESIDE POLITICS: RADIO AND POLITICAL CULTURE IN THE UNITED STATES, 1920–1940 (Baltimore: Johns Hopkins University Press, 2005), 49.

23. Silas Bent, *Radio Squatters*, INDEPENDENT 389 (Oct. 2, 1926).

24. *Tribune Co. v. Oak Leaves Broadcasting Station*, Cook County, Ill. Circuit Court (Nov. 17, 1926), reprinted in 68 CONGRESSIONAL RECORD—SENATE (Dec. 10, 1926), 215–219.

25. Ibid.

26. Philip T. Rosen, THE MODERN STENTORS: RADIO BROADCASTERS AND THE FEDERAL GOVERNMENT, 1920–1934 (Westport, CT: Praeger, 1980), 103.

27. CONGRESSIONAL RECORD—HOUSE (Jan. 29, 1927), 2579.

28. Regulators were conscious of the continuing tension. "The Commission, fearful that licensees would assert property interests in their coverage to the public, has inserted elaborate provisions in application forms precluding the assertion of any such right." Paul M. Segal & Harry P. Warner, *"Ownership" of Broadcasting "Frequencies": A Review*, 19 ROCKY MTN. L. REV. 111 (1947), 111.

29. Clarence C. Dill, RADIO LAW, PRACTICE, AND PROCEDURE (Washington, DC: National Book Co., 1938), 80–81. The term *free radio communication* is used here in its politically correct sense, a euphemism for *licensed and regulated*.

30. Federal Radio Commission, ANNUAL REPORT (1927), 10–11. This passage was from a speech given by Commissioner O. H. Caldwell. It should also be noted that KDKA began broadcasting in Pittsburgh on November 2, 1920, but was first licensed as a broadcasting station (only radio telegraphy licenses had previously been issued) in September 1921.

31. Quotation from *Red Lion* (1969).

32. *Federal Control of Radio Broadcasting*, 29 YALE LAW JOURNAL 247, footnote omitted (1929).

33. Quotation from *NBC v. United States*, 319 U.S. 190 (1943).

34. Dill (1938), 78.

35. Ibid.

36. *Ruling the Radio Waves*, OUTLOOK 463 (Nov. 25, 1925).

37. Barnouw (1966), 95.

Chapter 3. Protection by Subtraction

Epigraph: Erwin G. Krasnow, Lawrence D. Longley & Herbert A. Terry, THE POLITICS OF BROADCAST REGULATION, 3rd ed. (New York: St. Martin's, 1982), 16 (footnote omitted).

1. *Broadcasting Licenses*, 64 SCIENCE (Supplement, Dec. 17, 1926), xiv–xv.

2. As reported in Martin R. Bensman, THE BEGINNING OF BROADCAST REGULATION IN THE TWENTIETH CENTURY (Jefferson, NC: McFarland, 2000), 154.

3. Harry P. Warner, RADIO AND TELEVISION LAW, 1952 Cumulative Supplement (Albany: Matthew Bender, 1953), 766.

4. Estimates per a Department of Commerce study cited in Bensman (2000), 155.

5. Philip T. Rosen, THE MODERN STENTORS: RADIO BROADCASTERS AND THE FEDERAL GOVERNMENT, 1920–1934 (Westport, CT: Praeger, 1980), 125.

6. Federal Radio Commission, ANNUAL REPORT (1927), 13.

7. *Radio Welcomes Government Control*, LITERARY DIGEST 21 (April 9, 1927).

8. Robert McChesney, TELECOMMUNICATIONS, MASS MEDIA & DEMOCRACY: THE BATTLE FOR THE CONTROL OF U.S. BROADCASTING, 1928–1935 (New York: Oxford University Press, 1995), 19.

9. *Welcome to the Radio Commission*, RADIO BROADCAST 555 (April 1927).

10. *Stabilizing the Broadcast Situation*, RADIO BROADCAST 79 (June 1927).

11. McChesney (1995), 26.

12. Ibid., 30–31.

13. Using 1500–2000 KHz for radio broadcasting would have required some relocation of amateur radio operations. In fact, abundant idle bands were available for this.

14. McChesney (1995), 254. McChesney has described his own politics as socialist. Mark Karlin, *Robert McChesney: We Need to Advocate Radical Solutions to Systemic Problems*, Truth-out.org (Jan. 4, 2015).

15. Quoted in Erik Barnouw, A TOWER IN BABEL (New York: Oxford University Press, 1966), 219.

16. Thomas G. Krattenmaker & Lucas A. Powe, Jr., REGULATING BROADCAST PROGRAM-MING (Cambridge, MA: MIT Press, 1994), 20.

17. Federal Radio Commission, ANNUAL REPORT (1929).

18. Note, *Concepts of the Broadcast Media Under the First Amendment: A Reevaluation and a Proposal*, 47 N.Y.U. LAW REVIEW 84 (April 1972), 87 (footnote omitted).

19. Ithiel de Sola Pool, TECHNOLOGIES OF FREEDOM (Cambridge, MA: Harvard University Press, 1983), 124.

20. McChesney (1995), 28.

21. Lucas A. Powe, Jr., AMERICAN BROADCASTING AND THE FIRST AMENDMENT (Berkeley: University of California Press, 1987), 23.

22. Ibid., 13.

23. Ibid., 14–15.

24. Ibid., 15.

25. Ibid., 16.

26. Quoted in Powe (1987), 16.

27. Federal Radio Commission opinion in *Great Lakes Broadcasting Co.* (1929), quoted in Jonathan W. Emord, FREEDOM, TECHNOLOGY, AND THE FIRST AMENDMENT (San Francisco: Pacific Research Institute, 1991), 181.

28. Peter Huber, LAW AND DISORDER IN CYBERSPACE (New York: Oxford University Press, 1997), 43.

29. Edward Jay Epstein, NEWS FROM NOWHERE: TELEVISION AND THE NEWS (Chicago: Ivan R. Dee, 1973), vii.

30. Pool (1983), 122.

31. Quoted in Elizabeth Fones-Wolf, WAVES OF OPPOSITION: LABOR AND THE STRUGGLE FOR DEMOCRATIC RADIO (Urbana: University of Illinois Press, 2006), 20.

32. Ibid., 20–21.

33. Federal Radio Commission, ANNUAL REPORT (1929), 34.

34. Ibid., 32.

35. Ibid., 36.

36. Quoted in McChesney (1995), 71.

37. Ibid., 79.

39. Fones-Wolf (2006), 21.

39. Ibid., 242.

40. Federal Radio Commission, ANNUAL REPORT (1928), 155, 160.

41. *Debs Fund Buys Radio Station: Broadcasting Is to Begin Soon: Socialist, Labor, and Liberal Group Is Now Owner of WSOM*, 4 THE NEW LEADER (Aug. 6, 1927), 3.

42. Nathan Godfried & Ronald C. Kent, *Legitimizing the Mass Media Structure: The Socialists and American Broadcasting, 1926–1932*, in Ronald C. Kent et al., eds., CULTURE, GENDER, RACE AND U.S. LABOR HISTORY (Westport, CT: Greenwood, 1993), 135. See also Paul F. Gullifor & Brady Carlson, *Defining the Public Interest: Socialist Radio and the Case of WEVD*, 4 JOURNAL OF RADIO STUDIES (1997), 210.

43. Nathan Godfried, *Struggling over Politics and Culture: Organized Labor and Radio Station WEVD During the 1930s*, 42 LABOR HISTORY 347 (2001), 349.

44. Godfried & Kent (1993), 130.

45. Quoted in Brian Craig Dolber, *Sweating for Democracy: Working-Class Media and the Struggle for Hegemonic Jewishness, 1919–1941*, Doctoral Thesis, University of Illinois Department of Communications (2011), 276 (footnote omitted).

Chapter 4. Myth Calculation

Epigraph: Quoted in Les Brown, *Broadcast Regulation: Plan Makes Waves*, N.Y. TIMES (June 12, 1978).

1. Ithiel de Sola Pool, TECHNOLOGIES OF FREEDOM (Cambridge, MA: Harvard University Press, 1983).
2. Zhenzhi Guo, *A Chronicle of Private Radio in Shanghai*, 30 JOURNAL OF BROADCASTING & ELECTRONIC MEDIA 379 (1986).
3. Adrian Johns, DEATH OF A PIRATE: BRITISH RADIO AND THE MAKING OF THE INFORMATION AGE (New York: Norton, 2011).
4. Bernard Cros, *Why South Africa's Television Is Only Twenty Years Old: Debating Civilisation, 1958–1969*, O.R.A.C.L.E. (1996).
5. CAPE TIMES (May 4, 1967), quoted in *Television in South Africa* 10 CONTACT (1967). The quotation is attributed to Dr. Albert Hertzog, South African minister for posts and telegraphs at the time.
6. Ryan Nakashima, *Hollywood in China? Country's New Foreign Film Quotas Make the Industry Optimistic*, HUFFINGTON POST (April 17, 2012).
7. For an illuminating world tour, see Jack Goldschmidt & Tim Wu, WHO CONTROLS THE INTERNET? ILLUSIONS OF A BORDERLESS WORLD (New York: Oxford University Press, 2006).
8. *Schneider v. Smith*, 390 U.S. 17 (1968), 25.
9. David Sarnoff, *Uncensored and Uncontrolled*, 119 NATION 90 (1924).
10. Mary S. Mander, *The Public Debate About Broadcasting in the Twenties: An Interpretive History*, 28 JOURNAL OF BROADCASTING 167 (1984), 183–184 (citations omitted).
11. Clarence C. Dill, RADIO LAW, PRACTICE, AND PROCEDURE (Washington, DC: National Book Co., 1938), 127.
12. Ibid., 93.
13. *Red Lion Broadcasting Co. v. FCC*, 395 U.S. 367 (1969), 375–377 (citations omitted).
14. David L. Bazelon, *FCC Regulation of the Telecommunications Press*, 1975 DUKE LAW JOURNAL 213 (May 1975).
15. *Telecomm. Research & Action Ctr. v. FCC*, 801 F.2d 501 (D.C. Cir. 1986).
16. He continues: "It seems to me that were the [Supreme] Court to revisit the *Red Lion* issue . . . the odds are overwhelming that the current court would recognize the illogic of *Red Lion*." Laurence Tribe, *Plenary Address—Freedom of Speech and Press in the 21st Century: New Technology Meets Old Constitutionalism*, Progress & Freedom Foundation, Aspen Summit (Aug. 20, 2007), at 14: 45–18: 04.
17. Quoted in Erik Barnouw, A TOWER IN BABEL (New York: Oxford University Press, 1966), 197.

18. Bernard Schwartz, THE ECONOMIC REGULATION OF BUSINESS AND INDUSTRY: A LEGISLATIVE HISTORY OF U.S. REGULATORY AGENCIES, VOLUME 3 (New York: Chelsea House, 1973), 2078.

19. William Mayton, *The Illegitimacy of the Public Interest Standard at the FCC*, 38 EMORY LAW JOURNAL 715 (1989).

20. Clarence C. Dill, *A Traffic Cop for the Air*, 75 REVIEW OF REVIEWS 181 (1927), 181.

21. Columbia 250 Celebrates Columbians Ahead of Their Time, COLUMBIA 250 (Columbia University website visited Sept. 21, 2016, http://c250.columbia.edu/c250_celebrates/remarkable_columbians/edwin_howard_armstrong.html).

22. Lawrence Lessing, MAN OF HIGH FIDELITY: EDWIN HOWARD ARMSTRONG (New York: Bantam, 1954).

23. Yannis Tsividis, *Edwin Armstrong: Pioneer of the Airwaves*, COLUMBIA MAGAZINE (Spring 2002).

24. Christopher H. Sterling & Michael C. Keith, SOUNDS OF CHANGE: A HISTORY OF FM BROADCASTING IN AMERICA (Chapel Hill: University of North Carolina Press, 2008), 18–19.

25. Hugh Richard Slotten, *"Rainbow in the Sky": FM Radio, Technical Superiority, and Regulatory Decision-Making*, 37 TECHNOLOGY AND CULTURE 686 (Oct. 1996), 686.

26. Sterling & Keith (2008), 22.

27. Ibid., 184.

28. Hugh R. Slotten, RADIO AND TELEVISION REGULATION: BROADCAST TECHNOLOGY IN THE UNITED STATES, 1920–1960 (Baltimore: Johns Hopkins University Press, 2000), 117.

29. Slotten (1996), 692.

30. Lessing (1954), 196–197.

31. Ibid., 196–197, 200.

32. Slotten (1996), 693.

33. Sterling & Keith (2008), 54, 65.

34. Lessing (1954), 207, 211.

35. Sterling & Keith (2008), 57.

36. "The plain dishonesty of this order was promptly demonstrated when the FCC turned about and assigned the band it had just ordered FM to vacate to television, a service about twenty-five times more sensitive to any kind of interference than FM and which, moreover, was still required to use FM on its sound channel. Later the same band of frequencies was assigned to government safety and emergency radio services, in which interference of any kind could be tolerated even less than in commercial broadcasting or television. The fact is that none of the 'ionospheric interference' predicted for this band ever materialized." Lessing (1954), 213.

37. Quoted in Lawrence D. Longley, *The FM Shift in 1945*, 12 JOURNAL OF BROADCASTING 353 (1968), 355.

38. Ibid., 359.

39. Lessing (1954), 211.

40. Ibid., 223.

41. Vincent Ditingo, THE REMAKING OF RADIO (Boston: Focal, 1995), 18, 60.
42. Thomas W. Hazlett & David W. Sosa, *Was the Fairness Doctrine a "Chilling Effect"? Lessons from the Post-Deregulation Radio Market*, 26 JOURNAL OF LEGAL STUDIES 307 (Jan. 1997).
43. Sterling & Keith (2008), 11.
44. Armstrong's widow would receive more than ten million dollars in judgments and settlements from RCA, Emerson, Motorola, and nearly twenty other firms when the litigation finally ended in 1967.
45. Lessing (1954), 225–226. Armstrong's heroic struggle to launch FM landed the inventor in a star-crossed pantheon. He is one of twenty-two people chronicled in Ken Smith, RAW DEAL: HORRIBLE AND IRONIC STORIES OF FORGOTTEN AMERICANS (New York: Blast, 1998).

Chapter 5. Eureka-nomics

Epigraph: Mark S. Fowler & Daniel L. Brenner, *A Marketplace Approach to Broadcast Regulation*, 60 TEXAS LAW REVIEW 207 (1982), 221 (footnote omitted). Fowler was then FCC chairman and Brenner his chief of staff.

1. The phrase is from Thomas Sowell, INTELLECTUALS IN SOCIETY (New York: Basic, 2011).
2. Email to the author (April 6, 2016). Quoted with permission.
3. Telecommunications Science Panel, Commercial Technical Advisory Board, *Electromagnetic Spectrum Utilization: The Silent Crisis* (Washington, DC: U.S. Government Printing Office, 1966).
4. Thomas W. Hazlett, *Looking for Results: An Interview with Ronald Coase*, REASON (Jan. 1997).
5. Ronald H. Coase, *Biographical*, Nobel Prize website.
6. Ronald H. Coase, *Law and Economics at Chicago*, 36 JOURNAL OF LAW & ECONOMICS 239 (April 1993), 248.
7. *NBC v. United States*, 319 U.S. 190 (1943), 225.
8. Ibid., 228.
9. Ronald Coase, *The Federal Communications Commission*, 2 JOURNAL OF LAW & ECONOMICS 1 (1959), 14.
10. Ibid., 40.
11. Leo Herzel, *"Public Interest" and the Market in Color Television Regulation*, 18 UNIVERSITY OF CHICAGO LAW REVIEW 802 (1951).
12. Dallas W. Smythe, *Facing Facts About the Broadcasting Business*, 20 UNIVERSITY OF CHICAGO LAW REVIEW 96 (1952), 96, 98.
13. *Television Economics*, 13 FEDERAL COMMUNICATIONS BAR JOURNAL 51 (1953), 89. The exchange included Leo Herzel, *Rejoinder*, 20 UNIVERSITY OF CHICAGO LAW REVIEW 106 (1952).
14. Smythe (1952), 100.
15. Herzel (1952), 106.

16. Jerry Brito, *Sending Out an S.O.S.: Public Safety Communications Interoperability as a Collective Action Problem*, FEDERAL COMMUNICATIONS LAW JOURNAL 457 (2007).

17. President's Task Force on Communications Policy, *Final Report* (U.S. Government Printing Office, 1968), Chapter VIII, 16–18. Quoted in Harvey J. Levin, THE INVISIBLE RESOURCE: USE AND REGULATION OF THE RADIO SPECTRUM (Washington, DC: Resources for the Future, 1971), 2.

18. Coase (1993), 249.

19. Coase (1959), 16.

20. George J. Stigler, MEMOIRS OF AN UNREGULATED ECONOMIST (New York: Basic, 1988), 75.

21. Coase (1993), 250.

22. Stigler (1988), 73.

23. Quoted in George H. Douglas, H. L. MENCKEN, CRITIC OF AMERICAN LIFE (Hamden, CT: Archon, 1978), 118.

24. R. H. Coase, *The Problem of Social Cost*, 3 JOURNAL OF LAW & ECONOMICS 1 (1960).

25. Robert W. Hahn & Robert N. Stavins, *The Effect of Allowance Allocations on Cap-and-Trade System Performance*, 54 JOURNAL OF LAW AND ECONOMICS S267 (Nov. 2011).

26. Thomas W. Hazlett, David Porter & Vernon Smith, *Radio Spectrum and the Disruptive Clarity of Ronald Coase*, 54 JOURNAL OF LAW & ECONOMICS S125-65 (Nov. 2011).

27. Arthur M. Diamond, Jr., *Most-Cited Economics Papers and Current Research Fronts*, CURRENT COMMENTS (Jan. 9, 1989).

28. Coase (1993), 250.

29. Coase (1959), 28.

30. R. H. Coase, *Comment on Thomas W. Hazlett: Assigning Property Rights to Radio Spectrum Users—Why Did FCC License Auctions Take 67 Years?* 41 JOURNAL OF LAW & ECONOMICS 577 (Oct. 1998), 579.

31. H. H. Goldin & R. H. Coase, *Discussion of "Evaluation of Public Policy Relating to Radio and Television Broadcasting: Social and Economic Issues" (Coase)*, 41 LAND ECONOMICS 161 (May 1965), 168.

32. Coase (1998), 579.

33. Ibid., 580.

34. Ronald Coase, William Meckling & Jora Minasian, *Problems of Radio Frequency Allocation*, RAND Corporation DRU-1219-RC (1995).

35. Levin (1971).

36. Paul Baran, *Is the UHF Frequency Shortage a Self Made Problem?* Paper delivered to the Marconi Centennial Symposium, Bologna, Italy (June 23, 1995), 1.

37. Andrew J. Viterbi, *The History of Multiple Access and the Future of Multiple Services through Wireless Communication*, Address at Marconi Centennial and Rutgers WINLAB Decennial (Sept. 30, 1999).

38. The taxonomy owes to Levin (1971).

39. John Robinson, *Spectrum Management Policy in the United States: An Historical*

Account, OPP Working Paper No. 15 (Washington, DC: Federal Communications Commission, April 1985), 10, B-5, B-8, B-12.

40. Ithiel de Sola Pool, TECHNOLOGIES WITHOUT BORDERS (Cambridge: Harvard University Press, 1990), 28–29.

41. Cisco Global Policy and Government Affairs, *High Tech Policy Guide—Wireless Spectrum Management,* Cisco website (Jan. 2005).

42. Base stations, often perched atop cell towers, typically connect to the network via high-capacity fiber-optic wires. In some cases this "backhaul" is achieved through point-to-point wireless links.

43. Peter Huber, LAW AND DISORDER IN CYBERSPACE (New York: Oxford University Press, 1997), 67–68.

44. Figure 4-F in FCC, *National Broadband Plan* (March 2010), 41. See also: Peter Rysavy, *General Packet Radio Services (GPRS),* GSM DATA TODAY (Sept. 30, 1998); *The State of LTE,* OPEN SIGNAL (Feb. 2013).

45. Sacha Segan, *Fastest Mobile Networks in 2012,* PC MAGAZINE (June 18, 2012).

46. *The Cell Phone: Marty Cooper's Big Idea,* CBS NEWS (May 21, 2010).

47. Ibid.

48. D. W. Everitt, dean of the University of Illinois School of Engineering, and a member of the President's Communications Policy Board, as quoted in Coase et al. (1995), 46.

49. David Talbot, *What 5G Will Be: Crazy-Fast Wireless Tested in New York City,* MIT TECHNOLOGY REVIEW (May 22, 2013).

50. Tom Simonite, *Qualcomm Proposes a Cell-Phone Network by the People, for the People,* MIT TECHNOLOGY REVIEW (May 2, 2013).

51. Martin Cave, Chris Doyle & William Webb, ESSENTIALS OF MODERN SPECTRUM MANAGEMENT (Cambridge: Cambridge University Press, 2007), 4.

52. Ibid., 107.

53. Coase et al. (1995), 23.

54. Michael Marcus, *Does Today's FCC Have Sufficient Decision Making Throughput to Handle the 21st Century Spectrum Policy Workload?* paper delivered at the Telecommunications Policy Research Conference (Sept. 25, 2015) (comment taken from oral presentation).

55. Coase (1959), 18.

Part II. Silence of the Entrants

Epigraph: Quoted in an interview with Richard Heffner, *The Open Mind* (Dec. 7, 1975).

Chapter 6. The Death of DuMont

Epigraph: Milton Friedman, with Rose D. Friedman, CAPITALISM AND FREEDOM (Chicago: University of Chicago Press, 1962).

1. Robert W. Crandall, *Regulation of Television Broadcasting: How Costly Is the "Public Interest"?* REGULATION 31 (Jan.–Feb. 1978), 31.

2. The TV dial, encompassing VHF and UHF bands, stretched from channels 2 through 83. Channel 37, however, was reserved for radio astronomy, an ad hoc regulatory choice detailed in Harvey J. Levin, THE INVISIBLE RESOURCE: USE AND REGULATION OF THE RADIO SPECTRUM (Washington, DC: Resources for the Future, 1971), 163–167.

3. Bruce M. Owen, Jack H. Beebe & Willard G. Manning, TELEVISION ECONOMICS (Lexington, MA: Lexington, 1974), 12.

4. Cecilia Rothenberger, *The UHF Discount: Shortchanging The Public Interest,* 53 AMERICAN UNIVERSITY LAW REVIEW 689 (Dec. 2003), 696.

5. *Sixth Report & Order,* 41 FCC 148 (1952).

6. Note, *The Darkened Channels: UHF Television and the FCC,* 75 HARVARD LAW REVIEW 1578 (June 1962), 1579–1580.

7. Indeed, economists found that six national broadcast networks could have been successful under this approach. Roger G. Noll, Merton J. Peck & John J. McGowan, ECONOMIC ASPECTS OF TELEVISION REGULATION (Washington, DC: Brookings Institution, 1973), 316.

8. Owen et al. (1974), 124.

9. Philip J. Auter & Douglas A. Boyd, *DuMont: The Original Fourth Television Network,* 29 JOURNAL OF POPULAR CULTURE 63 (Winter 1995), 73.

10. Thomas L. Schuessler, *Structural Barriers to the Entry of Additional Television Networks: The Federal Communications Commission's Spectrum Management Policies,* 54 SOUTHERN CALIFORNIA LAW REVIEW 875 (1981), 929 (Table 10).

11. Note, *The Darkened Channels* (1962), 1581–1582.

12. Schuessler (1981), 955, 964.

13. *Ad Hoc Advisory Committee on Allocations to the Senate Committee on Interstate and Foreign Commerce, 85th Congress, 2nd Session, Allocation of Television Channels* (1958), 39.

14. Gary M. Fournier & Ellen S. Campbell, *Shifts in Broadcast Policies and the Value of Television Licenses,* 5 INFORMATION ECONOMICS & POLICY 87 (1993), 94 (footnote 13).

15. Laurie Thomas & Barry R. Litman, *Fox Broadcasting Company, Why Now? An Economic Study of the Rise of the Fourth Broadcast "Network,"* 35 JOURNAL OF BROADCASTING & ELECTRONIC MEDIA 139 (1991), 139.

16. Noll, Peck & McGowan (1973), 113–114; italics in original.

17. Matthew L. Spitzer, *The Constitutionality of Licensing Broadcasters,* 64 NEW YORK UNIVERSITY LAW. REV. (Nov. 1989), 990, 1053 (footnotes omitted).

18. Schuessler (1981), 1000.

19. Douglas Webbink, *The Impact of UHF Promotion: The All-Channel Television Receiver Law,* 11 LAW AND CONTEMPORARY PROBLEMS 538 (Summer 1969).

20. Minow & LaMay (1995), 96.

21. Senate Subcommittee on State, Justice and Commerce, S. Rep. No. 1043, 95th Congress, 2d Sess. 7 (1978). Cited in Schuessler (1981), 978–979.

22. Federal Communications Commission Staff Report, *Comparability for UHF Television* (1980), 235, quoted in Schuessler (1981), 980.

23. Kerri Smith, *The FCC Under Attack,* 2 Duke Law & Technology Review 1 (2003).

24. Schuessler (1981), 954–955.

25. Ibid., 935.

26. Ibid., 956.

27. Quoted in Lawrence D. Longley, *The FCC and the All-Channel Receiver Bill of 1962,* 13 Journal of Broadcasting 293 (Summer 1969), 297–298.

28. Webbink (1969).

29. Schuessler (1981), 961, citing 1961 FCC Proposed Rule Making.

30. Webbink (1969), 547.

31. Leonard Chazen & Leonard Ross, *Federal Regulation of Cable TV: The Visible Hand,* 83 Harvard Law Review 1820 (June 1970), 1824–1825.

32. Rolla Edward Park, *Cable Television, UHF Broadcasting, and FCC Regulatory Policy,* 15 Journal of Law and Economics 207 (April 1972), 208. Park cites *The Economics of the TV-CATV Interface, Staff Report to the F.C.C.* (July 15, 1970), 9.

33. Henry Geller, *A Modest Proposal for Modest Reform of the Federal Communications Commission,* 63 Georgetown Law Journal 705 (Feb. 1975), 709.

34. Webbink (1969), 552. The "estimate of the FCC was evidently incorrect and based on a misunderstanding of the way the price index for television receivers was constructed."

35. Ibid., 561.

Chapter 7. "Thank God for C-SPAN!"

Epigraph: Bruce Owen, The Internet Challenge to Television (Cambridge, MA: Harvard University Press, 1999).

1. Brian Lockman and Don Sarvey, Pioneers of Cable Television: The Pennsylvania Founders of an Industry (Jefferson, NC: McFarland, 2005), 10–13; Thomas R. Eisenmann, *Cable TV: From Community Antennas to Wired Cities,* Harvard Business Review (July 10, 2000).

2. William Emmons, *Public Policy and the Evolution of Cable Television: 1950–1990,* 21 Business and Economic History 182 (1992), 184.

3. *Carter Mountain Transmission Corporation v. FCC,* 321 F. 2d 359—Court of Appeals, Dist. of Columbia Circuit 1963.

4. Federal Communications Commission, *21st Annual Report for the Fiscal Year Ending June 30, 1965,* 82.

5. *Second Report and Order,* 2 FCC 2d 725 (1966), 785. See also J. Clay Smith, Jr., *Primer on the Regulatory Development of CATV (1950–72),* 18 Howard Law Journal 729 (1973–1975).

6. FCC (1966), 770–771.

7. 38 FCC 683 (1965), 701.

8. FCC (1966), 785.

9. Ibid., 776.
10. FCC (1965), 736, 745.
11. Patrick R. Parsons, BLUE SKIES: A HISTORY OF CABLE TELEVISION (Philadelphia: Temple University Press, 2008), 210.
12. Franklin M. Fisher, Victor E. Ferrall, Jr., David Belsey & Bridger Mitchell, *Community Antenna Television Systems and Local Television Station Audience*, 80 QUARTERLY JOURNAL OF ECONOMICS 227 (May 1966).
13. Stanley Besen & Robert W. Crandall, *The Deregulation of Cable Television*, 44 LAW & CONTEMPORARY PROBLEMS 81 (Winter 1981), 88.
14. Fisher et al. (1966), 250.
15. Rolla Edward Park, *Potential Impact of Cable Growth on Television Broadcasting*, Rand Corp. R-587-FF (Oct. 1970).
16. Ithiel de Sola Pool, TECHNOLOGIES OF FREEDOM (Cambridge, MA: Harvard University Press, 1983), 156.
17. *United States v. Midwest Video Corp.*, 406 US 649 (1972), 668–669.
18. Pool (1983), 156.
19. Emmons (1992), 185.
20. Besen & Crandall (1981), 93.
21. Park (1972), 231.
22. Monroe Price, *Requiem for the Wired Nation: Cable Policymaking at the FCC*, 61 VIRGINIA LAW REVIEW 541 (April 1975), 545 (footnote omitted).
23. Ibid., 552.
24. Cited in Thomas W. Hazlett, *Wiring the Constitution for Cable*, REGULATION, no. 1 (1988).
25. Price (1975), 543–544.
26. Besen & Crandall (1981), 95.
27. Ibid., 122.
28. For a brutal, first-person account of the wheeling and dealing, see Clinton Galloway, ANATOMY OF A HUSTLE: CABLE COMES TO SOUTH CENTRAL (n.p.: Phoenix, 2012). The volume details the ordeal Galloway's firm went through in seeking a cable franchise in Los Angeles. Despite winning a 1986 Supreme Court case establishing their First Amendment rights to compete, *Preferred Communications v. City of Los Angeles*, the company was excluded from the market due to bargains cut by City Hall and wealthy, well-connected political insiders. The author served in the case as an expert witness retained by Preferred.
29. Thomas W. Hazlett, *Private Monopoly and the Public Interest: An Economic Analysis of the Cable Television Franchise*, 134 UNIVERSITY OF PENNSYLVANIA LAW REVIEW 1335 (July 1986).
30. Website of the National Cable and Telecommunications Association.
31. Jeff Baumgartner, *7% of U.S. Homes Rely on Over-the-Air TV: CEA Study*, MULTICHANNEL NEWS (July 30, 2013).
32. Bruce Springsteen, *57 Channels (and Nothin' On)*, Columbia Records (released July 1992).

33. *Cable Dominated Emmy Awards*, RADIO + TELEVISION BUSINESS REPORT (Aug. 30, 2010).

34. News Release, *Writers Choose the 101 Best Written TV Series of All Time*, Writers' Guild of America, West (June 2, 2013).

35. Josef Adalian, *What Show Had the Richest Audience? What Do Dog Owners Watch? A Deep Dive into 2013's TV Ratings*, VULTURE (Dec. 19, 2013).

36. *Top 10 TV Shows on Social Media: New Fall Shows*, THE HOLLYWOOD REPORTER (Dec. 21, 2015).

37. The networks began to expand nightly national news programs to thirty minutes (minus commercials) in 1963. Steven Waldman & the Working Group on the Information Needs of Communities, *The Information Needs of Communities: The Changing Media Landscape in a Broadband Age*, Federal Communications Commission, Office of Strategic Planning & Policy Analysis (July 2011), 72.

38. Gabriel Sherman, *Chasing Fox: The Loud, Cartoonish Blood Sport That's Engorged MSNBC, Exhausted CNN—and Is Making Our Body Politic Delirious*, NEW YORK (Oct. 3, 2010).

39. Clay Shirky, *We Are Indeed Less Willing to Agree on What Constitutes Truth*, POYNTER .ORG (Oct. 17, 2012).

40. *Terminiello v. Chicago*, 337 U.S. 1 (1949), 4.

41. Dissenting Opinion of Mr. Justice Oliver Wendell Holmes in *Gitlow v. People*, 268 U.S. 652 (1925), 673.

42. Stephen Collinson, *How Jon Stewart Changed Politics*, CNN (Feb. 11, 2015).

43. Brian C. Anderson, SOUTH PARK CONSERVATIVES: THE REVOLT AGAINST LIBERAL MEDIA BIAS (Washington, DC: Regnery, 2005).

44. Bill Curry, *Jon Stewart Changed Everything: How "The Daily Show" Revolutionized TV & Revitalized the Democratic Party*, SALON (Aug. 4, 2015).

45. David Corn, *Happy Birthday, C-SPAN!* NATION (March 12, 2004).

46. Michelle Malkin, *C-SPAN Wins Seattle War; Regulatory Fight Continues*, SEATTLE TIMES (Feb. 11, 1997).

47. *Editorial: C-SPAN Founder Going Out Like a Lamb*, USA TODAY (March 20, 2012).

48. Thomas W. Hazlett, *Changing Channels: C-SPAN's Brian Lamb on How Unfiltered Reporting and Media Competition Are Transforming American Politics*, REASON (March 1996).

49. Corn (2004).

50. Waldman & Working (2011), 176.

51. Ibid., 176–177.

52. Hazlett (1996).

53. Thomas Sowell, KNOWLEDGE AND DECISIONS (New York: Basic, 1980), 188.

Chapter 8. Lost in Space

Epigraph: John Micklethwait & Adrian Wooldridge, THE FOURTH REVOLUTION: THE GLOBAL RACE TO REINVENT THE STATE (New York: Penguin, 2014).

1. Jeff Magenau, *Digital Audio Radio Services: Boon or Bust to the Public Interest?* 21 ANNALS OF AIR AND SPACE LAW 219 (1996).

2. *Notice of Inquiry in the Matter of the Commission's Rules with Regard to the Establishment and Regulation of New Digital Audio Radio Satellite Service,* 5 F.C.C. REC. 5237 (Aug. 1, 1990).

3. *Notice of Proposed Rulemaking in the Matter of Establishment of Rules and Policies for the Digital Audio Radio Satellite Service in the 2310–2360 MHz Frequency Band,* FCC 95-229 (June 14, 1995), ¶¶ 2, 3, 4, 17.

4. *Response of NAB to American Mobile Radio Corporation's Reply and Opposition to Petitions to Deny in File,* Nos. 26/27-DSS-LA-93; IO/l I-DSS-P-93, Federal Communications Commission (June 25, 1993), 3.

5. Reply comments of the NAB, Federal Communications Commission Gen. Docket No. 90-357 (Oct. 20, 1995), 2.

6. FCC Public Notice, DA 97-656 (April 2, 1997), *FCC Announces Auction Winners for Digital Audio Radio Service: Auction Raises $173.2 Million for Two Licenses,* 1.

7. National Association of Broadcasters, *In the Matter of Establishment of Rules and Policies for the Digital Audio Radio Satellite Service in the 2310–2360 MHz Frequency Band, Petition for Declaratory Ruling,* Federal Communications Commission, IB Docket No. 95-91 (April 14, 2004), 17.

8. Thomas W. Hazlett, *Local Motives: Why the FCC Should Scrap Its Absurd Rules for Satellite Radio,* SLATE (March 16, 2004).

9. *National Association of Broadcasters' Reply Comments to NAB Petition for Declaratory Ruling,* Federal Communications Commission, MB Docket No. 04-160 (June 21, 2004), 15–16.

10. *NAB's Fritts Speaks Out on XM/WCS Deal,* FMQB.com (July 15, 2005).

11. Ibid.

12. The author was retained as an expert by the merging parties. Eric Savitz, *Satellite Radio Stocks Rallying on Former FCC Economist Hazlett's Report Supporting the Deal,* BARRON'S (June 15, 2007).

13. SiriusXM website (visited May 6, 2016).

Chapter 9. Baptists, Bootleggers, and LPFM

Epigraph: Eric Hoffer, THE TEMPER OF OUR TIME (New York: Harper & Row, 1967).

1. Bruce Yandle, *Bootleggers and Baptists: The Education of a Regulatory Economist,* 7 REGULATION 12 (1983).

2. Jeremy Smith and Howard Rosenfeld, RADIO FOR PEOPLE, NOT PROFIT (Dollars& Sense.org, 1999).

3. Jesse Walker, REBELS ON THE AIR: AN ALTERNATIVE HISTORY OF RADIO IN AMERICA (New York: NYU Press, 2001), 180.

4. Smith & Rosenfeld (1999).

5. Federal Communications Commission, *In the Matter of Creation of a Low Power Radio Service: Notice of Proposed Rulemaking,* MM Docket No. 99-25 (Rel. Feb. 3, 1999), ¶ 11.

6. Ralph Nader, *Ralph Nader Supports Low-Power FM,* Prometheus Radio Project website (July 9, 1999).

7. Thomas W. Hazlett, *Low-Power Radio: An Autopsy,* FINANCIAL TIMES (Feb. 5, 2003).

8. Kathy Chen, *FCC Set to Open Air Waves to Low Power Radio,* WALL STREET JOURNAL (Jan. 17, 2000); Stephen Labaton, *FCC to Open Airwaves,* N.Y. TIMES (Jan. 23, 2000).

9. Edward Lewine, *Radio Pirates Drop Anchor Together,* N.Y. TIMES (Jan. 10, 1999).

10. Editorial, *New Voices in the Air,* L.A. TIMES (Feb. 10, 1999).

11. Quoted in Eric Klinenberg, FIGHTING FOR THE AIR: THE BATTLE TO CONTROL AMERICA'S MEDIA (New York: Henry Holt, 2007), 259.

12. Federal Communications Commission, *Factsheet—Low Power FM Radio: Allegations and Facts,* posted on FCC.gov (March 29, 2000).

13. Frank Ahrens, *Political Static May Block Low-Power FM; FCC, Congress Battle over Radio Plan,* WASHINGTON POST (May 15, 2000).

14. Thomas W. Hazlett & Bruno E. Viani, *Legislators vs. Regulators: The Case of Low Power FM Radio,* 7 BUSINESS & POLITICS 1 (April 2005), 2.

15. pete tridish, *Radio Activists Win a Pathetic Sliver of the Airwaves for Neighborhood Radio!* Prometheus Radio Project website (Summer 2000).

16. Stephen Labaton, *FCC Offers Low Power FM Stations,* N.Y. TIMES (Jan. 29, 1999).

17. Thomas W. Hazlett, *Who Killed Micro Radio?* ZDNET (April 17, 2001).

18. tridish (2000).

19. Edward Lewine, *Seeing More Chance for F.C.C. Support, Advocates of Low-Power Stations Share Advice,* N.Y. TIMES (Jan. 10, 1999).

20. H.R. 6533, The Local Community Radio Act of 2010, signed by President Barack Obama on Jan. 4, 2011.

21. Federal Communications Commission, *Creation of a Low Power Radio Service,* 47 CFR Part 73, 3204.

22. *In Historic Victory for Community Radio, FCC Puts 1,000 Low Power FM Frequencies Up for Grabs,* video produced by Democracy Now! (Prometheus Radio Project website; visited Sept. 5, 2013).

23. News Release, *Broadcast Station Totals as of December 31, 2014,* Federal Communications Commission (Jan. 7, 2015).

24. Federal Communications Commission, *Digital Audio Broadcasting Systems and Their Impact on the Terrestrial Radio Broadcast Service: First Report & Order,* FCC 02-286, MM Docket No. 99-325 (adopted Oct. 10, 2002).

25. Donald R. Lockett, THE ROAD TO DIGITAL RADIO IN THE UNITED STATES (Washington, DC: National Association of Broadcasters, 2004), xvii. The book was published as an "NAB Executive Technology Briefing."

Part III. Adventures in Content Regulation

Epigraph: John J. Mearsheimer, R. Wendell Harrison Distinguished Service Professor of Political Science at the University of Chicago, *Review Essay,* FOREIGN POLICY (Nov. 1, 2010).

Chapter 10. Orwell's Revenge

Epigraph: *CBS v. Democratic Nat'l Committee*, 412 U.S. 94 (1973), 154 and fn. 17 (Douglas, J., concurring).

1. *Red Lion Broadcasting Co., Inc. v. FCC*, 395 U.S. 367 (1969).
2. *Report on Editorializing*, 13 F.C.C. 1246 (1949), 1251.
3. *Miami Herald Pub. Co. v. Tornillo*, 418 U.S. 241 (1974).
4. Fred W. Friendly, THE GOOD GUYS, THE BAD GUYS, AND THE FIRST AMENDMENT: FREE SPEECH VS. FAIRNESS IN BROADCASTING (New York: Random House, 1975), 31.
5. *Annual Report of the Federal Radio Commission* (1928), 155.
6. Lucas A. Powe, Jr., AMERICAN BROADCASTING AND THE FIRST AMENDMENT (Berkeley: University of California Press, 1987), 108–109.
7. Ibid., 69.
8. Quoted in Robert A. Caro, THE YEARS OF LYNDON JOHNSON: MEANS OF ASCENT (New York: Vintage, 1991), 91.
9. *FCC Investigating Committee Loses Its Chairman; Cox Resigns Under Pressure*, BILLBOARD (Oct. 9, 1943).
10. Quoted in Caro (1991), 90.
11. Well, not that humiliated. Cox resigned as head of the special committee but remained in Congress until his death in 1952.
12. Caro (1991), 93–95.
13. Glen O. Robinson, *The Federal Communications Commission: An Essay on Regulatory Watchdogs*, 64 VIRGINIA LAW REVIEW 169 (March 1978), 224.
14. It made no legal or financial difference to LBJ, as Texas was a community property state.
15. Dan Fletcher, *Who's Our Richest President Ever?* TIME (May 24, 2010).
16. Powe (1987), 109.
17. *Mayflower Broadcasting Co.*, 8 F.C.C. 333 (1940), 340.
18. 13 F.C.C. 1246.
19. *The General Fairness Doctrine Obligations of Broadcast Licensees*, 102 F.C.C. 2d 145, 146 (1985).
20. For a fascinating discussion, see Jonathan Haidt, THE RIGHTEOUS MIND (New York: Vintage, 2010), 104–105.
21. Edward Jay Epstein, NEWS FROM NOWHERE: TELEVISION AND THE NEWS (Chicago: Ivan R. Dee, 1973), 64–65.
22. *Letter to Cullman Broadcasting Co.*, 40 F.C.C. 576 (1963).
23. Hargis is quoted in Friendly (1975), 5.
24. *NBC v. United States*, 319 U.S. 190 (1943), 215–216.
25. *Red Lion* (1969), 392–393.
26. Ibid., fn. 19. The source was given as: "Keynote Address, Sigma Delta Chi National Convention, Atlanta, Georgia, November 21, 1968."
27. Quoted in Epstein (1973), 72.
28. Quoted in Friendly (1975), 33.

29. Quoted ibid., 32. The "committee" referenced appears to be the National Committee for a Sane Nuclear Policy, which worked closely with the DNC.

30. Quoted ibid., 35.

31. Ibid., 35.

32. Quoted ibid., 37.

33. Ibid., 36.

34. Fred J. Cook, *Radio Right: Hate Clubs of the Air,* NATION (May 25, 1964), 523.

35. Friendly (1975), 39, 41, 42.

36. David L. Bazelon, *FCC Regulation of the Telecommunications Press,* 1975 DUKE LAW JOURNAL 213 (May 1975).

37. Nicholas Johnson, *With Due Regard for the Opinions of Others,* 8 CALIFORNIA LAWYER 52 (Aug. 1988), 52.

38. The FCC required all stations to offer some news or public affairs programming, deemed "nonentertainment" content by the Commission. Per rules crafted during the Carter administration, AM radio stations were supposed to use at least 8 percent of their program time for nonentertainment, FM stations 6 percent. Compliance on music-oriented stations was perfunctory, with "rip and read" news on the hour and half-hour (wherein an announcer would relay national wire service stories for two or three minutes).

39. Thomas W. Hazlett & David W. Sosa, *Was the Fairness Doctrine a "Chilling Effect"? Evidence from the Post-Deregulation Radio Market,* 26 JOURNAL OF LEGAL STUDIES 279 (Jan. 1997).

40. "The repeal of the Fairness Doctrine opened the door to talk radio. . . . Before the repeal, radio station license holders were very hesitant to talk about anything in politics because they were afraid they would be fined, shut down, or be forced to put things on that were not interesting to their audiences." Michael Harrison, publisher of TALKERS MAGAZINE, quoted in Keach Hagey, *Fairness Doctrine Fight Goes On,* POLITICO (Jan. 16, 2011).

41. Deborah Mesce, *Nader, Schlafly Team Up in Support of Fairness Rule,* Associated Press (April 7, 1987).

42. Sean Motley, *Newt Gingrich Co-Sponsored the 1987 Pro-Fairness Doctrine Bill,* PJ MEDIA (April 29, 2011).

43. *Merits of Fairness Doctrine Debated by Broadcasters,* N.Y. TIMES (March 20, 1987).

44. Enver Masud, *Broadcasting Fairness Doctrine Promised Balanced Coverage,* Wisdom Fund (July 25, 1997).

45. *The Fairness Doctrine and Claims of Systematic Imbalance in Television News Broadcasting: American Security Council Education Foundation v. FCC,* 93 HARVARD LAW REVIEW 1028 (March 1980), 1028.

46. Lexington, *Let the Blowhards Blow: The Case for Keeping the Airwaves Unfair and Unbalanced,* ECONOMIST (July 19, 2007).

47. Ibid.

48. Federal Communications Commission, *In the Matter of Inquiry into Section 73.1910 of the Commission's Rules and Regulations Concerning the General Fairness Obliga-*

tions of Broadcast Licensees: Report, Gen. Docket No. 84-282 (released Aug. 23, 1985), 171.

49. This section relies heavily on Thomas W. Hazlett & David Sosa, *"Chilling" the Internet? Lessons from FCC Regulation of Radio Broadcasting,* 4 MICHIGAN TELECOMMUNICATIONS AND TECHNOLOGY LAW REVIEW (1997).

50. *Columbia Broadcasting Sys. Inc. v. Federal Communications Comm'n,* 454 F.2d 1018 (D.C. Cir. 1971), 1020.

51. Quoted in Powe (1987), 124.

52. Daniel Schorr, CLEARING THE AIR (Boston: Houghton Mifflin, 1977), 48.

53. The FCC licenses broadcast outlets—radio and television stations—but not networks that supply programming. However, because each of the networks owned several TV stations (in the largest markets) the administration was able to exert, or at least threaten, leverage over them.

54. Schorr (1977), 139.

55. Senator John Pastore (D-RI), head of the Senate Communications Subcommittee, quoted ibid., 62.

56. CBS newsman Roger Mudd wrote a commentary critical of the network's decision, which was to be aired on CBS Radio the day after the announcement. It too was eliminated. Only after a memo outlining the meeting between White House staffers and Paley was leaked four and a half months later did CBS resume the practice of instant analysis of presidential speeches. Powe (1987), 139.

57. Quoted in Powe (1987), 131–132.

58. Anthony R. Carrozza, WILLIAM D. PAWLEY: THE EXTRAORDINARY LIFE OF THE ADVENTURER, ENTREPRENEUR, AND DIPLOMAT WHO FOUNDED THE FLYING TIGERS (Washington, DC: Potomac, 2012), 317–318.

59. "There is no doubt that the Nixon partisans were attacking the Post." Bill Mitchell, *A Remembrance of Courage,* Poynter.org (Aug. 20, 2002).

60. Bazelon (1975), Appendix.

61. Ibid.

Chapter 11. Must Carry This, Shall Not Carry That

Epigraph: Hernan Galperin, NEW TELEVISION, OLD POLITICS: THE TRANSITION TO DIGITAL TV IN THE UNITED STATES AND BRITAIN (Cambridge: Cambridge University Press, 2004), 69.

1. *Turner Broadcasting System, Inc. v. F.C.C.,* 520 U.S. 180 (1997), 189.

2. Charles Lubinsky, *Reconsidering Retransmission Consent: An Examination of the Retransmission Consent Provision (47 U.S.C. 325 (b)) of the 1992 Cable Act,* 49 FEDERAL COMMUNICATIONS LAW JOURNAL 99 (1996–1997), 146.

3. Mike Farrell, *Kagan: Retrans Fees Rise to $9.3B by 2020,* MULTICHANNEL NEWS (Oct. 27, 2014).

4. Cable TV operators that owned programming and then denied competing cable networks channel space constituted a somewhat different issue, one governed by

the antitrust laws. A related situation arose where incumbent cable operators used exclusive rights to create barriers for cable system entrants, a problem governed both by antitrust and "program access" rules in the 1992 Cable Act. See Thomas W. Hazlett, *Predation in Local Cable Television Markets*, 40 ANTITRUST BULLETIN 609 (Fall 1995).

5. George H. Shapiro, Philip B. Kurland & James P. Mercurio, "CABLESPEECH": THE CASE FOR FIRST AMENDMENT PROTECTION (New York: Harcourt Brace Jovanovich, 1983), 139.

6. Peter Huber, LAW AND DISORDER IN CYBERSPACE (New York: Oxford University Press, 1997), 149.

7. Robert B. Hobbs Jr., *Cable TV's "Must Carry" Rules: The Most Restrictive Alternative—Quincy Cable TV, Inc. v. FCC*, 8 CAMPBELL LAW REVIEW 339 (Spring 1986), 357.

8. Shapiro et al. (1983), 138.

9. *Home Box Office, Inc., v. FCC*, 567 F.2d 9 (D.C. Cir.), *cert. denied*, 434 U.S. 829 (1977), 36.

10. *Quincy Cable TV, Inc. v. F.C.C.*, 768 F.2d 1434 (D.C. Cir. 1985), *cert. denied*, 476 U.S. 1169 (1986); *Century Communications Corp. v. F.C.C.*, 835 F.2d 292 (D.C. Cir. 1987), *cert. denied*, 486 U.S. 1032 (1988).

11. Pub. L. 102-385, 102 Stat. 1460 (47 U.S.C. §§ 534, 535).

12. *Definition of Markets for Purposes of the Cable Television Mandatory Television Broadcast Signal Rules, Report and Order and Further Notice of Proposed Rulemaking*, 11 FCC Rcd 6201 (1996).

13. Bruce Owen, THE INTERNET CHALLENGE TO TELEVISION (Cambridge: Harvard University Press, 1999), 114.

14. *Turner Broadcasting System, Inc. v. FCC*, 512 U.S. 622 (1994) [hereafter *Turner I*].

15. The Court created a legal standard for evaluating the constitutionality of must-carry called "intermediate scrutiny," not so constraining to policy makers as the "strict scrutiny" routinely applied to cases involving the regulation of newspapers, books, and other traditional organs of the "free press."

16. *Turner Broadcasting System, Inc. v. F.C.C.*, 520 U.S. 180 (1997) [hereafter *Turner II*].

17. *Turner II*, citing *Turner I*, 190 (citations omitted).

18. *Turner II*, 191 (quoting congressional findings in the 1992 Cable Act's legislative history).

19. This was the figure in *Turner II*, but according to the Federal Communications Commission, 74.61 percent of U.S. homes subscribed to a multichannel video programming distributor (cable or satellite TV system) in December 1996. Federal Communications Commission, *In the Matter of Annual Assessment of the Status of Competition in the Market for the Delivery of Video Programming*, CS Docket No. 98-102 (Dec. 23, 1998), C1.

20. *Turner II*, 227 (Breyer concurring opinion).

21. *Turner II*, 257 (O'Connor dissent).

22. *Turner II*, 230 (O'Connor dissent) (footnotes omitted).

23. *Minneapolis Star & Tribune Co. v. Minnesota Comm'r of Revenue*, 460 U.S. 575 (1983).

24. *Miami Herald Pub. Co. v. Tornillo*, 418 U.S. 241 (1974).

25. Petition for Rehearing for Appellant Daniels Cablevision, Inc., *Turner Broadcasting System, Inc., et al., v. Federal Communications Commission*, No. 93-44, Supreme Court of the United States, October Term, 1994 (July 22, 1994), at 2.

26. *Carriage of Television Broadcast Signals by Cable Television Systems*, Federal Trade Commission (Nov. 26, 1991).

27. Michael Vita, *Must Carry Regulations for Cable Television Systems: An Empirical Analysis*, 12 JOURNAL OF REGULATORY ECONOMICS 159 (1997), 159.

28. Linda Greenhouse, *High Court Rules Cable Must Carry Local TV Stations*, N.Y. TIMES (April 1, 1997).

29. Thomas W. Hazlett, *Digitizing Must Carry Under Turner Broadcasting v. FCC (1997)*, 8 SUPREME COURT ECONOMIC REVIEW 141 (2000), 165–173.

30. Stephen McClellan, *Must Carry Upheld: Champagne, Real Pain*, BROADCASTING & CABLE (April 7, 1997).

31. ValueVision International, Inc., *Annual Report on Form 10-K for the Fiscal Year Ended January 31, 1997*, Securities and Exchange Commission, 32.

32. Home Shopping Network, Inc., *Annual Report, Form 10-K for the Year Ended Dec. 31, 1993*, Securities and Exchange Commission, 4.

33. Thomas W. Hazlett, *How Home Shopping Became King of Cable*, WALL STREET JOURNAL (July 14, 1994).

34. Stephen Frantzich & John Sullivan, THE C-SPAN REVOLUTION (Norman: University of Oklahoma Press, 1996), 70.

35. *Turner II*, Breyer Concurring, 82.

36. Reply Comments of C-SPAN Networks, *In the Matter of Carriage of the Transmissions of Digital Television Broadcast Stations, Amendments to Part 76 of the Commission's Rules*, Federal Communications Commission, CS Docket No. 98-120 (Dec. 22, 1998), footnote 16.

37. "Seven-Hundred and Fifty Commercial Television Stations in 211 Metropolitan Markets Across the U.S. Make Up the Local TV News Industry." Rocky Mountain Media Watch, *Local News on Cable: 1999* (Sept. 9, 1999), 1.

38. Lamb quoted in Frantzich & Sullivan (1996), 69–71 (footnote omitted).

39. Dean Alger, MEGAMEDIA: HOW GIANT CORPORATIONS DOMINATE MASS MEDIA, DISTORT COMPETITION AND ENDANGER DEMOCRACY (Lanham, MD: Rowman & Littlefield, 1998), 187.

40. Cass R. Sunstein, DEMOCRACY AND THE PROBLEM OF FREE SPEECH (New York: Free Press, 1995), 261.

41. Ibid., xviii.

42. *Columbia Broadcasting System, Inc. v Democratic Nat'l Comm*, 412 US 94, 160–161 (Douglas, J., concurring).

43. Sunstein (1995), xix.

44. Ibid., 5–6.

45. Ibid., 54–55.

46. Ibid., 17.

47. Quoted in Steven Waldman & the Working Group on the Information Needs of Communities, *The Information Needs of Communities: The Changing Media Landscape in a Broadband Age*, Federal Communications Commission, Office of Strategic Planning & Policy Analysis (July 2011), 280.

48. Sunstein (1995), 54.

49. Thomas G. Krattenmaker & Lucas A. Powe, Jr., REGULATING BROADCAST PROGRAMMING (Cambridge, MA: MIT Press, 1994), 232.

50. Sunstein (1995), 259–260.

51. R. H. Coase, *Evaluation of Public Policy Relating to Radio and Television Broadcasting: Social and Economic Issues*, 41 LAND ECONOMICS 161 (May 1965), 166–167.

Chapter 12. Indecent Exposure

Epigraph: Ithiel de Sola Pool, TECHNOLOGIES WITHOUT BOUNDARIES (Cambridge, MA: Harvard University Press, 1990), 251.

1. Lynee Joyrich, *Epistemology of the Console,* 27 CRITICAL INQUIRY (Spring 2001), 443–444, fn. 9.

2. Kaiser Family Foundation, *Parents, Media, and Public Policy: A Kaiser Family Foundation Survey,* Program for the Study of Media and Health, Publication No. 7156 (Sept. 23, 2004), 7.

3. William J. Clinton, PUBLIC PAPERS OF THE PRESIDENTS OF THE UNITED STATES: 1995 (Ann Arbor: University of Michigan Press, 1995), 345.

4. Robert H. Bork, SLOUCHING TOWARDS GOMORRAH: MODERN LIBERALISM AND THE AMERICAN DECLINE (New York: HarperCollins, 1996), 141–153.

5. *Miller v. California,* 413 U.S. 15 (1973).

6. Paul Gewirtz, On "I Know It When I See It," 105 YALE LAW JOURNAL 1023 (1996), 1024.

7. *FCC v. Pacifica Foundation,* 438 U.S. 726 (1978).

8. *Congress Asserts Its Dominion over FCC,* 117 BROADCASTING (Aug. 7, 1989), 28.

9. *FCC v. Fox Television Stations, Inc.,* 132 S. Ct. 2307 (2012).

10. Bork (1996), 147.

11. Krysten Crawford, *Howard Stern Jumps to Satellite,* CNN MONEY (Oct. 6, 2004).

12. *Viacom-Plaints Uncovered,* BROADCASTING & CABLE (April 17, 2005).

Part IV. Slouching Toward Freedom

Epigraph: William J. Baumol & Dorothy Robyn, TOWARD AN EVOLUTIONARY REGIME FOR SPECTRUM GOVERNANCE: LICENSING OR UNRESTRICTED ENTRY? (Washington, DC: AEI-Brookings Joint Center for Regulatory Studies, 2006), 65.

1. Peter W. Huber, Michael K. Kellogg & John Thorne, FEDERAL TELECOMMUNICATIONS LAW, 2nd ed. (Gaithersburg, MD: Aspen Law & Business, 1999), 874.

Chapter 13. The Thirty Years' War

Epigraph: Douglas W. Webbink, *Frequency Spectrum Deregulation Alternatives,* Federal Communications Commission, Office of Plans & Policies Working Paper No. 2 (Oct. 1980), 10.

1. President Bill Clinton, *Acceptance Speech,* Democratic National Convention (Aug. 29, 1996); Vice President Al Gore, *Acceptance Speech,* Democratic National Convention (Aug. 28, 1996).

2. *Phone Me by Air,* SATURDAY EVENING POST (July 28, 1945). See also Tom Farley, *Mobile Telephone History,* TELEKTRONIKK 24 (March 4, 2005).

3. A more technical description of cellular was provided in 1947. D. H. Ring, *Mobile Telephony—Wide Area Coverage,* Bell Laboratories Technical Memorandum (Dec. 11, 1947).

4. These units were not rushed to market by regulators. One-way police car radios appeared in 1921. Two-way radios were first deployed in Bayonne, New Jersey, in 1933. James B. Murray, Jr., WIRELESS NATION: THE FRENZIED LAUNCH OF THE CELLULAR REVOLUTION IN AMERICA (Cambridge, MA: Perseus, 2001), 16.

5. Ibid., 25.

6. Dan Steinbock, THE MOBILE REVOLUTION (London: Kogan Page, 2005), 41–42.

7. Tom Farley, *The Cell Phone Revolution,* 22 AMERICAN HERITAGE INVENTION & TECHNOLOGY (Jan. 2007).

8. *General Mobile Radio Service,* Docket No. 8976, 13 FCC 1190 (1949).

9. Christopher W. Mines, *Regulation and Re-Invention of Cellular Telephone Service in the United States and Great Britain,* Program on Information Resources Policy, Kennedy School of Government, Harvard University (Rev. 2, Feb. 9, 1992), 19.

10. George C. Calhoun, DIGITAL CELLULAR (Norwood, MA: Artech House, 1988), 46.

11. FCC TV license allocations are discussed in Chapter 6.

12. Kenneth E. Hardman, *A Primer on Cellular Mobile Telephone Systems,* FEDERAL BAR NEWS & JOURNAL 385 (Nov. 1982), 387.

13. By the mid-1990s there were one million PLMRS licenses, with about twelve phones operated per license. Michelle C. Farquhar, Chief, Wireless Telecommunications Bureau, *Private Land Mobile Radio Services: Background,* FCC Staff Paper (Dec. 18, 1996), Executive Summary.

14. Hardman (1982), 386–387.

15. Farquhar (1996), 1.

16. *Wireless Quick Facts,* CTIA, Wireless Association website.

17. Hardman (1982), 387.

18. John W. Berresford, *The Impact of Law and Regulation on Technology: The Case History of Cellular Radio,* 44 BUSINESS LAWYER 721 (May 1989), 724.

19. Robert W. Crandall, *Surprises from Telephone Deregulation and the AT&T Divestiture,* 78 AMERICAN ECONOMIC REVIEW 323 (May 1988), 325.

20. Jeffrey H. Rohlfs, Charles L. Jackson & Tracey E. Kelly, *Estimate of the Loss to the United States Caused by the FCC's Delay in Licensing Cellular Telecommunications,* National Economic Research Associates (Nov. 8, 1991, revised).

21. William C. Jakes, Jr., ed., MICROWAVE MOBILE COMMUNICATIONS (New York: Wiley, 1974), 1.

22. *Gartner Says Worldwide Mobile Phone Sales Increased 16 Per Cent in 2007*, Gartner (Feb. 27, 2008).

23. Motorola website, *About: What We Do.*

24. Zachary M. Seward, *The First Mobile Phone Call Was Made 40 Years Ago Today on April 3, 1973: Motorola Employee Martin Cooper Stood in Midtown Manhattan and Placed a Call to the Headquarters of Bell Labs in New Jersey*, ATLANTIC (April 3, 2013).

25. Hardman (1982), 385, 387.

26. *Land Mobile Radio Service*, SECOND REPORT AND ORDER, Docket 18262, FCC 2d (1974), 760.

27. Berresford (1989), 726.

28. FCC (1974), ¶¶ 12, 21.

29. 46 FCC 2d 760.

30. Joel West, *Institutional Standards in the Initial Deployment of Cellular Telephone Service on Three Continents*, CENTER FOR RESEARCH ON INFORMATION TECHNOLOGY AND ORGANIZATIONS (June 30, 1999), 11.

31. Rohlfs et al. (1991).

32. FCC, National Broadband Plan (2010), 79.

33. Berresford (1989), 726.

34. Farley (2007).

35. FCC (1974), ¶ 8.

36. Which became a theme among advocates of regulatory reform. Adam Thierer, PERMISSIONLESS INNOVATION: THE CONTINUING CASE FOR COMPREHENSIVE TECHNOLOGICAL FREEDOM (Arlington, VA: Mercatus Center at George Mason University, 2014).

37. FCC (1981), ¶ 96.

38. Murray (2001), 50.

39. Calhoun (1988), 15.

40. Quoted in Erwin G. Krasnow, Lawrence D. Longley & Herbert A. Terry, THE POLITICS OF BROADCAST REGULATION, 3rd ed. (New York: St. Martin's, 1982), 242.

41. Ibid., 243, 251.

42. Ibid., 249, 253, 255. Goldwater had support from other conservatives, including Senators Harrison Schmitt (R-NM), Larry Pressler (R-SD), and Ted Stevens (R-AK). *Senate Beats Van Deerlin to the Draw on 1934 Law*, BROADCASTING (March 19, 1979), 35.

43. Quoted in Krasnow et al. (1982), 252.

44. Ibid., 257, quoting the Office of Communication from the United Church of Christ.

45. Ibid., quoting FCC member Nicholas Johnson.

46. Ibid., 260.

Chapter 14. Deal of the Decade

Epigraph: James B. Murray, Jr., WIRELESS NATION: THE FRENZIED LAUNCH OF THE CELLULAR REVOLUTION IN AMERICA (Cambridge, MA: Perseus, 2001), 30.

1. Bernard Schwartz, The Professor and the Commissions (New York: Knopf, 1959), 164.
2. William B. Ray, Ups and Downs of Radio-TV Regulation (Ames: Iowa State University Press, 1990), 45.
3. Henry Goldin, *Discussion of "Evaluation of Public Policy Relating to Radio and Television Broadcasting: Social and Economic Issues" (Coase)*, 41 Land Economics 167 (May 1965), 168.
4. Erwin G. Krasnow, Lawrence D. Longley & Herbert A. Terry, The Politics of Broadcast Regulation, 3rd ed. (New York: St. Martin's, 1982), 17.
5. Ronald Coase, *The Federal Communications Commission*, 2 Journal of Law & Economics 1 (1959), 35–36.
6. Edward J. Markey, *The Fairness Doctrine, Congress, and the FCC*, 6 Comm. Law 1 (Summer 1988), 26–27. In 2013 Markey was elected to the Senate.
7. Jora Minasian, *Property Rights in Radiation: An Alternative Approach to Radio Frequency Allocation*, 18 Journal of Law & Economics 221 (1975), 268.
8. Ronald H. Coase, *Evaluation of Public Policy Relating to Radio and Television Broadcasting: Social and Economic Issues*, 41 Land Economics 161 (1965), 165, 167.
9. David T. Bazelon, *The First Amendment and the "New Media"—New Directions in Regulating Telecommunications*, in D. Brenner & W. Rivers, eds., Free but Regulated: Conflicting Traditions in Media Law 52 (1982), 55.
10. Quoted in Thomas W. Hazlett, *Making Money Out of the Air*, N.Y. Times (Dec. 2, 1987).
11. Nick Allard, *The New Spectrum Auction Law*, 18 Seton Hall Legislative Journal 13 (1994), 120.
12. Congressman John D. Dingell (D-MI), Chair of the House Energy & Commerce Committee, Legislation to codify the Fairness Doctrine: Hearings on H.R. 1934, Subcommittee on Telecommunications and Finance of the House Energy and Commerce Committee, 100th Congress, 1st Session (April 7, 1987), 10.
13. *Citizens Committee to Save WEFM v. FCC*, 506 F.2d 246 (1974), 268.
14. William Mayton, *The Illegitimacy of the Public Interest Standard at the FCC*, 38 Emory Law Journal 715 (1989), 759 (footnotes omitted).
15. The family viewing hour was eventually found illegal by U.S. courts—on antitrust grounds. In setting up one hour each evening (7–8 P.M.) in which TV programs would avoid certain types of content, the networks were found to have conspired to restrain trade.
16. Victor P. Goldberg, *Toward an Expanded Economic Theory of Contract*, 10 Journal of Economic Issues 45 (March 1976); Victor P. Goldberg, *Regulation and Administered Contracts*, 7 Bell Journal of Economics 426 (Autumn 1976).
17. David Henderson, *Armen Alchian*, The Concise Encyclopedia of Economics (Indianapolis: Liberty Fund, 2008), 522.
18. Andrei Schleifer & Robert W. Vishny, *Pervasive Shortages Under Socialism*, 23 Rand Journal of Economics 237 (Summer 1992).
19. Murray (2001), 37, 45–46.

20. Peter Huber, LAW AND DISORDER IN CYBERSPACE (New York: Oxford University Press, 1997), 106.
21. Murray (2001), 51.
22. Evan Kwerel & Alex D. Felker, *The Use of Auctions to Select FCC Licensees,* Federal Communications Commission, Office of Plans & Policies Working Paper No. 16 (May 1985).
23. Murray (2001), 45–47.
24. Dissenting Statement of Commissioner Glen O. Robinson in *Cowles Florida Broadcasting, Inc. et. al.,* 60 FCC 2d 372 (1976).
25. Peter Leeson, *Ordeals,* 55 JOURNAL OF LAW & ECONOMICS 691 (Aug. 2012).
26. Murray (2001), 65. In the event, the coverage variable was bypassed when the FCC chose to. Other "public interest" criteria were employed to deny Graphic Scanning, not an FCC favorite, from the Chicago franchise and award it to the only other applicant, John Kluge's Metromedia.
27. Murray (2001), 61.
28. Kwerel & Felker (1985), 12.
29. Thomas W. Hazlett, *Assigning Property Rights to Radio Spectrum Users: Why Did FCC License Auctions Take 67 Years?* 41 JOURNAL OF LAW & ECONOMICS 529 (Oct. 1998), 534.
30. Webbink (1980), 30, 32 (emphasis original).
31. Murray (2001), 68.
32. Quoted ibid., 136.
33. Thomas W. Hazlett & Robert J. Michaels, *The Cost of Rent Seeking: Evidence from the Cellular Telephone License Lotteries,* 39 SOUTHERN ECONOMIC JOURNAL 425 (Jan. 1993), 433.
34. Round 1: 190; round 2: 353; round 3: 567. Murray (2001), 88.
35. Kwerel & Felker (1985), 4.
36. Allard (1994), 115 (footnote omitted).
37. One notable measure proved perverse. The Commission thought it would be smart, when drawing lottery winners, to choose ten per market, a winner and then a succession of runners-up. Whenever a winner should fail to qualify (if its application were to be deemed deficient), the next in line would move up, no new drawing needed. It was intended to save time. In fact, the idea was a disaster. Now it was a certainty that "winners" two through ten would litigate in sequence.
38. About 100,000 applications were received for the 306 MSAs, and 280,000 for the 428 RSAs. Hazlett & Michaels (1993), 428, 431.
39. Thomas W. Hazlett, *"Spectrum Auctions" Only a First Step,* WALL STREET JOURNAL (Dec. 20, 1994).
40. Hazlett & Michaels (1993), 430.
41. Actual examples cited in Murray (2001), 162–178.
42. Fleming Meeks, *Cellular Suckers,* FORBES (July 1988), 41.
43. As shown in Hazlett & Michaels (1993). Gains here were calculated using *ex ante* values—what lottery winners would expect to get by selling licenses immediately.

44. David Ellen, *The Phone Flushaway: Uncle Sam Wants You to Win $20 Billion*, NEW REPUBLIC (Oct. 9, 1989), 13.

45. Department of Commerce, National Telecommunications and Information Administration (NTIA), *U.S. Spectrum Management Policy: Agenda for the Future* (Feb. 1991), D6. The NTIA estimates for MSA license values are here adjusted to include RSAs.

46. Carla Lazzareschi & Jube Shiver, *AT&T Will Buy McCaw Cellular for $12.6 Billion*, L.A. TIMES (Aug. 17, 1993).

47. William Kummel, *Spectrum Bids, Bets, and Budgets: Seeking an Optimal Allocation and Assignment Process for Domestic Commercial Electromagnetic Spectrum Products, Services, and Technology*, 48 FEDERAL COMMUNICATIONS LAW JOURNAL 511 (1996), 526–527.

48. Robert A. Caro, THE YEARS OF LYNDON JOHNSON: MEANS OF ASCENT (New York: Vintage, 1991), 94.

49. David Porter & Vernon Smith, *FCC License Auction Design: A 12-Year Experiment*, 3 JOURNAL OF LAW, ECONOMICS & POLICY 63 (2006), 64.

50. Mark Lewyn, *The Case for Auctioning Off the Airwaves*, BUSINESS WEEK (Oct. 18, 1992).

51. Federal Communications Commission, *FCC Report to Congress on Spectrum Auctions* (Oct. 9, 1997), 8.

52. The FCC estimated comparative hearings to take an average of eighteen months to assign a license, lotteries twelve; Kwerel & Felker (1985).

53. *Sell the Dial*, N.Y. TIMES (May 9, 1985), editorial page.

54. Thomas W. Hazlett, *Making Money Out of the Air*, N.Y. TIMES (Dec. 2, 1987); Thomas W. Hazlett, *Dial "G" for Giveaway*, BARRON'S (June 4, 1990).

55. Lewyn (1992).

56. Harry Kalven, *Broadcasting, Public Policy, and the First Amendment*, 10 JOURNAL OF LAW & ECONOMICS 15 (Oct. 1967), 30, 32.

57. William H. Melody, *Radio Spectrum Allocation: Role of the Market*, 70 AMERICAN ECONOMIC REVIEW 393 (May 1980), 393.

58. *Broadcast Renewal Applicant*, 66 F.C.C.2d 419, 434 n.2 (1977) [Commissioners Hooks and Fogarty, separate statement].

59. *Dead Before Arrival: Administration's Spectrum Auction Bill Gains Hill Backing*, 8 COMMUNICATIONS DAILY, No. 115 (June 15, 1988), 2. "The death of the auction proposal is at hand," Markey said, "and the principle of assigning spectrum by the public interest standard will again rise phoenix-like in its wake."

60. Cindy Skrzycki, *Congress Mulls New Ways for FCC to Divide Broadcast Spectrum*, WASH. POST F1 (June 26, 1991), F3.

61. A Bill to Establish Procedures to Improve the Allocation and Assignment to the Electromagnetic Spectrum, Serial No. 102-2 Hearings on H.R. 531 before the Subcomm. on Telecomm. and Fin. of the Comm. on Energy and Commerce, 102d Cong. 1st Sess. 89 (Feb. 21 and March 12, 1991).

62. Quoted in Janice Obuchowski, *The Unfinished Task of Spectrum Policy Reform*, 47 FEDERAL COMMUNICATIONS LAW JOURNAL 325 (1994).

63. Mike Mills, *Auction of Frequencies Sets Up a 21st Century Marketplace*, CONG. Q. (May 8, 1993), 1137.

64. Cited in Evan R. Kwerel & Gregory L. Rosston, *An Insider's View of FCC Spectrum Auctions*, 17 JOURNAL OF REGULATORY ECONOMICS 253 (2000), 258.

65. Robert J. Samuelson, *The Quiet Giveaway*, WASHINGTON POST (May 8, 1991).

66. Commissioner Gloria Tristani, *Broadcast Views*, Speech to the Federal Communications Bar Association (May 21, 1998).

67. Steven H. Wildstron & Mark Lewyn, *Airwaves for Sale: Contact Bill Clinton*, BUSINESS WEEK (May 9, 1993), 37.

68. Eva Kalman, THE ECONOMICS OF RADIO FREQUENCY ALLOCATION (Paris: Organization for Economic Cooperation & Development, 1993), 86.

69. Kurt A. Wimmer & Lee J. Tiedrich, *Competitive Bidding and Personal Communications Services: A New Paradigm for FCC Licensing*, 3 COMMLAW CONSPECTUS 17 (1994), 20.

70. Congressional Budget Office (CBO), *Auctioning Radio Spectrum Licenses* (March 1992), 21–22.

71. Thomas W. Hazlett & Matthew L. Spitzer, PUBLIC POLICY TOWARD CABLE TELEVISION: THE ECONOMICS OF RATE CONTROLS (Cambridge, MA: MIT Press, 1997).

72. Allard (1994), 34.

73. Ibid., 13.

74. Reed Hundt, YOU SAY YOU WANT A REVOLUTION (New Haven: Yale University Press, 2000), 93.

75. For the moment, the FCC abandoned the MSAs and RSAs that, in cellular, had resulted in 734 franchise areas. In the next round of PCS license auctions, held in 1996, a map with 493 basic trading areas (BTAs) would be used.

76. John McMillan, *Selling Spectrum Rights*, 8 JOURNAL OF ECONOMIC PERSPECTIVES 145 (Summer 1994), 146.

77. Peter C. Cramton, *The FCC Spectrum Auctions: An Early Assessment*, 6 JOURNAL OF ECONOMICS & MANAGEMENT STRATEGY 431 (1997), 433.

78. Federal Communications Commission, *Semiannual Report of the Inspector General* (Oct. 1994–March 1995), 2.

79. Hundt (2000), 96.

80. John W. Berresford, *The Impact of Law and Regulation on Technology: The Case History of Cellular Radio*, 44 BUSINESS LAWYER 721 (May 1989), 721.

81. Quoted in Allard (1994), 125.

Chapter 15. The Toaster Tsunami

Epigraph: Reed Hundt, YOU SAY YOU WANT A REVOLUTION: A STORY OF INFORMATION AGE POLITICS (New Haven: Yale University Press, 2000), 98.

1. Robert Pepper, *Spectrum Allocation for Personal Communication*, MIT Communications Forum (Feb. 25, 1993), 2–3. Pepper, a former professor of communications, was the influential longtime chief of the Office of Plans and Policies.

2. Jon Gertner, *How Bell Labs Invented the World We Live in Today,* Time (March 21, 2012). At its zenith, Bell Labs, the renowned AT&T research arm spun off as Lucent in 1996, had twenty-five thousand employees, of whom thirty-three hundred were scientists holding doctorates.

3. Evan Kwerel & John Williams, *A Proposal for a Rapid Transition to Market Allocation of Spectrum,* Federal Communications Commission OPP Working Paper No. 38 (Nov. 2002), 1.

4. Federal Communications Commission, *National Broadband Plan* (March 2010), 85, Exhibit 5-F.

5. Federal Communications Commission, *National Broadband Plan,* Chapter 5 ("Spectrum"), 75 (footnote omitted).

6. Louis Galambos & Eric John Abrahamson, Anytime Anywhere: Entrepreneurship and the Creation of a Wireless World (Cambridge: Cambridge University Press, 2002), 3.

7. Jerry Hausman, *Cellular Telephone, New Products, and the CPI,* 17 Journal of Business & Economic Statistics 188 (April 1999), 189.

8. Ibid., 189.

9. Galambos & Abrahamson (2002), 2.

10. Hausman (1999), 189.

11. It is noteworthy that such high retail prices were, until 1994, regulated at the option of state public utility commissions. Elimination of the rate controls, by federal statute, was followed by a rapid decline in consumer prices.

12. California Public Utilities Commission, Decision 90-6-025 (June 6, 1990), 8.

13. Federal Communications Commission, *Annual Report and Analysis of Competitive Market Conditions with Respect to Commercial Mobile Services* (July 28, 1995), ¶ 13 (footnote omitted).

14. *The FCC Encyclopedia: Broadband Personal Communications Services* (PCS), FCC .gov.

15. Robert G. Harris & C. Jeffrey Kraft, *Meddling Through: Regulating Local Telephone Competition in the United States,* 11 Journal of Economic Perspectives 93 (Fall 1997), 102.

16. Federal Communications Commission, *FCC Announces 99 Licenses for Broadband Personal Communications Services in Major Trading Areas,* News Release (June 23, 1995).

17. John Leslie King & Joel West, *Ma Bell's Orphan: U.S. Cellular Telephony, 1947–1996,* 26 Telecommunications Policy 189 (2002), 197.

18. O. Casey Corr, Money from Thin Air: The Story of Craig McCaw, the Visionary who Invented the Cell Phone Industry, and His Next Billion-Dollar Idea (New York: Crown Business, 2000), 154–155.

19. *History of McCaw Cellular Communications, Inc.,* Reference for Business, http://www.referenceforbusiness.com/history2/90/MCCAW-CELLULAR-COMMUNI CATIONS-INC.html#ixzz3AINxNNML.

20. *AT&T Launches Flat-Rate Service,* Los Angeles Times (May 8, 1998).

21. Walter Mossberg, *AT&T's One-Rate Plan Falls Short of Promise,* WALL STREET JOURNAL (July 8, 1999).

22. CTIA—The Wireless Association, *CTIA's Annualized Wireless Industry Results— December 1985 to December 2013,* CELLULAR TELECOMMUNICATIONS & INTERNET ASSOCIATION (2014), 2.

23. Ibid., 181. Annual data calculated by doubling six-month year-end interval in CTIA data.

24. Ibid.

25. Thomas W. Hazlett, *Is Federal Preemption Efficient in Cellular Phone Regulation?* 56 FEDERAL COMMUNICATIONS LAW JOURNAL 155 (Dec. 2003), Table 3.

26. Bank of America/Merrill Lynch, *Global Wireless Matrix 2Q2014* (Sept. 29, 2014), 301.

27. The calculation assumes a nominal per minute rate drop from fifty cents to three cents, and adjusts for inflation. The Consumer Price Index (CPI-U) averaged 130.7 in 1991, and 224.9 in 2011.

28. Galambos & Abrahamson (2002), 191.

29. Quoted ibid., 189.

30. Robert W. Crandall, COMPETITION AND CHAOS: U.S. TELECOMMUNICATIONS SINCE THE 1996 ACT (Washington, DC: Brookings Institution, 2005), 89.

31. This includes voice and texting service but excludes broadband.

32. *New Tech to Drive CE Industry Growth in 2015, Projects CEA's Midyear Sales and Forecasts Report,* Consumer Electronics Association (July 15, 2015).

33. *Jeepers, Creepers, Where'd You Get All Those Beepers,* BUSINESS WEEK (Sept. 29, 1991).

34. Reily Gregson, *Consumers Direct Paging Industry Growth,* RCR WIRELESS (Nov. 10, 1997).

35. FCC website.

36. Federal Communications Commission, *Annual Report and Analysis of Competitive Market Conditions with Respect to Commercial Mobile Services, Third Report,* FCC 98-91 (June 11, 1998), 58.

37. Ibid., 61, footnote 317. The prices were those given for GTE's CDPD service.

38. Nippon Telephone & Telegraph was the state monopoly telecommunications provider in Japan until its privatization in 1985. The government of Japan continues to own a minority stake in the company.

39. James Quintana Pearce, *3G—Hitting the Mass Market,* GIGAOM (Dec. 11, 2005).

40. John Beck & Mitchell Wade, DOCOMO: JAPAN'S WIRELESS TSUNAMI (New York: Amacom, 2003), 156.

41. Frank Rose, *Pocket Monster,* WIRED (Sept. 1, 2001).

42. Joseph A. Schumpeter, CAPITALISM, SOCIALISM, AND DEMOCRACY (New York: Harcourt, 1942), 83.

43. For a beautifully written biography of the noted intellectual, see Thomas K. McCraw, PROPHET OF INNOVATION (Cambridge, MA: Belknap Press of Harvard University Press, 2007).

44. Fred Vogelstein, Dogfight: How Apple and Google Went to War and Started a Revolution (New York: Farrar, Straus and Giroux, 2013), 120.

45. The wireless company was then known as Cingular. Its parent, SBC, acquired AT&T Wireless in 2004 and then AT&T (the long distance operator) in late 2005. SBC changed its name (and Cingular's) to AT&T in January 2007. Lloyd Vries, *From AT&T to Cingular and Back Again,* CBS News (Jan. 12, 2007).

46. Vogelstein (2013), 19.

47. Sarah Perez, *iTunes App Store Now Has 1.2 Million Apps, Has Seen 75 Billion Downloads to Date,* Tech Crunch (June 2, 2014).

48. Vogelstein (2013), 138.

49. Jay Yarow, *It's Official: Apple Is Just a Niche Player in Smartphones Now,* Business Insider (Nov. 2, 2012).

50. Gregg Keizer, *Microsoft Writes Off $7.6B, Admits Failure of Nokia Acquisition,* Computerworld (July 8, 2015).

51. Benzinga Editorial, *Apple Now Most Valuable Company in History,* Forbes (Aug. 21, 2012).

52. Initially, mobile carriers received 25 percent and Google 5 percent. Chris Smith, *Google Wants to Increase Its Share of Play Store Android App Revenue, New Report Says,* Android Authority (June 28, 2013).

53. Quoted in Walter Isaacson, Steve Jobs (New York: Simon & Schuster, 2011), 512. For better or worse, Jobs died without spending the $40 billion or destroying Android.

54. Thomas W. Hazlett, *Gravitational Shift: Competition in Mobile Phones Has Supplanted Competition Among Networks,* Barron's (March 31, 2012).

55. Jim Edwards, *Apple's Manufacturing Costs Reveal the Profits It Will Make on iPhone 6,* Business Insider (Sept. 24, 2014).

56. Comments of CTIA—The Wireless Association, submitted to the Federal Communications Commission in WT Docket No. 15-125 (June 29, 2015), 11.

57. Federal Communications Commission, *In the Matter of Digital Audio Broadcasting Systems and Their Impact on the Terrestrial Radio Broadcast Service: First Report and Order,* MM Docket No. 99-325 (released Oct. 11, 2002).

58. Scott R. Flick & Christine A. Reilly, *FCC Approves Digital Power Increase for Most FM IBOC Stations,* Pillsbury Advisory (Feb. 2010), 1.

59. *Declaration of James Martinek* (Nov. 28, 2005), 2; attached to Comments of T-Mobile USA, Inc., submitted to the Federal Communications Commission WT Docket No. 05-265 (Nov. 28, 2005).

60. More than fifty U.S. MVNOs were available in September 2016, as listed on Best-MVNO.com.

61. John W. Mayo & Scott Wallsten, *Enabling Efficient Wireless Communications: The Role of Secondary Spectrum Markets,* 22 Information Economics and Policy 61 (March 2010), Table 1.

62. Rolfe Winkler, *Mobile's Path to Glory,* Wall Street Journal (Dec. 30, 2012).

63. Robert Jensen, *The Digital Provide: Information (Technology), Market Performance,*

and Welfare in the South Indian Fisheries Sector, 122 QUARTERLY JOURNAL OF ECONOM-ICS 879 (Aug. 2007), 879.

64. Quoted in Kevin Sullivan, *For India's Traditional Fishermen, Cellphones Deliver a Sea Change*, WASHINGTON POST (Oct. 15, 2006).

65. Ibid.

66. Lars-Hendrik Röller & Leonard Waverman, *Telecommunications Infrastructure and Economic Development: A Simultaneous Approach*, 91 AMERICAN ECONOMIC REVIEW 909 (2001), 919.

67. Leonard Waverman, Meloria Meschi & Eugene Fuss, *The Impact of Telecoms on Economic Growth in Developing Countries*, Vodafone Policy Paper Series No. 2, 10 (March 2005), 10–11.

68. Roger G. Noll, *Telecommunications Reform in Developing Countries*, in Anne O. Krueger, ed., ECONOMIC POLICY REFORM: THE SECOND STAGE 83 (Chicago: University of Chicago Press, 2000), Table 6.1.

69. Sara Corbett, *Can the Cellphone Help End Global Poverty?* N.Y. TIMES MAGAZINE (April 13, 2008).

70. Ibid.

71. Russell Southwood, LESS WALK, MORE TALK: HOW CELTEL AND THE MOBILE PHONE CHANGED AFRICA (West Sussex, UK: Wiley, 2008).

72. Andy Hull, Nick McClellan & Troy K. Schneider, *Which Countries Have the Most Mobile Phones?* SLATE (Feb. 12, 2012).

73. Philip N. Howard, Aiden Duffy, Deen Freelon, Muzammil Hussain & Will Mari, *Opening Closed Regimes: What Was the Role of Social Media During the Arab Spring?* Working Paper 2011.1, Project on Information Technology & Political Islam, University of Washington (2011), 8.

74. Cory Doctorow, *Mobile Malware Infections Race Through Hong Kong's Umbrella Revolution*, BOING BOING (Oct. 3, 2014).

75. Adam Powell, *Cell Phones Combat Agricultural Pests in Asia*, USC Annenberg Center on Communication Leadership & Policy (March 10, 2014).

76. Waverman, Meschi & Fuss (2005), 11.

77. *Nobel Peace Prize for 2006 to Muhammad Yunus and Grameen Bank—Press Release*, Nobelprize.org (visited Sept. 25, 2016).

78. Waverman, Meschi & Fuss (2005), 10–11.

79. The author had a modest role in consulting with policy makers in Guatemala and El Salvador, in 1996 and 1997, via grants provided by the World Bank and the Bank for Inter-American Development. Pablo Spiller from the University of California, Berkeley, also advised these governments.

80. Pablo T. Spiller & Carlo G. Cardilli, *The Frontier of Telecommunications Deregulation: Small Countries Leading the Pack*, 11 JOURNAL OF ECONOMIC PERSPECTIVES 127 (Autumn 1997); Giancarlo Ibárgüen, *Liberating the Radio Spectrum in Guatemala*, 27 TELECOMMUNICATIONS POLICY 543 (Aug. 2003).

81. Thomas W. Hazlett, Giancarlo Ibárgüen & Wayne A. Leighton, *Property Rights to Radio Spectrum in Guatemala and El Salvador: An Experiment in Liberalization*, 3

REVIEW OF LAW & ECONOMICS 437 (Dec. 2007). See also Thomas W. Hazlett & Roberto E. Muñoz, *Spectrum Allocation in Latin America: An Economic Analysis*, 21 INFORMATION ECONOMICS AND POLICY 261 (June 2009).

82. The reform process is well described in Carlos Sabino & Wayne Leighton, *Privatization of Telecommunications in Guatemala: A Tale Worth Telling*, Amazon Kindle Edition (2013).

Chapter 16. *Dirigiste* Backlash

Epigraph: Robert W. Lucky, *Wired and Wireless Networks Compete—Cooperatively*, IEEE SPECTRUM (Oct. 31, 2012).

1. Gregory L. Rosston & Thomas W. Hazlett, *Comments of 37 Concerned Economists*, AEI-Brookings Joint Center for Regulatory Studies Related Publication (Feb. 2001), 6. The statement was submitted to the Federal Communications Commission, *In the Matter of Promoting Efficient Use of Spectrum Through Elimination of Barriers to the Development of Secondary Markets*, WT Docket No. 00-230 (Feb 7, 2001). Signatories included Martin Neil Baily (International Institute for Economics), Jonathan Baker (American University), Timothy Bresnahan (Stanford), Ronald Coase (University of Chicago), Peter Cramton (University of Maryland), Robert W. Crandall (Brookings Institution), Richard Gilbert (U.C. Berkeley), Shane Greenstein (Northwestern), Robert W. Hahn (American Enterprise Institute), Robert Hall (Stanford), Barry Harris (Economists, Inc.), Robert Harris (U.C. Berkeley), Jerry A. Hausman (MIT), Thomas W. Hazlett (American Enterprise Institute), Andrew Joskow (National Economic Research Associates), Alfred E. Kahn (Cornell), Michael Katz (U.C. Berkeley), Robert E. Litan (Brookings Institution), Paul Milgrom (Stanford), Roger G. Noll (Stanford), Janusz Ordover (NYU), Bruce M. Owen (Economists, Inc.), Michael Riordan (Columbia), William Rogerson (Northwestern), Gregory L. Rosston (Stanford), Daniel L. Rubinfeld (U.C. Berkeley), David Salant (National Economic Research Associates), Richard L. Schmalensee (MIT), Marius Schwartz (Georgetown), Howard Shelanski (U.C. Berkeley), J. Gregory Sidak (American Enterprise Institute), Pablo Spiller (U.C. Berkeley), David Teece (U.C. Berkeley), Michael Topper (Cornerstone Research), Hal R. Varian (U.C. Berkeley), Leonard Waverman (London Business School), and Lawrence J. White (NYU).

2. In Eli Noam's dissenting take, *Spectrum Auctions: Yesterday's Heresy, Today's Orthodoxy, Tomorrow's Anachronism. Taking the Next Step to Open Spectrum Access*, 41 JOURNAL OF LAW & ECONOMICS 765 (Oct. 1998). See also Timothy J. Brennan, *The Spectrum as Commons: Tomorrow's Vision, Not Today's Prescription*, 41 JOURNAL OF LAW & ECONOMICS 791 (Oct. 1998); and Thomas W. Hazlett, *Spectrum Flash Dance: Eli Noam's Proposal for "Open Access" to Radio Waves*, 41 JOURNAL OF LAW & ECONOMICS 805 (Oct. 1998).

3. Reed Hundt, *Spectrum Policy and Auctions: What's Right, What's Left*, Speech to Citizens for a Sound Economy (June 18, 1997).

4. Thomas W. Hazlett, *Property Rights and the Value of Wireless Values*, 51 JOURNAL OF LAW & ECONOMICS 563 (Aug. 2008).

5. *Review of Radio Spectrum Management by Professor Martin Cave for Department of Trade and Industry Her Majesty's Treasury* (March 2002), 16.

6. Ibid.

7. *German Phone Auction Ends*, CNN MONEY (Aug. 17, 2000).

8. Russell Carlberg, *The Persistence of the Dirigiste Model: Wireless Spectrum Allocation in Europe, à la Française*, 54 FEDERAL COMMUNICATIONS LAW JOURNAL 129 (2001), 132.

9. Tomaso Duso & Jo Seldeslachts, *The Political Economy of Mobile Telecommunications Liberalization: Evidence from the OECD Countries*, 38 JOURNAL OF COMPARATIVE ECONOMICS 199 (2010), 211.

10. *Spectrum-Starved US Prepares to Feast: Broadband Floodgates Open*, REGISTER (March 30, 2005).

11. Comments of Evan Kwerel, Ronald Coase's Anniversary lectures at George Mason, Convergence Law Institute (Notes) (Oct. 30, 2009).

12. Commissioner Furchtgott-Roth made the comment at a Washington, DC, conference, circa 2002. He confirmed the accuracy of the quotation in an email to the author Oct. 19, 2015.

13. Thomas W. Hazlett & Babette E. L. Boliek, *Use of Designated Entity Preferences in Assigning Wireless Licenses*, 51 FEDERAL COMMUNICATIONS LAW JOURNAL 639 (May 1999), 640.

14. Three categories were established. Very small companies had less than $15 million in annual revenues (previous three-year average); small companies had $15 million to $40 million. Entrepreneurial companies could have between $40 million and $125 million, and no more than $500 million in assets. See ibid., 641.

15. Prices calculated in Thomas W. Hazlett, David Porter & Vernon L. Smith, *Radio Spectrum and the Disruptive Clarity of Ronald Coase*, 54 JOURNAL OF LAW & ECONOMICS S125 (Nov. 2011), Appendix 1A.

16. Ibid. See also Robert W. Crandall, COMPETITION AND CHAOS: U.S. TELECOMMUNICATIONS SINCE THE 1996 ACT (Washington, DC: Brookings Institution, 2005), 105.

17. *FCC v. Nextwave Personal Communications, Inc.*, 254 F.3d 130, 133 (2001). In 2003, the Supreme Court upheld the decision in *FCC v. Nextwave Personal Communications, Inc.*, 537 U.S. 293 (2003), 254 F.3d 130, affirmed.

18. The long, bloody path, described as "The Miracle of Compound Interests," is painfully detailed in Harold Furchtgott-Roth, A TOUGH ACT TO FOLLOW: THE TELECOMMUNICATIONS ACT OF 1996 AND THE SEPARATION OF POWERS (Washington, DC: AEI Press, 2006), 125–144.

19. Federal Communications Commission, *In the Matter of Advanced Television Systems and Their Impact upon the Existing Television Broadcast Service: Sixth Further Notice of Proposed Rule Making*, MM Docket No. 87-268 (Released Aug. 14, 1996).

20. Mary Greczyn, *FCC Wireless Bureau Delays 700 MHz Auction Until Sept.*, COMMUNICATIONS DAILY (Feb. 1, 2001).

21. Ibid.

22. Patrick Ross, *Bush Wants to Delay Airwaves Auction*, News.com (April 9, 2001).

23. *Bush Administration Hopes to Sell Congress on 700 MHz Sale Delays, Analog TV Lease Fees*, TELECOMMUNICATIONS REPORTS (March 5, 2001), 3.

24. Lynette Luna, *Spectrum Quandary Puts 3G at Risk*, TELEPHONY ONLINE (July 23, 2001).

25. The name is instructive. The company's receiver dishes would point north, picking up signals broadcast from a terrestrial transmitter in the northern edge of each local market. This traffic would peacefully coexist with existing satellite services, which (like DirecTV and DISH) deploy receiver dishes facing south. That orientation helps receive signals from satellites on geosynchronous orbits over the Earth's equator. Northpoint planned to feed its transmitters through conventional satellite or fiber links, and to then use the north-to-south broadcasts for last-mile connections to homes and businesses, market by market. The north-to-south emissions would hit existing satellite dishes in the back (or the back of the surface they were mounted on) and leave their line of sight to the equator unaffected.

26. "'This will be the Southwest Airlines of subscription television,' said Sophia Collier, the president of Northpoint Technology, the small company that . . . said it could offer 96 digital channels and high-speed Internet access for a total of $40 a month—versus the $80 to $100 that cable companies typically charge now"; Jennifer Lee, *Many Bidders May Pursue New Method to Carry TV*, N.Y. TIMES (April 30, 2002).

27. Stephen Labaton, *An Earthly Idea for Doubling the Airwaves*, N.Y. TIMES (April 8, 2001). I note that I served as an economic expert for Northpoint Technology in 2001–2002.

28. Federal Communications Commission, *In the Matter of Amendment of Parts 2 and 25 of the Commission's Rules to Permit Operation of NGSO FSS Systems Co-Frequency with GSO and Terrestrial Systems in the Ku-Band Frequency Range: Memorandum Opinion and Order and Second Report and Order*, ET Docket No. 98-206 (Released May 23, 2002), Joint Statement of Chairman Michael Powell and Commissioner Kathleen Q. Abernathy, 6–7.

29. Northpoint had patented the technology, but this did not protect the company from appropriation. First, whatever value the licenses held were substantially discounted by the fact that rival operators held the spectrum rights necessary to deploy them. Second, intellectual property rights are costly to enforce, as seen in the litigation that did ensue when a Mexican operator attempted to use Northpoint's network design; Northpoint was unable to defend its asserted claims in court. Third, Northpoint's socially useful contribution at the regulatory agency involved nonpatentable innovations in developing a spectrum allocation.

30. Joint Statement of Chairman Michael Powell and Commissioner Kathleen Q. Abernathy (2002), 6–7.

31. Mark Lewis, *FCC Gives Northpoint Half a Loaf*, FORBES (April 24, 2002).

32. Dwight Lee, *In Defense of Excessive Government*, 65 SOUTHERN ECONOMIC JOURNAL (1999).

33. Geller & Lampert prepared a monograph on the subject cited in 5 FCC Rcd. 2766 (1990).

34. Anne West, *Pioneer Preferences: Analysis Through Five Models*, 45 FEDERAL COMMUNICATIONS LAW JOURNAL 149 (Dec. 1992), 152.

35. *Qualcomm, Inc. v. FCC, et al.*, U.S. Court of Appeals for the District of Columbia Circuit, No. 98-1246 (July 23, 1999).

36. Federal Communications Commission, *Review of the Pioneer's Preference Rules: Order*, ET Docket No. 93-266 (Aug. 29, 1997).

37. John McMillan, *Selling Spectrum Rights*, 8 JOURNAL OF ECONOMIC PERSPECTIVES 145 (Summer 1994), 147.

38. Paul Klemperer, *How (Not) to Run Auctions: The European 3G Telecom Auctions*, 46 EUROPEAN ECONOMIC REVIEW 829 (2002), 840.

39. As shown empirically in Thomas W. Hazlett & Roberto E. Muñoz, *A Welfare Analysis of Spectrum Allocation Policies*, 40 RAND JOURNAL ON ECONOMICS 424 (Autumn 2009).

40. Ian Ayres & Peter Cramton, *Deficit Reduction Through Diversity: How Affirmative Action at the FCC Increased Auction Competition*, 48 STANFORD LAW REVIEW 761 (1996).

41. Paul Milgrom, PUTTING AUCTION THEORY TO WORK (Cambridge: Cambridge University Press, 2004), 21.

42. Michael H. Rothkopf & Coleman Bazelon, *Interlicense Competition: Spectrum Deregulation Without Confiscation or Giveaways*, New America Foundation Spectrum Policy Program, Working Paper No. 8 (2003).

43. The proof is provided in Thomas W. Hazlett, Roberto Muñoz & Diego Avanzini, *What Really Matters in Spectrum Allocation Design*, 10 NORTHWESTERN JOURNAL OF TECHNOLOGY & INTELLECTUAL PROPERTY 93 (Jan. 2012), 117–120.

44. George Gilder, *Auctioning the Airwaves*, FORBES ASAP (April 1, 1994).

45. George Gilder, MICROCOSM: THE QUANTUM REVOLUTION IN ECONOMICS AND TECHNOLOGY (New York: Free Press, 2000), 94.

46. www.Wi-Fi.org, cited in Federal Communications Commission, *In the Matter of Revision of Part 15 of the Commission's Rules to Permit Unlicensed National Information Infrastructure (U-NII) Devices in the 5 GHz Band, First Report and Order*, ET Docket No. 13-49 (April 1, 2014), 4.

47. The FCC has made the spectrum for unlicensed devices using spread spectrum techniques available in increments: 300 MHz in 1997, 255 MHz in 2003. See FCC, *Revision of Part 15*.

48. Michael J. Marcus, *Wi-Fi and Bluetooth: The Path from Carter and Reagan-Era Faith in Deregulation to Widespread Products Impacting Our World*, 11 INFO 19 (2009). The story has been retold in Richard Rhodes, HEDY'S FOLLY: THE LIFE AND BREAKTHROUGH INVENTIONS OF HEDY LAMARR, THE MOST BEAUTIFUL WOMAN IN THE WORLD (New York: Doubleday, 2011).

49. *Inductees: Hedy Lamarr*, National Inventors Hall of Fame website (visited Oct. 21, 2015).

50. David Mock, THE QUALCOMM EQUATION: HOW A FLEDGLING TELECOM COMPANY FORGED

A New Path to Big Profits and Market Dominance (New York: AMACOM, 2005), 19, 70.

51. Rhodes (2011), 206.

52. Ibid.

53. Rules for unlicensed devices using spread spectrum technology were adopted in many countries, allowing for important economies of scale in producing and adopting wi-fi (and, later, Bluetooth) in the 2.4 GHz and 5 GHz bands. Because these frequencies (dubbed ISM for industrial, scientific, and medical applications) were commonly reserved for electronic devices that, like certain types of hospital equipment and microwave ovens, emitted energy for noncommunications purposes, regulators abroad did not confront insurmountable resistance in adding new local area network radios to the list. Michael Marcus believes, however, that most other regimes "were not as flexible as the FCC's 1985 rules . . . as most countries fear 'permissionless innovation' in spectrum"; Michael Marcus, emails to the author (Oct. 21–22, 2015).

54. IEEE Communications Society, *Award for Public Service in the Field of Telecommunications Winner Biographies* (2013).

55. *And the Winners Were,* Economist (Dec. 4, 2003).

56. *Spectrum Policy: On the Same Wavelength,* Economist (Aug. 12, 2004).

57. Lawrence Lessig, *Spectrum for All,* CIO Insight (March 14, 2003).

58. Yochai Benkler, *Some Economics of Wireless Communications,* 16 Harvard Journal of Law & Technology 25 (Fall 2002), 30.

59. Kevin Werbach, *SuperCommons: Toward a United Theory of Wireless Communication,* 82 Texas Law Review 863 (March 2004), 867.

60. Lawrence Lessig, *Technology Over Ideology: FCC Chair Michael Powell Has Confounded Liberals and Free Market Purists,* Wired (Dec. 2004).

61. Evan Kwerel & John Williams, *A Proposal for a Rapid Transition to Market Allocation of Spectrum,* Federal Communications Commission, Office of Plans & Policy Working Paper No. 38 (Nov. 15, 2002), 31.

62. The idea of selling bandwidth rights to accommodate demands for unlicensed allocations was further explored by Commission economists Mark Bykowsky, Mark Olson, and William Sharkey in a 2008 paper. Responding to the "mobile data tsunami" and the resulting "spectrum crunch," the FCC would adopt a spinoff of the Kwerel-Williams "big bang" in the 2010 National Broadband Plan. Both approaches are discussed below.

63. BrainyQuote.com (visited Nov. 4, 2014).

64. News Release, *Consumer Electronics Industry Revenues to Reach All-Time High in 2014, Projects CEA's Semi-Annual Sales and Forecasts Report,* Consumer Technology Association (July 15, 2014).

65. *Wireless LAN Gear Nears $5 Billion in 2014; 802.11ac Access Points Take Off,* Infonetics (April 2, 2015).

66. Thomas W. Hazlett & Michael Honig, *Valuing Spectrum Allocations,* 23 Michigan Telecommunications & Technology Law Review (Dec. 2016).

67. Charles Jackson, Raymond Pickholtz & Dale Hatfield, *Spread Spectrum Is Good—*

But it Does Not Obsolete NBC v. U.S.! 58 FEDERAL COMMUNICATIONS LAW JOURNAL 245 (April 2006), 251. The authors were engineering professors at George Washington University and (Dale Hatfield) the University of Colorado, Boulder.

68. Ibid., 260.

69. John M. Peha, *Wireless Communications and Coexistence for Smart Environments*, 7 IEEE PERSONAL COMMUNICATIONS 66 (Oct. 2000), 66. Peha is now professor of electrical engineering at Carnegie Mellon University.

70. Federal Communications Commission, *Revision of Part 15 of the Commission's Rules Regarding Ultra-Wideband Transmission Systems, First Report and Order*, 17 F.C.C.R. 7435, ¶ 71 (2002).

71. Federal Communications Commission, *In the Matter of Additional Spectrum for Unlicensed Devices Below 900 MHz and in the 3 GHz Band*, ET Docket No. 02-380 (Dec. 20, 2002), ¶¶ 6, 7 (footnote omitted).

72. Federal Communications Commission, *In the Matter of Revision of Parts 2 and 15 of the Commission's Rules to Permit Unlicensed National Information Infrastructure (U-NII) Devices in the 5 GHz Band, Report and Order*, ET Docket No. 03-122, 18 FCC Rcd 24484 (2003).

73. Federal Communications Commission, *In the Matter of Wireless Operations in the 3650–3700 MHz Band, Rules for Wireless Broadband Services in the 3650–3700 MHz Band*, ET Docket No. 04-151 (March 16, 2005).

74. Ibid., ¶¶ 14, 28.

75. Werbach (2004), 878 (footnotes omitted).

76. Federal Communications Commission, *In the Matter of Unlicensed Operation in the TV Broadcast Bands, Report and Order and Further Notice of Proposed Rule Making*, ET Docket No. 04-186 (Oct. 18, 2006), ¶ 13.

77. *The Difference Engine: Bigger than Wi-Fi*, ECONOMIST (Sept. 23, 2010).

78. Federal Communications Commission, *In the Matter of Protecting and Promoting the Open Internet: Comments of the Wireless Internet Service Providers Association*, GN Docket No. 14-28 (July 16, 2014), 3.

79. Federal Communications Commission *WISPA Comments on TV Whitespaces*, ET Docket No. 04-186 (Feb. 20, 2007), 2 (emphasis in original; footnote omitted).

80. Patrick S. Ryan, *Wireless Spectrum Allocation and New Technologies: Reviewing Old and New Paradigms Through a Case Study of the U.S. Ultra Wideband Proceeding*, German Working Papers in Law and Economics (2002), 11.

81. Rafe Neeldeman, editor, RED HERRING (March 17, 1999); David G. Leeper, *Wireless Data Blaster*, SCIENTIFIC AMERICAN (May 2002).

82. Craig Mathias Follow, *Whatever Happened to UWB?* NETWORK WORLD (March 17, 2009).

83. Federal Communications Commission, *In the Matter of Wireless Operations in the 3650–3700 MHz Band, Rules for Wireless Broadband Services in the 3650–3700 MHz Band*, ET Docket No. 04-151 (March 16, 2005), ¶ 15.

84. As the assistant secretary of commerce, John Kneuer, triumphantly noted in a 2004 policy round table: "As for the Administration, we have doubled the amount

of spectrum available for unlicensed services in the five-gigahertz band: It is 255-megahertz of additional spectrum." Heritage Lectures No. 852, Broadband by 2007: A Look at the President's Initiative, Heritage Foundation (May 13, 2004), 3.

85. Jeff Berolucci, *Six Things That Block Your Wi-Fi, and How to Fix Them*, PC WORLD (May 16, 2011).

86. The FCC citing comments of the Center for Technology and Democracy. Federal Communications Commission, *In the Matter of Amendment of the Commission's Rules to Provide for Unlicensed NII/SUPERNet Operations in the 5 GHz Frequency Range: Notice of Proposed Rule Making*, ET Docket No. 96-102 (Released May 6, 1996), ¶ 20.

87. "T-Mobile's disruptive spirit has benefited consumers"; Kevin Kelleher, *You Should Thank This Company for Better Phone Plans*, TIME (Sept. 28, 2015).

88. Thomas W. Hazlett, *Spectrum Tragedies*, 22 YALE JOURNAL ON REGULATION 242 (Summer 2005).

89. 70 MHz was allocated to 700 MHz licenses auctioned in 2003, 2004, and 2008; 90 MHz was allocated to AWS-1 licenses sold in 2006. See Hazlett, Porter & Smith (2011).

90. Gregory L. Rosston, *The Long and Winding Road: The FCC Paves the Path with Good Intentions*, 27 TELECOMMUNICATIONS POLICY 501 (2003); Hazlett, Muñoz & Avanzini (2012), 100.

91. Shane Greenstein, HOW THE INTERNET BECAME COMMERCIAL: INNOVATION, PRIVATIZATION & THE BIRTH OF A NEW NETWORK (Princeton: Princeton University Press, 2015), 397.

92. For extremely local uses of radios, which involve power levels so low that they very rarely disturb neighbors, it generally makes sense to let well enough alone and avoid either property rights or regulation. Thomas W. Hazlett & Sarah Oh, *Exactitude v. Economics: Radio Spectrum and the "Harmful Interference" Conundrum*, 26 BERKELEY TECHNOLOGY LAW JOURNAL 227 (Spring 2013).

93. Even with estimates nearly two decades old. Thomas W. Hazlett, *The U.S. Digital TV Transition: Time to Toss the Negroponte Switch*, AEI-Brookings Joint Center Working Paper No. 01-15 (Nov. 2001).

94. *In the Matter of Spectrum for Broadband: A National Broadband Plan for Our Future, to the Broadband Task Force*, Comments—NBP Public Notice #26, the Association for Maximum Service Television, Inc. and the National Association of Broadcasters (NAB), Federal Communications Commission, GN Docket Nos. 09-47, 09-137 (Dec. 22).

95. National Association of Broadcasters, *114th Congress: Broadcasters Policy Agenda* (Jan. 26, 2015).

96. Hazlett & Honig (2016).

97. *Innovatio IP Ventures, LLC, Patent Litigation*, U.S. (N. Ill., Eastern Div.), MDL Docket No. 2303, Case No. 11 C 9308 (Sept. 27, 2013), 32.

98. Joann Muller, *Should Talking Cars Share Coveted Airwaves with Wifi Providers?* FORBES (Feb. 22, 2014).

99. Mark M. Bykowsky, Mark A. Olson & William W. Sharkey, *A Market-Based Approach to Establishing Licensing Rules: Licensed Versus Unlicensed Use of Spectrum*, Federal Communications Commission OSP Working Paper No. 43 (Feb. 2008).

100. *Your Phone on Steroids: With 5G Mobile, Wireless Will Go Even Faster than Fibre*, Economist (March 23, 2015).

101. Gregory Staple & Kevin Werbach, *The End of Spectrum Scarcity*, IEEE Spectrum (March 2004).

102. Thomas W. Hazlett, *Property Rights and the Value of Wireless Licenses*, 51 Journal of Law & Economics 563 (Aug. 2008).

103. Before the AWS-3 auction, where licenses were allotted 65 MHz, there was an estimated 580 MHz available for mobile services.

Chapter 17. What Would Coase Do?

Epigraphs: Friedrich A. von Hayek, acceptance speech, Nobel Prize in Economics (1974); *It's Complicated* (2009), written and directed by Nancy Meyers.

1. The total cost of the study came to $20 million, with $7 million coming from the general FCC budget. Cecilia Kang, *National Broadband Plan Cost $20 Million*, Washington Post (March 25, 2010).

2. Thomas W. Hazlett, *Broadband Miracle*, Wall Street Journal (Aug. 26, 2004).

3. FCC, National Broadband Plan (2010), 81.

4. Cited in Bruce M. Owen, *Lies, Damn Lies, and the Legal Fictions of American Broadcast Regulation*, American Enterprise Institute Conference on the Role of Government in the Transition to Digital Television (Oct. 26, 2001), 4. The idea of an FCC two-sided auction to reallocate TV band spectrum was also suggested in James Murray, Jr., Wireless Nation: The Frenzied Launch of the Cellular Revolution in America (Cambridge, MA: Perseus, 2002), 321–322, and was elaborately detailed as a "big bang" release of new spectrum rights in Evan Kwerel & John Williams, *A Proposal for a Rapid Transition to Market Allocation of Spectrum*, Federal Communications Commission, Office of Plans & Policy Working Paper No. 38 (Nov. 15, 2002).

5. Ronald Coase, William Meckling & Jora Minasian, Problems of Radio Frequency Allocation, RAND Corporation DRU-1219-RC (1995), 85.

6. "One of the most important ideas that needs 'softening up' is the response to the claim that spectrum licensees that receive flexibility or that receive the rights to sell their spectrum are getting a 'windfall' "; James Speta, *Making Spectrum Reform "Thinkable,"* 4 Journal on Telecommunications & High-Tech Law 183 (2006), 211–212.

7. Grant Gross, *Study Disputes Predictions of Coming Spectrum Crunch*, Computerworld (Aug. 22, 2014).

8. Video at http://marconisociety.org/.

9. 73 Federal Register (Jan. 15, 2008), 2437.

10. Federal Communications Commission, *In the Matter of Service Rules for Advanced*

Wireless Services H Block—Implementing Section 6401 of the Middle Class Tax Relief and Job Creation Act of 2012 Related to the 1915–1920 MHz and 1995–2000 MHz Bands, Report and Order (June 27, 2013), ¶ 3.

11. Federal Communications Commission, ET Docket No. 258, *Notice of Proposed Rulemaking & Order,* 16 FCC Record 596 (2001), 23194, fn. 3.

12. Alfred Kahn, Letting Go—Deregulating the Process of Deregulation: Temptation of the Kleptocrats and the Political Economy of Regulatory Disingenuousness (East Lansing: Michigan State University Press, 1998).

13. Joel Brinkley, Defining Vision: The Battle for the Future of Television (New York: Harcourt, 1997), 4 (emphasis original).

14. Fritts quoted ibid., 18–19.

15. Abel quoted ibid., 30.

16. Thomas W. Hazlett, *Analog Television Dies with a Whimper,* RealClearMarkets (June 30, 2009).

17. Representative Ed Markey, *June 11, 2009: Markey Statement on DTV Transition,* press release.

18. Hazlett (2009).

19. *DTV Transition Impact,* Consumer Electronics Association (Aug. 2009), 2. A random CEA survey of 3,030 U.S. adults found just 9 percent of households connected by TV "antenna only." Another 2 percent were not aware how they were connected. Eighty-nine percent indicated that they subscribed to a video service through cable, satellite, or fiber connection.

20. Mark Huffman, *Senate Approves Delay in Digital TV Conversion,* Consumer Affairs (Jan. 27, 2009).

21. John Curran, *Kennard: DTV Transition May Be "Train Wreck" Waiting to Happen,* Communications Reports Daily (Sept. 16, 2008).

22. U.S. Department of Commerce, National Telecommunications and Information Administration, TV Converter Box Coupon Program (Dec. 9, 2009).

23. *Analog to Digital Transition Required by the FCC,* Accuval.net (Aug. 2009).

24. Gerald R. Faulhaber, *The Future of Wireless Telecommunications: Spectrum as a Critical Resource,* 18 Information Economics and Policy 256 (2006), 262.

25. The story was relayed to the author during a phone conversation with Lex Felker, who had retired from the Commission, on June 17, 2009.

26. 142 Congressional Record—Senate 10672, 10672–10676 (1996). Discussion of the Pressler draft follows Thomas W. Hazlett, *The Wireless Craze, the Unlimited Bandwidth Myth, the Spectrum Auction Faux Pas, and the Punchline to Ronald Coase's "Big Joke": An Essay on Airwave Allocation Policy,* 15 Harvard Journal of Law & Technology 335 (Spring 2001), 442–443. I note that I consulted with Senator Pressler's staff in developing this proposal.

27. Jonathan Klick, John McDonald & Thomas Stratmann, *Mobile Phones and Crime Deterrence: An Underappreciated Link,* in Alon Harel & Keith N. Hylton, eds., Research Handbook on the Economics of Criminal Law (Northampton, MA: Edward Elgar, 2012).

28. I am indebted to Steve Sharkey and Christopher Wieczorek of T-Mobile USA for explaining the inner workings of the reallocation under the CSEA to me; telephone conversation Dec. 18, 2014.

29. National Telecommunications & Information Administration, U.S. Department of Commerce, *Relocation of Federal Radio Systems from the 1710–1755 MHz Spectrum Band: Seventh Annual Progress Report* (March 2014), 1–3.

30. FCC NBP (2010), 82.

31. Which he attributes to another Nobel laureate, the late Leo Hurwicz; Eric Maskin, *Leonid Hurwicz,* Guardian (July 20, 2008).

32. Hazlett (2009).

33. A sprightly description of the proposal is given by Richard Thaler, *The Buried Treasure in Your TV Dial,* N.Y. Times (Feb. 27, 2010).

34. *TV Licensing and the Law,* TVLicensing.co.uk.

35. Cable, satellite, telco, and broadband providers could compete for the contract. The marginal cost of each satellite TV customer, not counting customer acquisition, is less than $300 per household. Economies of scale in a bulk contract, and bidding competition, would tend to push this down.

36. FCC NBP (2010), 82.

37. Greenhill & Co., *Incentive Auction Opportunities for Broadcasters,* prepared for the Federal Communications Commission (Oct. 2014), 2 (emphasis original).

38. Abel M. Winn & Mathew W. McCarter, *Who's Holding Out? An Experimental Study of the Benefits and Burdens of Eminent Domain,* Working Paper (May 15, 2014).

39. FCC NBP (2010), 82.

40. Quoted in Grant Gross, *FCC Chairman Aims for TV Spectrum Auction in Mid-2015: The FCC Will Conduct the Complex Auction When Its Software and Systems Are Ready, Wheeler Says,* Network World (Dec. 6, 2013).

41. Quoted in John Eggerton, *Verveer: First Two Auctions Should Cover Most of Statutory Financial Obligations,* Broadcasting & Cable (March 18, 2014).

42. The 2015 FCC request for funding asks for $375 million, of which $106 million is "to support the timely implementation of the Incentive Auctions program." *Fiscal Year 2015 Budget in Brief* (March 2014), 2.

43. I am, like many economists, consulting with a participant in the FCC auction.

44. Paul de Sa, *Weekend Media Blast: 2015 Year in (P)review,* Bernstein Research (Dec. 5, 2014), 3. The possibility exists, however, that once the flexible-use licenses are assigned to high bidders, things may speed up. At that point, the new rights constitute overlays yielding prospective bandwidth claimants incentives to push the process forward.

45. The necessary legislation, which became law on Feb. 22, 2012, was Title VI of the Middle Class Tax Relief and Job Creation Act of 2012 (this portion known as the Spectrum Act). Not only did this legislative detour delay implementation of the incentive auction, it included provisions (at the urging of incumbent broadcasters) that FCC officials condemned as unhelpful to the reallocation process.

46. Deborah D. McAdams, *FCC Aims to Clear 84 MHz of TV Spectrum,* Radio World (Dec. 15, 2014).

47. Federal Communications Commission, *FCC Establishes Bidding Procedures for 2016 Incentive Auction: Auction Scheduled to Begin March 29, 2016,* News Release (Aug. 6, 2015).

48. FCC NBP (2010), 92.

Chapter 18. Hoarders Anonymous

Epigraph: Ithiel de Sola Pool, Technologies Without Boundaries (Cambridge, MA: Harvard University Press, 1990), 42.

1. Michael A. Heller, The Gridlock Economy (New York: Basic, 2008), 83 (footnote omitted). Following the collapse of the World Trade Center South Tower due to a terrorist attack on September 11, 2001, New York City issued orders to evacuate the North Tower. Firefighters never received the call. "Clear warnings . . . were transmitted 21 minutes before the building fell, and officials say they were relayed to police officers, most of whom managed to escape. Yet most firefighters never heard those warnings, or earlier orders to get out. Their radio system failed frequently that morning. Even if the radio network had been reliable, it was not linked to the police system"; Jim Dwyer, Kevin Flynn & Ford Fessenden, *Fatal Confusion: A Troubled Emergency Response; 9/11 Exposed Deadly Flaws in Rescue Plan,* N.Y. Times (July 7, 2002).

2. Wireless industry executive quoted in Dorothy Robyn, *Buildings and Bandwidth: Lessons for Spectrum Policy from Federal Property Management,* Brookings Institution Economic Studies (Sept. 2014), 19.

3. President's Council of Advisors on Science and Technology (PCAST), *Report to the President: Realizing the Full Potential of Government-Held Spectrum to Spur Economic Growth* (July 20, 2012), ix.

4. Leslie Marx, *Federal Spectrum According to the PCAST Report,* Duke University (Aug. 12, 2012).

5. Gregory L. Rosston, *Increasing Wireless Value: Technology, Spectrum and Incentives,* 12 Journal on Telecommunications & High-Tech Law 89 (Spring 2014), 94.

6. Martin Cave & Adrian Foster, *Solving Spectrum Gridlock: Reforms to Liberalize Radio Spectrum Management in Canada in the Face of Growing Scarcity,* C. D. Howe Institute Commentary No. 303 (May 2010), 9 (footnotes omitted).

7. *Policy Evaluation Report: AIP,* Ofcom (July 3, 2009).

8. Valerie Strauss, *Mega Millions: Do Lotteries Really Benefit Public Schools?* Washington Post (March 30, 2012); Rodney E. Stanley & P. Edward French, *Can Students Truly Benefit from State Lotteries: A Look at Lottery Expenditures Towards Education in the American States,* Social Science Journal (2003).

9. Dorothy Robyn, Making Waves: Alternative Paths to Flexible Use Spectrum (Washington, DC: Aspen Institute, Communications and Society Program, 2015), 5.

10. T. Randolph Beard, George S. Ford, Lawrence J. Spiwak & Michael Stern, *Market Mechanisms and the Efficient Use and Management of Scarce Spectrum Resources,* 66 Fed. Comm. Law Journal 263 (April 2014), 290.

11. PCAST (2012), vii.

12. General Survey of Radio Frequency Bands—30 MHz to 3 GHz Version 2.0, Shared Spectrum (Sept. 23, 2010), 8.

13. Published on NTIA website.

14. Remarks of Commissioner Jessica Rosenworcel, CTIA 2013—*The Mobile Marketplace,* Las Vegas, Nevada (May 22, 2013), 4. The Obama White House also endorsed "an auction for the right to negotiate with a federal agency for spectrum. . . . Bidders would also pay for system improvements the agency makes to assist in freeing up the spectrum"; Stephanie Kanowitz, *White House Officials Propose Spectrum Allocation Strategies,* FIERCE MOBILE GOVERNMENT (Oct. 8, 2014), reporting on comments of Tom Power, White House Office of Science and Technology Policy.

15. Peter W. Huber, Michael K. Kellogg & John Thorne, FEDERAL TELECOMMUNICATIONS LAW, 2nd ed. (Gaithersburg, MD: Aspen Law & Business), 17.

16. Paul R. Milgrom, Jonathan Levin & Assaf Eilat, *The Case for Unlicensed Spectrum,* http://dx.doi.org/10.2139/ssrn.1948257 (Oct. 23, 2011), 13 (footnote omitted, emphasis original).

17. Heller (2008), 85–86.

18. Thomas W. Hazlett & Brent Skorup, *Tragedy of the Regulatory Commons: LightSquared and the Missing Spectrum Rights,* 13 DUKE LAW & TECHNOLOGY REVIEW (Dec. 2014).

19. Federal Communications Commission, *Report & Order and Notice of Proposed Rulemaking* (Released Feb. 10, 2003) in IB Docket No. 01-185 and IB Docket No. 02-364, ¶ 1, fn. 2; ¶28, fn. 61.

20. *The LightSquared Network: An Investigation of the FCC's Role: Hearing Before the Subcommittee on Oversight & Investigations of the House Committee on Energy & Commerce,* 112th Cong. (Sept. 21, 2012), 32, questions from Chairman Cliff Stearns, Sr.

21. *LightSquared Subsidiary LLC Request for Modification of Its Authority for an Ancillary Terrestrial Component, Order and Authorization,* 26 F.C.C. Rcd. 566, 569 (2011).

22. Ibid., 581 (footnote omitted).

Part V. Beyond

Epigraph: H. L. Mencken, *Clinical Notes,* AMERICAN MERCURY (Jan. 1924), 75.

Chapter 19. The Abolitionists

Epigraph: Herbert Stein, *Herb Stein's Unfamiliar Quotations on Money, Madness, and Making Mistakes,* SLATE (May 15, 1997).

1. James M. Landis, *Report on Regulatory Agencies to the President-Elect* (1960), 53; quoted in John F. Duffy, Randolph J. May, Wayne T. Brough, Braden Cox, James L. Gattuso, Solveig Singleton & Adam Thierer, *Report from the Working Group on Institutional Reform Release 1.0,* Progress and Freedom Foundation (Nov. 2006), 2.

2. Newton Minow, *Suggestions for Improvement of the Administrative Process: Federal Communications Commission,* 15 ADMINISTRATIVE LAW REVIEW 146 (1963), 146. Note that the Commission was then composed of seven members. It was reduced to five during the Nixon administration.

3. National Telecommunications & Information Administration, U.S. Department of Commerce, *Telecom 2000: Charting the Course for a New Century* (Oct. 1988).

4. National Telecommunications & Information Administration, U.S. Department of Commerce, *U.S. Spectrum Management Policy: An Agenda for the Future,* Spectrum SP 91-23 (Feb. 1991).

5. William E. Kennard, *A New FCC for the 21st Century: Draft Strategic Plan* (Aug. 1999), 1.

6. Dorothy Robyn, MAKING WAVES: ALTERNATIVE PATHS TO FLEXIBLE USE SPECTRUM (Washington, DC: Aspen Institute, Communications and Society Program, 2015), 46.

7. There are, however, advantages to using general courts. In particular, they may be less vulnerable to capture by vested interests.

8. Robyn (2015), 46.

9. Declan McCullagh, *Abolish the FCC,* ZDNET (June 7, 2004).

10. Declan McCullagh, *Letters in Response: Abolish the FCC? You're Crazy,* ZDNET (June 14, 2004).

11. Jack Shafer, *New Wave: The Case for Killing the FCC and Selling Off Spectrum,* SLATE (Jan. 17, 2007).

12. Lawrence Lessig, *It's Time to Demolish the FCC,* NEWSWEEK (Dec. 22, 2008).

13. FCC, National Broadband Plan (2010), 79.

14. See, e.g., Kenneth Heyer, *Welfare Standards and Merger Analysis: Why Not the Best?* 2 COMPETITION POLICY INTERNATIONAL (Nov. 2006).

15. Robert D. Hershey, Jr., *Sophia Collier, Soda Entrepreneur, Uncorks a Money Fund,* N.Y. TIMES (Oct. 8, 1995).

16. Jonas Wessel (Head of Spectrum Analysis), *Challenges and Strategies Under the Current Framework, Development of a Long Term Spectrum Strategy for Sweden,* Swedish Post and Telecom Authority, Presentation in Lisbon (Sept. 20, 2013), Slide 6.

17. Karen Wrege, *Spectrum Liberalization: Approaches in Five Countries,* KB Enterprises White Paper commissioned by the GSM Association (Oct. 9, 2009).

18. Olga Kharif, *Hi-Wire's High Wire Act,* BUSINESS WEEK (Aug. 28, 2006).

19. Press Release, *Live in New York City: Crown Castle's Modeo Subsidiary Launches Live Mobile TV Beta Service in Nation's Largest Metro Area,* CROWN CASTLE (Jan. 8, 2007).

20. Sam Churchill, *HiWire: 24 Mobile TV Channels,* Wireless.org (July 18, 2007).

21. Sam Churchill, *Qualcomm Sells MediaFLO Spectrum for $1.93B,* DailyWireless.org (Dec. 20, 2010).

22. Thomas W. Merrill & Henry E. Smith, *What Happened to Property in Law and Economics?* 111 YALE LAW JOURNAL 357 (Nov. 2001).

23. Ronald Coase, *The Federal Communications Commission,* 2 JOURNAL OF LAW & ECONOMICS 1 (1959), 34.

24. This occurred in the United States courtesy of the Air Commerce Act of 1926.

The economic rationale for the unbundling is that land use and sky routes are not complements; indeed, it would thwart economic progress—via tragedy of the anticommons—to require airlines to buy fly-over rights from each and every parcel proprietor below. That is detailed nicely in David D. Friedman, LAW'S ORDER: WHAT ECONOMICS HAS TO DO WITH LAW AND WHY IT MATTERS (Princeton: Princeton University Press, 2000), 44, 112–113.

25. Milton Friedman, *The Conventional Wisdom of J. K. Galbraith,* in FROM GALBRAITH TO ECONOMIC FREEDOM, 12 IEA Occasional Paper, No. 49 (London: Institute of Economic Affairs, 1977).

Chapter 20. Spectrum Policy as if the Future Mattered

Epigraph: Lawrence Lessig, THE FUTURE OF IDEAS: THE FATE OF THE COMMONS IN A CONNECTED WORLD (New York: Random House, 2001), 84.

1. Mike Chartier, *Local Spectrum Sovereignty: An Inflection Point in Allocation,* Proceedings of the International Symposium on Advanced Radio Technologies 29, U.S. Department of Commerce (March 2004).

2. Gerald R. Faulhaber & David J. Farber, *Spectrum Management: Property Rights, Markets, and the Commons,* in Lorrie Faith Cranor & Steven S. Wildman, eds., RETHINKING RIGHTS AND REGULATIONS: INSTITUTIONAL RESPONSES TO NEW COMMUNICATIONS TECHNOLOGIES (Cambridge, MA: MIT Press, 2003), 199.

3. Thomas W. Hazlett & Sarah Oh, *Exactitude v. Economics: Radio Spectrum and the "Harmful Interference" Conundrum,* 28 BERKELEY TECHNOLOGY LAW JOURNAL 227 (Spring 2013).

4. Congressional Research Service, *Spectrum Policy: Public Safety and Wireless Communications Interference,* CRS Report for Congress RL32408 (2004), 12.

5. Ronald Coase, *The Federal Communications Commission,* 2 JOURNAL OF LAW & ECONOMICS 1 (1959), 27.

6. At an Aspen Institute forum, Verizon executive Charla Rath explained: "You can negotiate rights at the borders—that you can use your neighbor's spectrum, [that] they can use yours. . . . If you talk to our engineers, they're doing [this] constantly. . . . We rarely go to the FCC [for] help. . . . It's just part of a normal negotiation we do"; Dorothy Robyn, MAKING WAVES: ALTERNATIVE PATHS TO FLEXIBLE USE SPECTRUM (Washington, DC: Aspen Institute, Communications and Society Program, 2015), 14.

7. FCC website.

8. It may be objected that the FCC has published hundreds of additional pages of rules and regulations relating to wireless licenses, but the same—or more—could be said of the contract, property, and criminal laws governing eBay's inventory. In any event, existing flexible-use boilerplate allows new bandwidth to be expeditiously deployed. The standard approach is simply referenced as TAS: "A TAS package describes an exclusive package of spectrum rights in terms of time, space and frequency"; Timothy K. Forde & Linda E. Doyle, *Exclusivity, Externalities &*

Easements: Dynamic Spectrum Access and Local Coasean Bargaining, in New Frontiers in Dynamic Spectrum Access Networks 303 (2007), 306.

9. Federal Communications Commission, *Spectrum Policy Task Force Report* (Nov. 15, 2002), 18, 25–26.

10. ITU Development Sector, Study Groups, *Draft Final Report on Question 10-2/1: Regulatory Trends for Adapting Licensing Frameworks to a Converged Environment* (July 21, 2009), 2.

11. Jonathan M. Barnett, *The Host's Dilemma: Strategic Forfeiture in Platform Markets for Informational Goods,* 124 Harvard Law Review 1861 (2011).

12. CREDO advertisement in the Nation (Dec. 2010) (emphasis original).

13. Project Loon (April 17, 2015).

14. Coase (1959), 34.

15. Quoted in Clifford Winston, Government Failure versus Market Failure (Washington, DC: Brookings. 2006), 1.

16. Alfred Toombs, *The Radio Battle of 1941,* Radio News 7 (March 1941), 44.

17. Stuart Minor Benjamin, *Spectrum Abundance and the Choice Between Private and Public Control,* 78 N.Y.U. Law Review 2007 (2003), provides a nice discussion.

18. Martin Cave, *New Spectrum-Using Technologies and the Future of Spectrum Management: A European Policy Perspective,* in Ed Richards, Robin Foster & Tom Kiedrowski, eds., Communications: The Next Decade (London: Ofcom, Nov. 2006), 224.

19. Mark M. Bykowsky, Mark A. Olson & William W. Sharkey, *A Market-Based Approach to Establishing Licensing Rules: Licensed Versus Unlicensed Use of Spectrum,* Federal Communications Commission OSP Working Paper No. 43 (Feb. 2008). A slick slide presentation is also available.

20. Thomas W. Hazlett & Michael Honig, *Valuing Spectrum Allocations,* 23 Michigan Journal of Telecommunications and Technology Law (Dec. 2016), both endorses and critiques various aspects of the proposal.

21. Economists William Baumol and Dorothy Robyn described the analogy with gusto in Toward an Evolutionary Regime for Spectrum Governance: Licensing or Unrestricted Entry (Washington, DC: AEI-Brookings Joint Center for Regulatory Studies, 2006). See Appendix A: "An Analog Suggesting Viability of a Private Commons: Voluntary Licensing of Intellectual Property."

22. Stephen Gandel, *Will This Company Save Wi-Fi or Destroy It?* Fortune (Nov. 17, 2015).

23. Sophia Yan & Christopher Flavelle, *GE Healthcare Path to Patient-Monitoring Boom Impeded by Boeing,* Bloomberg Business (July 22, 2010).

24. Brooks Boliek & Kim Hart, *FCC Refereeing Airwaves Fight: New Skirmishes Are Developing on the Spectrum Front, and the FCC Is Caught in the Middle,* Politico (Aug. 5, 2011).

25. "The wireless LAN industry's big trade group, the Wi-Fi Alliance, worries that carriers will have an edge in the unlicensed bands because their networks are centrally managed. Wi-Fi networks, on the other hand, tend to be a patchwork of access points and routers all operating independently but miraculously managing

to cooperate. Introducing a centrally controlled . . . [LTE unlicensed] network into that mix could mess up that mojo"; Kevin Fitchard, *Wi-Fi Industry Is Worried About Mobile Invading Its Airwaves*, GIGAOM (Feb. 13, 2015).

26. Jon Gold, *LTE-U Supporters and Opponents of LTE-U Both Claim Victory After Collaborative Testing*, NETWORK WORLD (Nov. 11, 2015).

27. Paul R. Milgrom, Jonathan Levin & Assaf Eilat, *The Case for Unlicensed Spectrum*, http://dx.doi.org/10.2139/ssrn.1948257 (Oct. 23, 2011), 3.

28. Federal Communications Commission, *Spectrum Policy Task Force Report* (Nov. 15, 2002), 38.

29. R. H. Coase, *Evaluation of Public Policy Relating to Radio and Television Broadcasting: Social and Economic Issues*, 41 LAND ECONOMICS 161 (May 1965), 163.

30. Lawrence J. White, *"Propertyzing" the Electromagnetic Spectrum: Why It's Important, and How to Begin*, 9 MEDIA LAW & POLICY 19 (Fall 2000).

31. "Central Park is a commons: an extraordinary resource of peacefulness in the center of a city that is anything but; an escape and refuge, that anyone can take and use without the permission of anyone else"; Larry Lessig, *The Architecture of Innovation*, 51 DUKE LAW JOURNAL 1783 (2002), 1788 (footnotes omitted).

32. A further discussion occurs in Thomas W. Hazlett & Coleman Bazelon, *Market Allocation of Radio Spectrum*, International Telecommunications Union Workshop on Market Mechanisms for Spectrum Management (Jan. 2007).

33. White (2000), 3.

34. Brent Skorup, *Sweeten the Deal: Transfer of Federal Spectrum Through Overlay Licenses*, Mercatus Center Working Paper (Aug. 2015).

35. Jon Peha explains the issue adroitly: "We would like for [our] schools and libraries to have great broadband access, but if you create a band just for schools and libraries, their spectrum would be free but their equipment would be a hundred times as expensive"; quoted in Robyn (2015), 23.

36. Robyn (2015), 22.

37. Without endorsing each of the particular reform proposals made, the following are provocative. Dorothy Robyn, MAKING WAVES: ALTERNATIVE PATHS TO FLEXIBLE USE SPECTRUM (Washington, DC: Aspen Institute, Communications and Society Program, 2015); Dorothy Robyn, *Buildings and Bandwidth: Lessons for Spectrum Policy from Federal Property Management*, Brookings Institution Economic Studies (Sept. 2014); Brent Skorup, *Reclaiming Federal Spectrum: Proposals and Recommendations*, 15 COLUMBIA SCIENCE & TECHNOLOGY LAW REVIEW (Fall 2013); Federal Communications Commission, National Broadband Plan (March 2010), chapter 5; Thomas M. Lenard, Lawrence J. White & James L. Riso, *Increasing Spectrum for Broadband: What Are the Options?* Technology Policy Institute (Feb. 2010). An interesting review of liberalization across countries is found in Karen Wrege, *Spectrum Liberalization: Experience in Five Countries*, KB Enterprises White Paper for the GSM Association (Oct. 9, 2009). And for policy ideas that I generally endorse, see Thomas W. Hazlett, *The Wireless Craze, the Unlimited Bandwidth Myth, the Spectrum Auction Faux Pas, and the Punchline to Ronald Coase's "Big Joke": An*

Essay on Airwave Allocation Policy, 15 HARVARD JOURNAL OF LAW & TECHNOLOGY 335 (Spring 2001); Thomas W. Hazlett, *Optimal Abolition of FCC Allocation of Radio Spectrum,* 22 JOURNAL OF ECONOMIC PERSPECTIVES 103 (Winter 2008).

38. Bryan Walsh, *Cell-Phone Safety,* TIME (March 15, 2010).

39. Valeen Afualo & John McMillan, *Auction of Rights to Public Property,* THE NEW PALGRAVE DICTIONARY OF ECONOMICS AND THE LAW (Nov. 1996), 4–5; Paul Milgrom, PUTTING AUCTION THEORY TO WORK (Cambridge: Cambridge University Press, 2004).

40. Conversation with the author (Jan. 16, 2015).

41. "Everybody knew that Lyndon Johnson did not want to deal with communications because of his wife's ownership of the TV stations"; notes from Interview with Clay T. Whitehead (July 14, 2008), 15–16.

42. Bobby Baker, WHEELING AND DEALING: CONFESSIONS OF A CAPITOL HILL OPERATOR (New York: Norton, 1978), 82.

43. John Eggerton, *Wheeler: FCC Should Look into Consumers Held Hostage over Corporate Disputes—Chairman Candidate Says Might Be Time for Another Newton Minow Call for Better Programming,* MULTICHANNEL NEWS (June 18, 2013).

44. The key episode of airline deregulation was described in Stephen Breyer, REGULATION AND ITS REFORM (Cambridge, MA: Harvard University Press, 1982), 317–340.

45. Peter Huber, LAW AND DISORDER IN CYBERSPACE (New York: Oxford University Press, 1997), 70.

46. Vivek Wadhwa, *How Technology Will Eat Medicine,* WALL STREET JOURNAL (Jan. 9, 2015).

INDEX

Page numbers in *italics* refer to illustrations.